Advances in High Temperature Chemistry

Volume 2

CONTRIBUTORS TO THIS VOLUME

LEO BREWER

M. P. FREEMAN

FRANK T. GREENE

THOMAS A. MILNE

TETSUO NOGUCHI

GERD M. ROSENBLATT

G. A. SOMORJAI

W. WELTNER, JR.

ADVANCES IN HIGH TEMPERATURE CHEMISTRY

Edited by LEROY EYRING
DEPARTMENT OF CHEMISTRY
ARIZONA STATE UNIVERSITY
TEMPE, ARIZONA

VOLUME 2

ACADEMIC PRESS New York and London 1969

ACADEMIC PRESS, INC.
111 Fifth Avenue, New York, New York 10003

United Kingdom Edition published by
ACADEMIC PRESS, INC. (LONDON) LTD.
Berkeley Square House, London W1X 6BA

LIBRARY OF CONGRESS CATALOG CARD NUMBER: 67-22786

PRINTED IN THE UNITED STATES OF AMERICA

Contributors

Numbers in parentheses refer to the pages on which the authors' contributions begin.

LEO BREWER, *Inorganic Materials Research Division, Lawrence Radiation Laboratory and Department of Chemistry, University of California, Berkeley, California* (1)

M. P. FREEMAN, *Central Research Division, American Cyanamid Company, Stamford, Connecticut* (151)

FRANK T. GREENE, *Midwest Research Institute, Kansas City, Missouri* (107)

THOMAS A. MILNE, *Midwest Research Institute, Kansas City, Missouri* (107)

TETSUO NOGUCHI, *Solar Research Laboratory, Government Industrial Research Institute, Hirate-Machi, Kita-Ku, Nagoya, Japan* (235)

GERD M. ROSENBLATT, *Department of Chemistry, Pennsylvania State University, University Park, Pennsylvania* (1)

G. A. SOMORJAI, *Inorganic Materials Research Division, Lawrence Radiation Laboratory and Department of Chemistry, University of California, Berkeley, California* (203)

W. WELTNER, JR., *Department of Chemistry, University of Florida, Gainesville, Florida* (85)

Foreword

Most matter of the visible universe is at high temperature. The radiation from this hot matter, in particular the sun, is responsible for the occurrence and maintenance of life on earth. The sun has been increasingly supplemented in history to improve the quality of living on the relatively cold surface of the earth as we have emerged into what has been called a high energy society.

The incredible manifold of chemistry which is life occurs in a very narrow and rather low temperature range. Life processes are achieved at body temperature by means of enzyme systems which greatly alter the energy and entropy requirements for reactions which would otherwise need high temperatures. In the environment of living things such complex catalyst systems do not generally function, hence many desirable reactions must be accomplished by elevating the temperature.

High temperature chemistry must be used to forge the special materials required for the more efficient energy producing machines of the future. Each "blast-off" from Cape Kennedy is an immense energy producing high temperature reaction with vast, often unwanted, side reactions, all of which must be controlled. High temperature chemistry is an essential component of the cornucopia which shall yield the fullness of our high energy society.

Leo Brewer relates the usual reaction of persons first encountering high temperature chemistry, "Oh, it must be very difficult to produce these high temperatures, but, at least once you get them, the chemistry must be very simple," and quickly assures his audience that, in fact, just the opposite is true.*

To apprehend the province of high temperature chemistry is very difficult indeed. This difficulty springs in part from the fact that temperatures only a few degrees above absolute zero are "high" for certain magnetic interactions, while three orders of magnitude higher temperatures are required to qualify as "high" for chemical dissociation processes and one must go many orders of magnitude still higher for even the simplest nuclear reactions. While we shall be concerned here with advances in knowledge of the high temperature behavior of materials and the means of extending our range of experimentation, we shall emphasize the complex and unfamiliar characteristics of matter at high temperature.

* L. Brewer, Principles of High Temperature Chemistry, *Proc. Robert A. Welch Foundation Conf. Chem. Res.* Vol. VI, "Topics in Modern Inorganic Chemistry" (W. O. Milligan, ed.). Houston, Texas, 1963.

Since high temperature chemistry is then what high temperature chemists do and say, this serial publication will be a living definition of the field. Even so, as in the first volume, definitions of high temperature chemistry will be offered from time to time for the purpose of pointing up its special and changing characteristics.

These volumes shall serve as an aid to the advance of high temperature chemistry by interpreting the complexities of the field at propitious times and collecting these interpretations together in one place. These didactic articles are directed to persons actually working in high temperature chemistry, and others with deep interest in this work.

Tempe, Arizona
September 1969

L. EYRING

Contents

CONTRIBUTORS v

FOREWORD vii

CONTENTS OF VOLUME 1 xi

Dissociation Energies and Free Energy Functions of Gaseous Monoxides

Leo Brewer and Gerd M. Rosenblatt

I. Introduction 1
II. The Free Energy Functions 1
III. Sources and Evaluation of Data 37
References 74

The Matrix-Isolation Technique Applied to High Temperature Molecules

W. Weltner, Jr.

I. Introduction 85
II. Experimental Procedures 86
III. Optical Spectroscopy 87
IV. Electron-Paramagnetic-Resonance Spectroscopy 98
V. Orientation of Molecules in Matrices 101
References 103

Molecular Beams in High Temperature Chemistry

Thomas A. Milne and Frank T. Greene

I. Introduction 107
II. Beam Formation from High Pressures 108
III. Molecular Beam Techniques 113
IV. Molecular Beam Applications 120
References 140

Chemical Research in Streaming Thermal Plasmas

M. P. Freeman

I. Introduction 151
II. Recent Research 175
References 198

Studies of the Sublimation Mechanism of Solids

G. A. Somorjai

I. Introduction 204
II. Principles of Vaporization Studies 205
III. Experimental Considerations 207
IV. Studies of the Sublimation Mechanism of Sodium Chloride Single Crystals 209
V. Studies of the Sublimation Mechanism of Cadmium Sulfide Single Crystals 216
VI. Studies of the Sublimation Mechanism of Gallium Arsenide Single Crystals 221
VII. Low Energy Electron Diffraction (LEED) Study of the Vaporizing Silver (100) Surface 228
VIII. Conclusions 231
References 233

High Temperature Phase Studies with a Solar Furnace

Tetsuo Noguchi

I. Introduction 235
II. The Nature of a Solar Furnace 236
III. Temperature and Emissivity Measurement 243
IV. Freezing Point Determination of High Melting Refractory Oxides 244
V. Liquidus Curve Measurement 248
VI. Phase Studies at High Temperatures 250
VII. Summary and Conclusions 259
References 260

Author Index 263
Subject Index 278

Contents of Volume 1

A Definition of High Temperature Chemistry
Paul Goldfinger

The Spectroscopy of Diatomic Transition Element Molecules
C. J. Cheetham and R. F. Barrow

Electronic Structure of Molecules of High Temperature Interest
K. Douglas Carlson and Charles R. Claydon

Gaseous Ternary Compounds of the Alkali Metals
Alfred Büchler and Joan B. Berkowitz-Mattuck

Valence States in High Temperature Chemistry; Energic Structures in Transitional Elements
R. J. Thorn and G. H. Winslow

Bond Energy Relationships in Polyvalent Metal Fluorides
D. L. Hildenbrand

The Chemical Reactions of High Temperature Species
J. B. Ezell, J. C. Thompson, J. L. Margrave, and P. L. Timms

High Temperature Adiabatic Calorimetry
Edgar F. Westrum, Jr.

Plasmas for High Temperature Chemistry
Thomas B. Reed

AUTHOR INDEX—SUBJECT INDEX

Dissociation Energies and Free Energy Functions of Gaseous Monoxides

Leo Brewer

INORGANIC MATERIALS RESEARCH DIVISION, THE LAWRENCE RADIATION LABORATORY AND DEPARTMENT OF CHEMISTRY, UNIVERSITY OF CALIFORNIA, BERKELEY, CALIFORNIA

and Gerd M. Rosenblatt

DEPARTMENT OF CHEMISTRY, PENNSYLVANIA STATE UNIVERSITY, UNIVERSITY PARK, PENNSYLVANIA

I. Introduction 1
II. The Free Energy Functions 1
III. Sources and Evaluation of Data 37
References 74

I. Introduction

The major vaporization processes of a very large proportion of the oxides can be adequately described up to the atmospheric boiling point in terms of gaseous elemental products and gaseous MO and MO_2 species. Thermodynamic data are available for the calculation of the partial pressures of the elemental species in equilibrium with most oxides. Brewer and Rosenblatt (1961) have presented data to allow calculation of the equilibrium MO_2 partial pressures. The present paper provides a parallel set of tabulations to allow calculation of the MO partial pressures.

II. The Free Energy Functions

The calculation of free energy functions can be carried out with more confidence for the gaseous MO molecule than for the MO_2 molecule, since there is no question of bond angles nor any low-frequency bending vibration to be considered. The vibrational and rotational contributions to the free energy functions were calculated on the basis of a harmonic oscillator–rigid rotator model. The values of B_0, the rotational constant, and of the vibrational frequency, $\tilde{\nu}(1\text{–}0) = \omega_e - 2\omega_e x_e$, are tabulated in Table III along with references to the source of the data. Where data were lacking, vibrational frequencies and internuclear distances were estimated from consideration of position in the periodic table and strength of bonding. (Brewer and Chandrasekharaiah, 1960). The uncertainties due to these estimates are negligible compared to uncertainties in electronic contributions and in the experimental equilibria data. For many of the molecules, there are low-lying electronic states with vibrational frequencies and internuclear distances somewhat different from those of the ground state. The vibrational and rotational contributions of these states were not included. Since the vibrational frequencies and rotational constants of the excited states are usually smaller than those of the ground state, the inclusion of these terms would increase the partition function. This neglect as well as error due to use of a harmonic oscillator–rigid rotator model is offset by use of a procedure for estimating the electronic contribution that overestimates the electronic partition function as is described below.

The problem of adequately considering the electronic contribution to the partition function, particularly for the transition metal compounds, is as difficult for the diatomic as for the triatomic oxides. In the previous dioxide and dihalide compilations (Brewer and Rosenblatt, 1961; Brewer *et al.* 1963), the electronic portion of the free energy function of the transition metal compounds has been obtained from that of the free ion, and the same recipe is proposed for this compilation. Thus the electronic partition function of a gaseous diatomic oxide of a transition element is taken the same as the electronic partition function of the doubly charged gaseous ion as calculated from electronic levels listed by Moore (1949, 1952, 1958) and other sources as indicated in Table IV.

The free divalent ion must have the same number of electronic states as the gaseous dihalide or monoxide, including repulsive states, even though the approach of the anions perturbs the levels of the cation. The procedure used here assumes that the average energy shifts of the levels will be not much different for M^{2+}, MX_2, and MO. For the dihalides the doubly charged cation should be a reasonable model, but the calculations of Brewer and Mastick (1951) show that a M^+O^- model is more reasonable than $M^{2+}O^{2-}$ for gaseous oxides. An estimation based on M^{2+} should not be as good for MO as for MX_2, but procedures based on M^+ together with $O^-(^2P)$ would be unduly complicated.

The difficulties of fixing the electronic partition function Q_{el} will be illustrated first by the examples of the fourth-group oxides, TiO, ZrO, and HfO, and subsequently by the alkaline earths. The ground states of ZrO and HfO have been assigned as $^1\Sigma$ states (Weltner and Mcleod, 1965a, b), whereas TiO is reported by Phillips (1952, 1969) to have a $^3\Delta$ ground state. If only the ground states are considered, the electronic contribution to the free energy function, $[-(F^\circ - H_0^\circ)/T]_{el} = R \ln Q_{el}$, of TiO gas would be 3.56 cal/deg mole with zero contribution for ZrO and HfO. However, if one considers the distribution of the ten valence electrons among possible low-lying molecular orbitals, there will be eight electrons in the $\sigma^2\sigma^2\pi^4$ orbitals with the remaining two electrons in one of the combinations $\sigma\delta$ ($^3\Delta$, $^1\Delta$), σ^2 ($^1\Sigma^+$), $\sigma\pi$ ($^3\Pi$, $^1\Pi$), and δ^2 ($^3\Sigma^-$, $^1\Delta$). The resulting electronic states are given in parentheses. Of the seven states, $^3\Delta$, $^1\Delta$, $^1\Sigma$, and $^1\Pi$ have been observed for both TiO and ZrO. The low-lying states of HfO are not well-characterized experimentally. For TiO, the $^1\Sigma$ state is around 2000 cm^{-1} above the $^3\Delta$ and $^1\Delta$ states. For ZrO it appears that the $^1\Sigma$ state is just barely below the delta states (Brewer and Green, 1969).

The $^3\Delta$ state of ZrO has been estimated by Brewer and Green to be 1000 cm^{-1} and the $^1\Delta$ state 5000 cm^{-1} above the $^1\Sigma$ state. If these states are now included, the electronic contribution to $R \ln Q$ for ZrO at 2000°K, for example, is increased from zero to 2.7 cal/deg. mole. There are other excited states that could increase this value. The $^3\Pi$ state, which should be around 8000 cm^{-1}, would only make a small contribution, but it is possible that the $^3\Sigma^-$ state is low-lying. If it were as low as 2000 cm^{-1}, the electronic contribution for ZrO would be 3.1 cal/deg mole. For contrast, the electronic contribution for the free gaseous Zr^{2+} ion is 5.02 cal/deg mole.

For TiO the spectroscopic data are more complete. The energy of the $^1\Delta$ state above the ground $^3\Delta$ state has been roughly fixed at around 580 cm^{-1} (Phillips, 1952). The $^1\Sigma$ state is 2218.77 cm^{-1} above the $^1\Delta$ state or about 2800 cm^{-1} above the ground state. The $^3\Pi$ state has been estimated (Brewer and Green, 1969) to be around 9000 cm^{-1} The inclusion of these excited states would increase $R \ln Q_{el}$ at 2000°K from 3.56 to a value of 4.00 cal/deg mole, which would be a lower limit if the $^3\Sigma^-$ state is low-lying. The electronic contribution for free gaseous Ti^{2+} ion at 2000°K is 5.72 cal/deg mole. Even if the $^3\Sigma^-$ state were put at 2000 cm^{-1} above the ground state, the resulting $R \ln Q_{el}$ of TiO at 2000°K would not be above 4.2 cal/deg mole. Several calculations have been made of the energies of the low-lying levels of TiO. Two Hartree–Fock calculations (Carlson and Moser, 1963, 1967; Carlson and Nesbet, 1964) have yielded values of 4985 and 2151 cm^{-1} for the $^1\Delta$ state compared with the experimental 581 cm^{-1}, and 8743 and 6694 cm^{-1} for the $^1\Sigma$ state compared with the experimental 2289 cm^{-1}. Since these energies are higher than the experimental values, their use will yield smaller partition functions than the minimum value based on just the observed levels. For example, the Carlson–Nesbet value yields 3.58 cal/deg mole for $R \ln Q_{el}$ at 2000°K, and a slightly higher value is given by the Carlson and Moser (1967) energy values. Carlson and Nesbet report that their calculations predict a very high energy for the $^3\Sigma^-$ level, in contrast to the prediction of Berg and Sinanoglu (1960) from a ligand-field model that the $^3\Sigma^-$ state is the ground state.

For HfO (and ZrO as well), Hartree–Fock calculations are impractical at present. The HfO spectral data are inadequate to tabulate energy levels, and there are no intermediate calculations of $R \ln Q_{el}$ of HfO between the zero value based on the ground state alone and the 4.27 cal/deg mole based on the electronic entropy of free Hf^{2+}.

The comparison of the various procedures for obtaining the electronic contributions to the free energy functions of TiO, ZrO, and HfO is given in Table I. Table I indicates that the M^{2+} model overestimates $R \ln Q_{el}$ by

TABLE I. Electronic Contributions to $-(F^\circ - H_0^\circ)/T$ for Fourth Group Oxides[a]

Oxide	Ground state only	Known spectroscopic states	Inclusion of low-lying $^3\Sigma^-$	Hartree–Fock calculations	M^{2+}
TiO	3.56	4.0	4.2	3.6	5.72
ZrO	0.0	2.7	3.1	—	5.02
HfO	0.0	—	—	—	4.27

[a] Data given in cal/deg mole.

1 to 2 cal/deg mole, but is much preferable to calculations based only on the ground state or incomplete knowledge of very low-lying electronic states. As noted above, part of this error is canceled by the underestimation of the vibrational and rotational contributions. For most monoxides, there is no alternative to this recipe to provide an estimate of the electronic partition function, since not even the ground electronic states are clearly established for most transition metal monoxides. There are only a few theoretical calculations of the type available for TiO. It was considered more important to follow a consistent procedure applicable to all of the transition oxides, rather than to combine different methods. Although the uncertainty in the electronic contribution to the free energy functions may be as much as 2 to 3 cal/deg mole in some instances, a large part of the error cancels out in equilibrium calculations involving pairs of monoxides or the monoxides together with dihalides or dioxides if this consistent procedure is used.

The divalent ion model for electronic partition functions was used for monoxides of all the elements from Sc, Y, and La through Zn, Cd, and Hg and for the lanthanides and actinides. It must be emphasized that the combination of the free energy functions of this paper with free energy functions from other sources, where no attempt is made to estimate a complete electronic partition function, can result in abnormally large errors. Also the enthalpies tabulated in this paper should not be combined with enthalpies obtained through use of different free energy functions. If the dissociation energies tabulated here are used for purposes other than equilibrium calculations, such as spectroscopic purposes, for example, it should be recognized that the D_0 values for the transition metal oxides and the alkaline earths may be high by 0.1 to 0.2 eV.

For most nontransition oxides, the ground electronic state can be established without question, and contributions by excited electronic states are small at temperatures below 3000°K. The alkaline earths are an outstanding exception. Molecular orbital correlations based on C_2 and other diatomic molecules with eight valence electrons indicate that six electronic states corresponding to the three electronic configurations $\sigma^2\sigma^2\pi^4$ ($^1\Sigma^+$), $\sigma^2\sigma^2\sigma\pi^3$ ($^3\Pi$, $^1\Pi$), and $\sigma^2\sigma^2\sigma^2\pi^2$ ($^3\Sigma^-$, $^1\Delta$, $^1\Sigma^+$) could possibly lie below 20,000 cm^{-1} for all of the alkaline earth oxides. In contrast to C_2 and other nonpolar molecules, the lowest states of the alkaline earths should have strong ionic bonding. The states $^1\Sigma^+$, $^3\Pi$, and $^1\Pi$ can be correlated to $M^+ + O^-$ dissociation products and would be expected to be the lowest electronic states because of their ionic character. The $^1\Pi$ state has been observed for BeO and MgO at 9406 and 3563 cm^{-1}, respectively, above the $^1\Sigma^+$ state. By Hund's rule, the $^3\Pi$ state should lie below the $^1\Pi$ state. If the splitting between the $^3\Pi$ and $^1\Pi$ states were as large as the 7700 cm^{-1} observed for the corresponding states of C_2, the $^3\Pi$ state would be below or close to the $^1\Sigma^+$ state and would dominate the electronic partition function (Brewer, 1963). The triplet systems of the alkaline earths have been too complicated to resolve, and only for the related molecule BeS has a triplet band system been resolved (Cheetham *et al.*, 1965). In this instance, the triplet state is seen in absorption, indicating that it must be a low-lying state.

Several Hartree–Fock calculations have been carried out for BeO. Verhaegen and Richards (1966) calculate the $^1\Pi$ and $^3\Pi$ states to be 14,400 and 11,800 cm^{-1}, respectively, above the $^1\Sigma^+$ state with a splitting of 2600 cm^{-1}. A more complete calculation by Huo *et al.* (1967) places the $^1\Pi$ and $^3\Pi$ states at 7379 and 8302 cm^{-1}, respectively, below the $^1\Sigma$ state. Since it is known experimentally that the $^1\Pi$ is 9406 cm^{-1} above the $^1\Sigma$ state, the authors argue that the error in the calculations can be corrected by adding 16,785 cm^{-1} to both the $^1\Pi$ and $^3\Pi$ energies, but that the calculated splitting of 923 cm^{-1} should be accepted as accurate. This would place the $^3\Pi$ state at 8483 cm^{-1}. If one uses the Verhaegen and Richards splitting, the $^3\Pi$ would be at 5806 cm^{-1} above the $^1\Sigma^+$ state.

For MgO Richards *et al.* (1966) calculate $^3\Pi$ and $^3\Sigma^+$ at 2250 cm^{-1} and $^1\Pi$ at 4600 cm^{-1}, compared with the experimental value of 3563 cm^{-1} for the $^1\Pi$ above the $^1\Sigma$ state. It does not appear that the calculations are sufficiently accurate to fix the thermodynamic properties. If one considers the differences in vibrational and rotational contributions as well as the electronic degeneracies, the partition function based on the $^1\Sigma^+$ state alone would be increased by a factor of 10 if the $^3\Pi$ were at a comparable energy. Schofield (1967) has reviewed various experimental data for the alkaline earths. Most of the experimental data indicate large partition functions corresponding to the $^3\Pi$ state lying very close to the $^1\Sigma$ state. However, the experimental uncertainty of these data is large, and they do not provide a definitive answer to the question of the position of the triplet state. An indirect spectroscopic approach to the question of the location of triplet states may be provided by the $^1\Sigma$–$^1\Sigma$ transition of $Ca^{16}O$. Hultin and Lagerqvist (1951) have reported that the upper singlet state is perturbed by six distinct states or substates in agreement with the molecular orbital correlations mentioned above which would predict perturbations by only the six substates $^3\Sigma^-_{0+}$, $^3\Sigma^-_1$, $^1\Pi_1$, $^3\Pi_1$, $^3\Pi_{0^+}$, and $^1\Sigma_0{}^+$. Thus all of these states must be at 12,000 cm^{-1} or below. For the heavier alkaline earths, one can expect Hund's case (c) coupling to be approached to give a distribution of five lambda-doubled states and five single states between 0 and 12,000 cm^{-1}. However, until the shifts of the perturbations have been worked out for the isotopic molecule $Ca^{18}O$, it is not possible to say how the energies of these substates are distributed.

Only for BaO does there seem to be reasonable agreement between several different types of measurements. BaO is unique among the alkaline earths in that it vaporizes predominantly to diatomic BaO and total vapor pressure measurements can be equated to the BaO pressure. Electric and magnetic resonance measurements by Wharton *et al.* (1962) and Brooks and Kaufman (1965) have shown a 0^+ ground state for BaO. For such a heavy molecule, it is not clear whether the ground state should be assigned to $^1\Sigma_0{}^+$ or $^3\Pi_{0^+}$ for comparison with the other alkaline earths, although one would have to assume that the 0^-, 1, and 2 components of the $^3\Pi$ are displaced to higher energies. Kaufman *et al.* (1965) report similar results for SrO.

The above discussion illustrates the unsatisfactory state of our knowledge of the low-lying electronic states of the alkaline earths. We have no basis for estimating the energy levels of BaO. For the other alkaline earths, we have adopted a set of low-lying electronic states as a compromise among the various conflicting data. These values are shown in Table II. The energy values chosen for the alkaline earths may yield electronic entropies that are

TABLE II. LOW-LYING ELECTRONIC STATES OF THE ALKALINE EARTH DIATOMIC GASES[a]

	$^1\Sigma^+$	$^3\Pi$	$^1\Pi$	$^3\Sigma^-$
BeO	0	1000	9235	8000
MgO	0	0	3503	9000
CaO	0	0	3200	3000
SrO	0	0	1900	1000

[a] Data given in cm^{-1}.

high by 1 to 2 cal/deg mole and are thus consistent with the procedure used for the transition element oxides. Such errors in the entropies or free energy functions cancel out in most equilibrium calculations. Since the free energy functions evaluated by the procedures outlined above are used together with equilibrium measurements to obtain the enthalpies of formation that are tabulated later in this chapter, the experimental equilibrium constants can be recovered with no error due to the free energy functions. It is, of course, essential that the free energy functions tabulated in this chapter be used only with the enthalpies tabulated here or with enthalpies obtained through use of the same free energy functions. For calculations far removed from the temperature range of the equilibrium measurements, an error in the free energy function will not completely cancel. For example, if equilibrium measurements at 2000°K have been used to evaluate the enthalpy of formation of an oxide, an error of 2 cal/deg mole in the free energy function will result in an error of 0.4 cal/deg mole in $\Delta F^\circ/T$ at 2500°K and 0.7 cal/deg mole in $\Delta F^\circ/T$ at 3000°K.

In evaluation of high temperature equilibria data it was necessary in many instances to use the enthalpy of atomization of the elements. Table V tabulates the values used for the elemental atomization enthalpies at 298.15°K, $\Delta H^\circ_{298}f$(M, g), along with references to the sources of the values.

The enthalpy data for the gaseous diatomic oxides are presented in two forms. Table VI presents the enthalpy changes in kcal/mole for the reaction MO(g) = M(g) + O(g) at 0° and 298.15°K as D°_0 and D°_{298}, respectively. Table VII presents the enthalpy changes for the reaction M(condensed) + $\frac{1}{2}O_2$(g) = MO(g) as $\Delta H^\circ_{298}f$(MO). Table VIII presents values of the enthalpies of formation of condensed oxides, $\Delta H^\circ_{298}f$, that were used in evaluating the high temperature equilibria data.

The free energy functions, $-(F^\circ - H^\circ_{298})/T$ in cal/degree mole, are tabulated in Table IX at 100° intervals between 298° and 3000°K along with values of $H^\circ_{298} - H_0^\circ$ and $H^\circ_{2000} - H_0^\circ$ in cal/mole for each gaseous diatomic oxide, and will be found in alphabetical order of the elemental symbols. To allow modification of the free energy functions if complete spectroscopic data on the low-lying electronic states become available, values of the electronic contribution to $-(F^\circ - H_0^\circ)/T$ are also tabulated in Table IX for each diatomic oxide.

Section III refers to sources of the data used and presents a discussion of their treatment to obtain dissociation enthalpies of the gaseous diatomic oxides. The physical constants used in treating the data are based upon the NAS-NRC (1963) recommendations. As for the tables, the order of presentation is in alphabetical order of the elemental symbols.

TABLE III. ROTATIONAL AND VIBRATIONAL CONSTANTS OF DIATOMIC OXIDES[a,b]

Molecule	B_0	Ref.	$\tilde{\nu}(1-0) = \omega_e - 2\omega_e x_e$	Ref.
AgO	0.3015	Uhler (1953)	484.4	Uhler (1953)
AlO	0.63946	Lagerqvist *et al.* (1957)	965.29	Lagerqvist *et al.* (1957)
AsO	0.48164	Collomon and Morgan (1965)	957.37	Collomon and Morgan (1965)
AuO	(0.2632)	Brewer and Chandrasekharaiah (1960)	(420)	Brewer and Chandrasekharaiah (1960)
BO	1.7737	Lagerqvist *et al.* (1958)	1862.07	Lagerqvist *et al.* (1958)
BaO	0.31192	Wharton and Klemperer (1963)	665.70	Lagerqvist *et al.* (1950)
BeO	1.6415	Herzberg (1950)	1463.66	Herzberg (1950)
BiO	0.3023	Barrow *et al.* (1967)	683.7	Barrow *et al.* (1967)
BrO	0.463	Durie and Ramsey (1958)	699	Coleman and Gaydon (1947)
CO	1.9225	Rank *et al.* (1965)	2143.24	Rank *et al.* (1965)
CaO	0.4428	Hultin and Lagerqvist (1951)	722.5	Hultin and Lagerqvist (1951)
CdO	(0.3515)	Brewer and Chandrasekharaiah (1960)	(550)	Brewer and Chandrasekharaiah (1960)
CeO	0.3568	Ames and Barrow (1967)	830	Ames and Barrow (1967)
ClO	0.619	Carrington *et al.* (1967)	853	Porter (1950)
CoO	(0.4921)	Brewer and Chandrasekharaiah (1960)	840	Rosen (1951)
CrO	0.5261	Ninomiya (1955)	885.8	Ghosh (1932)
CsO	(0.2310)	Brewer and Chandrasekharaiah (1960)	(490)	Estimated
CuO	0.4429	Lagerqvist and Uhler (1967)	622	Rosen (1951)
FO	(1.138)	JANAF (1966a)	1028	Arkell *et al.* (1965)
FeO	0.3477	Dhumwad and Narasimham (1966)	861.9	Dhumwad and Narasimham (1966)
GaO	0.4271	Raziunas *et al.* (1965a)	755.01	Howell (1945)
GdO	(0.358)	Estimated	(850)	Estimated
GeO	0.4859	Toerring (1966)	976.9	Rowlinson and Barrow (1953)
HO	18.514	Haar and Friedman (1955)	3404.0	Haar and Friedman (1955)
HfO	(0.3792)	Estimated	968	Gatterer *et al.* (1957), Weltner and McLeod (1965b)
HgO	(0.299)	Brewer and Chandrasekharaiah (1960)	(480)	Brewer and Chandrasekharaiah (1960)

TABLE III (continued)

Molecule	B_0	Ref.	$\tilde{\nu}(1\text{–}0) = \omega_e - 2\omega_e x_e$	Ref.
IO	0.33891	Durie *et al.* (1960)	672.89	Durie *et al.* (1960)
InO	(0.3582)	Brewer and Chandrasekharaiah (1960)	695.63	Watson and Shambon (1936)
IrO	(0.3333)	Brewer and Chandrasekharaiah (1960)	(785)	Brewer and Chandrasekharaiah (1960)
KO	(0.3367)	Brewer and Chandrasekharaiah (1960)	(670)	Brewer and Chandrasekharaiah (1960)
LaO	0.3519	Akerlind (1962b)	807.1	Akerlind (1962b)
LiO	1.3272	White *et al.* (1963)	745	White *et al.* (1963)
MgO	0.5718	Lagerqvist and Uhler (1949)	774.70	Lagerqvist and Uhler (1949)
MnO	(0.4878)	Brewer and Chandrasekharaiah (1960)	829.97	Das Sarma (1959)
MoO	(0.4107)	Brewer and Chandrasekharaiah (1960)	950	Swaminathan and Krishnamurty (1954)
NO	1.67185	Hall and Dowling (1966)	1875.96	Meyer *et al.* (1965)
NaO	(0.5769)	Brewer and Chandrasekharaiah (1960)	(758)	Brewer and Chandrasekharaiah (1960)
NbO	0.4310	Uhler (1954)	981.37	Uhler (1954)
NiO	(0.4982)	Brewer and Chandrasekharaiah (1960)	615	Rosen (1951)
OsO	(0.3336)	Brewer and Chandrasekharaiah (1960)	(785)	Brewer and Chandrasekharaiah (1960)
PO	0.7321	Singh (1959)	1220.28	Rao (1958)
PbO	0.30631	Barrow *et al.* (1961); Toerring (1964)	714.4	Barrow *et al.* (1961)
PdO	(0.3826)	Brewer and Chandrasekharaiah (1960)	(810)	Estimated
PtO	(0.3293)	Brewer and Chandrasekharaiah (1960)	(785)	Brewer and Chandrasekharaiah (1960)
RbO	(0.2583)	Brewer and Chandrasekharaiah (1960)	(560)	Brewer and Chandrasekharaiah (1960)
ReO	(0.3452)	Brewer and Chandrasekharaiah (1960)	(760)	Brewer and Chandrasekharaiah (1960)
RhO	(0.3842)	Brewer and Chandrasekharaiah (1960)	(810)	Estimated
RuO	0.422	Raziunas *et al.* (1965b)	854.6	Raziunas *et al.* (1965b)
SO	0.71793	Powell and Lide (1964); Amano *et al.* (1967); Daniels and Dorain (1966)	1135.96	Norrish and Oldershaw (1959)
SbO	0.3490	Laksham (1960)	811.3	Laksham (1960)
ScO	0.5135	Akerlind (1962a)	963.7	Akerlind (1962a)
SeO	0.4638	Barrow and Deutsch (1963)	905.63	Barrow and Deutsch (1963)

SiO	0.7248	Lagerqvist and Uhler (1953)	1229.60	Lagerqvist and Uhler (1953)
SnO	0.3550	Lagerqvist *et al*. (1959)	814.9	Jevons (1938)
SrO	0.33688	Kaufman *et al*. (1965)	645.57	Lagerqvist and Selin (1956)
TaO	0.4019	Cheetham and Barrow (1967)	1021.87	Cheetham and Barrow (1967), Weltner and McLeod (1965c)
TeO	0.3548	Chandler *et al*. (1965)	789.69	Chandler *et al*. (1965)
ThO	0.33199	Edvinsson *et al*. (1965)	890.99	Edvinsson *et al*. (1965)
TiO	0.5340	Phillips (1951)	1000.00	Phillips (1951)
TlO	(0.3048)	Brewer and Chandrasekharaiah (1960)	(660)	Estimated
UO	0.2927	De Maria *et al*. (1960)	920	De Maria *et al*. (1960)
VO	0.5463	Lagerqvist and Selin (1957)	1001.62	Lagerqvist and Selin (1957)
WO	(0.3615)	Brewer and Chandrasekharaiah (1960)	1054.9	Gatterer and Krishnamurty (1952), Weltner and McLeod (1965d)
YO	0.3881	Uhler and Akerlind (1961)	847.6	Akerlind (1962b)
ZnO	(0.4332)	Brewer and Chandrasekharaiah (1960)	(680)	Brewer and Chandrasekharaiah (1960)
ZrO	0.4230	Uhler and Akerlind (1956); Weltner and McLeod (1965a,b)	969.76	Uhler and Akerlind (1956), Weltner and McLeod (1965a,b)

[a] Data given in cm^{-1}.

[b] Values in parentheses are estimated.

TABLE IV. ELECTRONIC ENERGY LEVELS AND DEGENERACIES USED IN COMPUTING GASEOUS OXIDE FREE ENERGY FUNCTIONS

Molecule	Energy levels (cm^{-1}) and degeneracies[a]			Ref.
AgO	0(6)	4607(4)		Ag^{2+} levels (Moore, 1958)
AlO	$^2\Sigma$ 0(2)	$^2\Pi$ 8660(4)		Shetlar (1965)
AsO	$^2\Pi_{1/2}$ 0(2)	$^2\Pi_{3/2}$ 1026(2)		Collomon and Morgan (1965)
AuO	0(6)	8420(4)	4787(10)	Pt^+ levels (Moore, 1958)
	9356(8)	13,329(6)	15,791(4)	
	16,821(6)	18,098(8)		
BO	$^2\Sigma$ 0(2)			Herzberg (1950)
BaO	$^1\Sigma$ 0(1)	$^1\Sigma$ 16,722(1)		Herzberg (1950)
BeO	$^1\Sigma$ 0(1)	$^3\Pi$ 1000(6)	$^3\Sigma$ 8000(3)	Estimated
	$^1\Pi$ 9235(2)			
BiO	$^2\Pi_{1/2}$ 0(2)	$^2\Pi_{3/2}$ 8000(2)	$^2\Pi_{1/2}$ 14,096(2)	Barrow *et al.* (1967)
BrO	$^2\Pi_{3/2}$ 0(2)	$^2\Pi_{1/2}$ 3685(2)		Durie and Ramsey (1958), Br levels (Moore, 1952), Durie *et al.* (1960)
CO	$^1\Sigma$ 0(1)			Herzberg (1950)
CaO	$^1\Sigma$ 0(1)	$^3\Pi$ 0(6)	$^3\Sigma$ 3000(3)	Estimated
	$^1\Pi$ 3200(2)	$^1\Delta$ 7500(2)	$^1\Sigma$ 11,549(1)	
CdO	0(1)			Cd^{2+} levels (Moore, 1958)
CeO	0(5)	1016(7)	1971(9)	La^+ levels (Moore, 1958)
	1394(5)	1895(3)	2592(5)	
	3250(7)	5250(1)	5718(3)	
	6227(5)	7395(1)	7473(9)	
	10,095(5)	15,000(28)	16,599(9)	
	18,000(59)			
ClO	$^2\Pi_{3/2}$ 0(2)	$^2\Pi_{1/2}$ 881(2)		Durie and Ramsey (1958), Cl levels (Moore, 1949), Durie *et al.* (1960)
CoO	0(10)	841(8)	1451(6)	Co^{2+} levels (Moore, 1952)
	1867(4)	15,202(6)	19,429(4)	
	15,811(2)	16,978(10)	17,766(8)	

CrO	0(1)	60(3)	182(5)	Cr^{2+} levels (Moore, 1952)
	356(7)	575(9)	16,771(1)	
	17,167(3)	17,850(5)	17,273(9)	
	17,395(11)	17,529(13)	18,451(5)	
	18,510(7)	18,582(9)		
CsO	0(4)	404(2)		F levels (Moore, 1949), Durie *et al.* (1960)
CuO	0(6)	2072(4)		Cu^{2+} levels (Moore, 1952)
FO	$^2\Pi_{3/2}$ 0(2)	$^2\Pi_{1/2}$ 404(2)		F levels (Moore, 1949)
FeO	0(9)	46(7)	739(5)	Fe^{2+} levels (Moore, 1952)
	932(3)	1027(1)	20,000(63)	
GaO	$^2\Sigma$ 0(2)	$^2\Pi$ 4000(4)		Raziunas *et al.* (1965a)
GdO	0(5)	279(7)	694(9)	Gd^{2+} levels (Callahan, 1963)
	1310(11)	2283(13)	9356(11)	
	9718(9)	10,015(7)	10,234(5)	
	10,387(3)			
GeO	$^1\Sigma$ 0(1)			Herzberg (1950)
OH	$^2\Pi_{3/2}$ 0(2)	$^2\Pi_{1/2}$ 140(2)		Haar and Friedman (1955), Papousek (1961)
HfO	0(5)	1500(7)	3000(9)	Hf^{2+} levels estimated
	5000(5)	8000(9)		
HgO	0(1)			Hg^{2+} levels (Moore, 1958)
IO	$^2\Pi_{3/2}$ 0(2)	$^2\Pi_{1/2}$ 7603(2)		Durie *et al.* (1960), I levels (Moore, 1958)
InO	$^2\Sigma$ 0(2)			Herzberg (1950)
IrO	0(10)	3593(8)	3929(6)	Os^{+} levels (Moore, 1958)
	5592(4)	6637(2)	7892(6)	
	11,500(14)	13,200(18)	15,600(10)	
	17,400(32)	20,000(24)		
KO	0(4)	404(2)		F levels (Moore, 1949)
LaO	0(4)	1603(6)	13,591(2)	La^{2+} levels (Moore, 1958)
LiO	0(4)	404(2)		F levels (Moore, 1949)
MgO	$^1\Sigma$ 0(1)	$^3\Pi$ 0(6)	$^1\Pi$ 3503(2)	Estimated
	$^3\Sigma$ 9000(3)			
MnO	0(6)			Mn^{2+} levels (Moore, 1952)

TABLE IV (continued)

Molecule	Energy levels (cm^{-1}) and degeneracies			Ref.
MoO	0(1)	159(3)	438(5)	Nb^+ levels (Moore, 1952)
	801(7)	1225(9)		
NO	$^2\Pi_{1/2}$ 0(2)	$^2\Pi_{3/2}$ 123(2)		James and Thibault (1964)
NaO	0(4)	404(2)		F levels (Moore, 1949)
NbO	0(4)	517(6)	1177(8)	Nb^{2+} levels (Moore, 1952)
	1939(10)			
NiO	0(9)	1361(7)	2270(5)	Ni^{2+} levels (Moore, 1952)
	16,662(5)	16,978(3)	17,231(1)	
OsO	0(7)	14,200(25)	19,000(50)	Re^+ levels (Moore, 1959)
PO	$^2\Pi_{1/2}$ 0(2)	$^2\Pi_{3/2}$ 224(2)		Rao (1958), Singh (1959), Santharam and Rao (1962)
PbO	$^1\Sigma$ 0(1)			Herzberg (1950)
PdO	0(9)	3230(7)	4687(5)	Pd^{2+} levels (Moore, 1959)
	10,230(5)	13,699(1)	13,470(3)	
	14,634(5)	17,880(9)		
PtO	0(9)	4159(7)	2740(5)	Os levels (Moore, 1959)
	5766(3)	6093(1)	5144(11)	
	8743(9)	10,166(5)	11,200(16)	
	13,000(13)	14,500(40)	16,000(26)	
	19,000(40)			
RbO	0(4)	404(2)		F levels (Moore, 1949)
ReO	0(2)	1519(4)	3172(6)	W^+ levels (Moore, 1959)
	4716(8)	6147(10)	7240(6)	
	8711(4)	8833(2)	10,592(4)	
	11,301(6)	13,400(16)	14,900(28)	
	16,500(24)	19,000(96)		
RhO	0(10)	2148(8)	3486(6)	Rh^{2+} levels (Iglesias, 1966)
	4322(4)	11,062(6)	10,997(4)	
	12,470(2)	13,030(10)	15,129(8)	

	18,000(38)			
RuO	0(9)	1159(7)	1826(5)	Ru^{2+} levels (Moore, 1959)
	2266(3)	2476(1)		
SO	$^3\Sigma$ 0(3)			Norrish and Oldershaw (1959)
SbO	$^2\Pi_{1/2}$ 0(2)	$^2\Pi_{3/2}$ 2276(2)		Laksham (1960)
ScO	0(4)	197(6)		Sc^{2+} levels (Moore, 1949)
SeO	$^3\Sigma_0$ 0(1)	$^3\Sigma_1$ 172(2)		Barrow and Deutsch (1963)
SiO	$^1\Sigma$ 0(1)			Lagerqvist and Uhler (1953)
SnO	$^1\Sigma$ 0(1)			Lagerqvist *et al.* (1959)
SrO	$^1\Sigma$ 0(1)	$^3\Pi$ 0(6)	$^3\Sigma$ 1000(3)	Estimated
	$^1\Pi$ 1900(2)	$^1\Delta$ 6000(2)	$^1\Sigma$ 10,872(1)	
TaO	0(4)	3051(6)	3645(4)	Hf^+ levels (Moore, 1959)
	4905(6)	6344(8)	8362(10)	
	11,952(2)	12,921(4)	13,486(6)	
	12,070(6)	15,084(8)	14,359(4)	
	15,254(2)	17,500(32)		
TeO	$^3\Sigma_0$ 0(1)	$^3\Sigma_1$ 2750(2)		Chandler *et al.* (1965)
ThO	0(5)	5027(1)	5097(7)	Th^{2+} levels (Charles, 1958)
	5300(11)	5461(3)	5871(7)	
	6475(9)	7113(19)	7813(3)	
	8310(7)	8731(3)	810(9)	
	1320(5)	3337(7)	3994(21)	
	4613(5)	8952(9)	9247(13)	
	9820(16)	11,000(37)		
TiO	0(5)	184(7)	422(9)	Ti^{2+} levels (Moore, 1949)
	8473(5)	10,536(1)	10,603(3)	
	10,721(5)	14,053(1)	14,398(9)	
TlO	0(2)			Tl^{2+} levels (Moore, 1959)
UO	0(5)	2869(7)	5094(9)	Th levels (Charles, 1958)
	5563(3)	6362(5)	7502(7)	
	8545(12)	8606(1)	8662(3)	
	10,807(3)	12,083(9)	13,297(9)	

TABLE IV (continued)

Molecule	Energy levels (cm^{-1}) and degeneracies			Ref.
VO	0(4)	145(6)	339(8)	V^{2+} levels (Moore, 1949)
	583(10)	11,207(2)	11,387(4)	
	11,513(2)	11,590(4)	11,771(6)	
	11,966(8)	12,187(10)	16,229(4)	
	16,376(6)	16,822(10)	16,977(12)	
WO	0(3)	1031(5)	2642(7)	Ta^{+} levels (Moore, 1959)
	4416(9)	6187(11)	3180(5)	
	6831(7)	9746(9)	4125(1)	
	5331(3)	5658(5)	9690(5)	
	11,300(15)	12,800(36)	14,400(33)	
	17,600(29)			
YO	0(4)	725(6)	7466(2)	Y^{2+} levels (Moore, 1952)
ZnO	0(1)			Zn^{2+} levels (Moore, 1952)
ZrO	0(5)	681(7)	1486(9)	Zr^{2+} levels (Moore, 1952)
	5742(5)	8062(1)	8326(3)	
	8838(5)	11,049(9)	13,832(1)	
	18,500(15)			

[a] Degeneracies are given in parentheses.

TABLE V. HEATS OF ATOMIZATION OF THE ELEMENTS AT 298.15°K[a]

Element	$\Delta H^\circ_{298} f$(M, g)		Ref.
Ag	67.9	±0.2	Hultgren *et al.* (1968)
Al	78.7	±0.5	Hultgren *et al.* (1968)
As*	72.3	±3	Wagman *et al.* (1968)
Au	88.0	±0.5	Hildenbrand and Hall (1962), Hultgren *et al.* (1963)
B	136.5	±3	Hultgren *et al.* (1964)
Ba	42.5	±1	Lewis *et al.* (1961)
Be	77.5	±1.5	Hultgren *et al.* (1967)
Bi	50.1	±0.5	Hultgren *et al.* (1968)
Br*	26.74		Wagman *et al.* (1968), from $\frac{1}{2}$ $Br_2(l)$
C	171.3	±0.5	Wagman *et al.* (1968)
Ca	42.6	±0.4	Hultgren *et al.* (1968)
Cd	26.72	±0.15	Hultgren *et al.* (1966)
Ce	101	±3	Hultgren *et al.* (1967)
Cl*	29.08		Wagman *et al.* (1968), from $\frac{1}{2}$ Cl_2(g)
Co	102.4	±1	Hultgren *et al.* (1966)
Cr	95	±1	Hultgren *et al.* (1966)
Cs	18.7	±0.1	Hultgren *et al.* (1963)
Cu	80.7	±0.3	Hultgren *et al.* (1968)
Dy	71	±1	Habermann and Daane (1964), Savage *et al.* (1959)
Er	75.8	±1	Hultgren *et al.* (1966)
Eu	42.4	±0.2	Revision of Hultgren *et al.* (1966) to agree with entropy of Gerstein *et al.* (1967)
F*	18.9		Wagman *et al.* (1968), from $\frac{1}{2}$ F_2(g)
Fe	99.3	±0.3	Hultgren *et al.* (1967)
Ga	65.4	±0.5	Munir and Searcy (1964), from Ga(s)
Gd	95.0	±0.5	Hultgren *et al.* (1967), Hoenig *et al.* (1967)
Ge	89.5	±0.5	Hultgren *et al.* (1965)
H*	52.095		Wagman *et al.* (1968), from $\frac{1}{2}$ H_2(g)
Hf	148	±1	Hultgren *et al.* (1966)
Hg	14.69	±0.03	Hultgren *et al.* (1963), from Hg(*l*)
Ho	71.9	±0.3	Hultgren *et al.* (1966)
I*	25.535		Wagman *et al.* (1968), from $\frac{1}{2}$ I_2(s)
In	58	±1	Wagman *et al.* (1968)
Ir	160	±1	Hampson and Walker (1961), Panish and Reif (1961), Paule and Margrave (1963), Hultgren *et al.* (1963)
K	21.42	±0.05	Hultgren *et al.* (1963)
La	103	±1	Hultgren *et al.* (1967)
Li	38.6	±0.4	Hultgren *et al.* (1963)
Lu	102.2	±0.4	Hultgren *et al.* (1966)
Mg	35.0	±0.3	Hultgren *et al.* (1966)
Mn	67.7	±1	Hultgren *et al.* (1967)
Mo	157.3	±0.5	Hultgren *et al.* (1964)
N*	112.98	±0.06	Douglas (1952), Wagman *et al.* (1968), from $\frac{1}{2}$ N_2(g)
Na	25.85	±0.15	Hultgren *et al.* (1963)
Nb	172.4	±1	Hultgren *et al.* (1966)
Nd	77	±1	Habermann and Daane (1964), White *et al.* (1961),

TABLE V (continued)

Element	$\Delta H^\circ_{298}f$(M, g)		Ref.
			Spedding and Daane (1954), Johnson *et al.* (1956)
Ni	102.8	$\pm$0.5	Hultgren *et al.* (1966)
O*	59.55	$\pm$0.02	Brix and Herzberg (1954), Wagman *et al.* (1968), from $\frac{1}{2}$ O_2(g)
Os	188	$\pm$1.5	Panish and Reif (1962), Carrera *et al.* (1964)
P*	79.4	$\pm$10	Wagman *et al.* (1968), from triclinic red P
Pb	46.62	$\pm$0.3	Hultgren *et al.* (1965), Wagman *et al.* (1968)
Pd	90.0	$\pm$0.5	Hultgren *et al.* (1968)
Pr	85.0	$\pm$0.5	Hultgren *et al.* (1967)
Pt	135.2	$\pm$0.3	Dreger and Margrave (1960), Hampson and Walker (1961)
Pu	84.1	$\pm$4	Hultgren *et al.* (1965)
Rb	19.6	$\pm$0.1	Hultgren *et al.* (1963)
Re	185	$\pm$1.5	Blackburn (1966), Plante and Szwarc (1966), Strassmair and Stark (1967), Sherwood *et al.* (1955)
Rh	133	$\pm$1	Dreger and Margrave (1961), Hampson and Walker (1961), Panish and Reif (1961), Strassmair and Stark (1967)
Ru	155.5	$\pm$1.5	Panish and Reif (1962), Paule and Margrave (1963), Carrera *et al.* (1964)
S*	66.6	$\pm$2	Wagman *et al.* (1968)
Sb*	63.2	$\pm$0.6	Hultgren *et al.* (1968)
Sc	90.3	$\pm$1	Hultgren *et al.* (1966)
Se*	54.3	$\pm$1	Wagman *et al.* (1968)
Si	108.9	$\pm$1	Wagman *et el.* (1968), Hultgren *et al.* (1965)
Sm	49.4	$\pm$0.5	Hultgren *et al.* (1966)
Sn	72.2	$\pm$0.5	Hultgren *et al.* (1963)
Sr	39.1	$\pm$0.5	Stull and Sinke (1956)
Ta	186.9	$\pm$0.6	Hultgren *et al.* (1965)
Tb	92.9	$\pm$0.5	Hultgren *et al.* (1967)
Te*	47.0	$\pm$0.5	Wagman *et al.* (1968), Hultgren *et al.* (1964)
Th	137.5	$\pm$0.5	Hultgren *et al.* (1967)
Ti	112.3	$\pm$0.5	Hultgren *et al.* (1966)
Tl	43.55	$\pm$0.1	Hultgren *et al.* (1963)
Tm	55.5	$\pm$1	Hultgren *et al.* (1967)
U	126	$\pm$3	De Maria *et al.* (1960), Drowart *et al.* (1964b, 1965c, 1967), Storms (1965, 1966), Alexander *et al.* (1969)
V	122.9	$\pm$0.3	Hultgren *et al.* (1966)
W	203.1	$\pm$1	Hultgren *et al.* (1965), Szwarc *et al.* (1965), JANAF (1966a)
Y	101.5	$\pm$0.5	Hultgren *et al.* (1967)
Yb	36.35	$\pm$0.2	Hultgren *et al.* (1966)
Zn	31.25	$\pm$0.1	Hultgren *et al.* (1963), Wagman *et al.* (1968)
Zr	145.5	$\pm$1	Hultgren *et al.* (1967)

[a] Corresponds to ΔH°_{vap} for all except elements marked with *.

TABLE VI. Dissociation Energies of Gaseous Diatomic Oxides

Molecule	D°_{298} (kcal/mole)	D_0° (kcal/mole)	Uncertainty
AgO	51	50	±20
AlO	116	115	±5
ArO	—	<1	—
AsO	115	114	±3
BO	192	191	±5
BaO	131	131	±6
BeO	98	97	±7
BiO	87	86	±3
BrO	56.2	55.3	±0.6
CO	257.26	256.16	±0.77
CaO	84	83	±7
CdO	≤67	≤66	—
CeO	188	187	±6
ClO	64.29	63.33	±0.03
CoO	88	87	±5
CrO	110	109	±10
CsO	67	66	±8
CuO	82	81	±15
DyO	146	145	±10
ErO	147	146	±10
EuO	130	129	±10
FO	56	55	±10
FeO	96	95	±5
GaO	68	67	±15
GdO	162	161	±6
GeO	158.2	156.9	±3
HO	102.34	101.36	±0.3
HeO	—	0	—
HfO	185	184	±10
HgO	—	(<65)	—
HoO	149	148	±10
IO	47	46	±7
InO	≤77	≤76	—
IrO	≤94	≤93	—
KO	57	56	±8
KrO	—	≤1	—
LaO	188	187	±5
LiO	78	77	±6
LuO	159	158	±8
MgO	79	78	±7
MnO	96	95	±8
MoO	115	114	±12
NO	150.8	149.9	±0.2
NaO	73	72	±12
NbO	189	187	±10
NdO	168	167	±8
NeO	—	<1	—

TABLE VI (continued)

Molecule	D°_{298} (kcal/mole)	D_0° (kcal/mole)	Uncertainty
NiO	89	88	±5
O_2	119.11	117.97	±0.04
OsO	<142	<141	—
PO	119.6	118.8	±3
PbO	89.4	88.4	±2
PdO	56	55	±7
PrO	180	179	±8
PtO	83	82	±8
PuO	163	162	±15
RbO	(61)	(60)	±20
RhO	90	89	±15
RuO	115	114	±15
SO	124.69	123.58	±0.03
SbO	89	88	±20
ScO	155	154	±5
SeO	101	100	±15
SiO	192.3	190.9	±2
SmO	134	133	±8
SnO	127	126	±2
SrO	93	92	±6
TaO	183	182	±15
TbO	165	164	±8
TeO	81	80	+10, −1
ThO	192	191	±10
TiO	158	157	±8
TlO	≪75	≪75	—
TmO	122	121	±15
UO	182	181	±8
VO	154	153	±5
WO	156	155	±6
XeO	9	8	±5
YO	162	161	±5
Yb	98	97	±15
ZnO	≤66	≤65	—
ZrO	181	180	±10

TABLE VII. HEATS OF FORMATION OF GASEOUS DIATOMIC OXIDES AT 298.15°K

Molecule	$\Delta H^\circ_{298}f(\text{MO})$	Molecule	$\Delta H^\circ_{298}f(\text{MO})$	Molecule	$\Delta H^\circ_{298}f(\text{MO})$
AgO	+76	HO	+9.31	RbO	+18
AlO	+21.8	HfO	+23	RhO	+103
AsO	+17	HoO	—18	RuO	+100
BO	+4	IO	+38	SO	+1.496
BaO	—29	InO	$\geq$41	SbO	+34
BeO	+39	IrO	$\geq$126	ScO	—5
BiO	+23	KO	+24	SeO	+13
BrO	+30.1	LaO	—26	SiO	—23.8
CO	—26.42	LiO	+20	SmO	—25
CaO	+18	LuO	+3	SnO	+5
CdO	$\geq$19	MgO	+17	SrO	+6
CeO	—17	MnO	+31	TaO	+63
ClO	+24.34	MoO	+102	TbO	—16
CoO	+74	NO	+21.7	TeO	+26
CrO	+45	NaO	+12	ThO	+5
CsO	+11	NbO	+43	TiO	+14
CuO	+58	NdO	—31	TlO	$\gg$28
DyO	—15	NiO	+74	TmO	(—7)
ErO	—12	O_2	0.0	UO	+4
EuO	—29	PO	+19.4	VO	+29
FO	+22	PbO	+16.8	WO	+107
FeO	+63	PdO	+93	YO	—1
GaO	+57	PrO	—35	YbO	(—2)
GdO	—8	PtO	+111	ZnO	$\geq$25
GeO	—9.2	PuO	—19	ZrO	+24

TABLE VIII. HEATS OF FORMATION OF SOME CONDENSED OXIDES AT 298.15°K

Oxide	$\Delta H^{\circ}_{298}f$ (kcal/mole)		Ref.
Ag_2O	−7.3	±0.1	Otto (1966)
$Al_2O_3(\alpha)$	−400.5	±0.3	Wagman *et al.* (1968), JANAF (1964)
As_4O_6 (octahedral)	−314	±2	Wagman *et al.* (1968)
As_4O_6 (monoclinic)	−313	±2	Wagman *et al.* (1968)
As_2O_4	−189.7		Wagman *et el.* (1968)
As_2O_5	−221.05		Wagman *et al.* (1968)
B_2O_3	−304.2	±0.1	Wagman *et al.* (1968)
BaO	−132.1	±2	Flidlider *et al.* (1966)
BaO_2	−151	±6	Brewer (1953)
BeO	−143.1	±4	Cosgrove and Snyder (1953)
Bi_2O_3	−137.16	±1	Wagman *et el.* (1968)
BrO_2	+11.6		Wagman *et al.* (1968)
CaO	−151.8	±0.3	Huber and Holley (1956)
CaO_2	−157.5	±1	Rossini *et al.* (1952)
CdO	−61.7	±0.5	Wagman *et al.* (1968)
Ce_2O_3 (hexagonal)	−429.3	±0.7	Baker and Holley (1968)
CeO_2	−260.2	±0.3	Huber and Holley (1953b)
CoO	−57.1	±0.3	Boyle *et al.* (1954)
Cr_2O_3	−272.7	±0.4	Mah (1954), Novokhatskii and Lenev (1966)
Cs_2O	−81	±2	Rossini *et al.* (1952), Gunn (1967)
CsO_2	−68.3	±0.5	D'Orazio and Wood (1965), Gunn (1967)
Cu_2O	−40.8	±0.3	Mah *et al.* (1967)
CuO	−37.2	±0.2	Mah *et al.* (1967)
Dy_2O_3	−445.8	±0.9	Huber *et al.* (1956a)
Er_2O_3	−453.6	±0.5	Huber *et al.* (1956b)
Eu_2O_3 (monoclinic)	−394	±1	Huber *et al.* (1964a)
$Fe_{0.95}O$	−63.8	±0.4	Lewis *et al.* (1961)
Fe_3O_4	−267.8	±1	Lewis *et al.* (1961)
Fe_2O_3	−196.8	±1	Lewis *et al.* (1961)
Ga_2O_3 (β)	−260.3	±1	Wagman *et al.* (1968)
Gd_2O_3 (B form)	−433.9	±0.9	Huber and Holley (1955)
GeO_2 (hexagonal)	−132.2	±1.2	Bills and Cotton (1964)
HfO_2	−273.6	±0.3	Huber and Holley (1968a)
HgO	−21.70	±0.02	JANAF (1962)
Ho_2O_3	−449.6	±1.2	Huber *et al.* (1957b)
I_2O_5	−37.78		Wagman *et al.* (1968)
In_2O_3	−221.3	±0.4	Wagman *et al.* (1968)
IrO_2	−57.4	±2	Bell *et al.* (1966)
K_2O	−86.9	±1	Rossini *et al.* (1952), Gunn (1967)
KO_2	−68.0	±0.4	Gilles and Margrave (1956), D'Orazio and Wood (1965), Gunn (1967)
La_2O_3 (hexagonal)	−428.6	±0.2	Huber and Holley (1953a), Fitzgibbon *et al.* (1965)
Li_2O	−143.0	±0.5	JANAF (1964)
Li_2O_2	−152	±2	Rossini *et al.* (1952)

TABLE VIII (continued)

Oxide	$\Delta H^\circ_{298}f$ (kcal/mole)		Ref.
Lu_2O_3	–448.9	±1.8	Huber *et al.* (1960b)
MgO	–143.8	±0.1	Holley and Huber (1951), Shomate and Huffman (1943), Pankratz and Kelley (1963b)
MgO_2	–148.9	±1	Rossini *et al.* (1952)
MnO	–92.05	±0.1	Lewis *et al.* (1961)
Mn_2O_3	–228.7	±0.3	Otto (1964)
Mn_3O_4	–331.3	±0.3	Mah (1960)
MnO_2	–124.4	±0.2	Mah (1960)
MoO_2	–140.8	±0.2	Mah (1957)
MoO_3	–178.1	±0.1	Mah (1957)
Na_2O	–99.6	±1	Rossini *et al.* (1952), Wagman *et al.* (1968)
Na_2O_2	–122.9	±1	Gilles and Margrave (1956), Wagman *et al.* (1968)
NaO_2	–62.2	±0.7	Gilles and Margrave (1956), Wagman *et al.* (1968)
NbO	–98	±2	Schäfer and Liedmeier (1964), Morozova and Stolyarova (1960)
NbO_2	–190.3	±0.5	Mah (1958), Morozova and Stolyarova (1960), Kusenko and Gel'd (1960a)
Nb_2O_5	–454.7	±1	Humphrey (1954), Kusenko and Gel'd (1960b), Morozova and Stolyarova (1960), Kornilov *et al.* (1962)
Nd_2O_3 (hexagonal)	–432.1	±0.2	Fitzgibbon *et al.* (1968)
NiO	–57.3	±0.1	Lewis *et al.* (1961), Coughlin (1954), Boyle *et al.* (1954), Hahn and Muan (1961)
NpO_2	–256.7	±0.6	Huber and Holley (1968b)
P_4O_6	–375	±5	Wagman *et al.* (1968) from triclinic red P
P_4O_{10} (hexagonal)	–696.4	±5	Wagman *et al.* (1968) from triclinic red P
PbO(α), red	–52.34	±0.4	Wagman *et al.* (1968)
PbO(β), yellow	–51.94	±0.3	Wagman *et al.* (1968)
Pb_3O_4	–171.7	±0.5	Wagman *et al.* (1968)
PbO_2	–66.3		Wagman *et al.* (1968)
PdO	–27.6	±0.8	Goldberg and Hepler (1968)
$PrO_{1.5}$ (hexagonal)	–217.9	±0.8	Stubblefield *et al.* (1956)
$PrO_{1.5}$ (cubic)	–218.4	±0.8	Stubblefield *et al.* (1956)
$PrO_{1.703}$	–223.5	±0.8	Stubblefield *et al.* (1956)
$PrO_{1.833}$	–227.6	±0.8	Stubblefield *et al.* (1956)
PuO_2	–252.9	±0.4	Oetting (1967)
Rb_2O	–81	±2	Rossini *et al.* (1952), Gunn (1967)
RbO_2	–66.7	±1	D'Orazio and Wood (1965), Gunn (1967)

TABLE VIII (continued)

Oxide	$\Delta H^\circ_{298}f$ (kcal/mole)		Ref.
ReO_2	—101.3	±4	Cobble *et al.* (1953)
ReO_3	—146	±3	Cobble *et al.* (1953)
Re_2O_7	—295.9	±2	Cobble *et al.* (1953)
RuO_2	—72	±1	Shchukarev and Ryabov (1960), Bell and Tagami (1963), Schäfer *et al.* (1963)
SO_2 (*l*)	—76.6	±0.1	Wagman *et al.* (1968)
Sb_4O_6 (cubic)	—344.3	±3	Wagman *et al.* (1968)
Sb_4O_6 (orthohombic)	—338.7	±1.4	Wagman *et al.* (1968)
Sb_2O_4	—216.9	±1.1	Wagman *et al.* (1968)
Sb_2O_5	—232.3		Wagman *et al.* (1968)
Sc_2O_3	—456.2	±0.5	Huber *et al.* (1963)
SeO_2	—53.86		Wagman *et al.* (1968)
Se_2O_5	—97.6		Wagman *et al.* (1968)
SeO_3	—39.9		Wagman *et al.* (1968)
SiO_2 (α crist.)	—217.37		Wagman *et al.* (1968)
SiO_2 (α quartz)	—217.72		Wagman *et al.* (1968)
SiO_2 (α tridymite)	—217.27		Wagman *et al.* (1968)
Sm_2O_3	—433.9	±0.5	Huber *et al.* (1955)
SnO	—68.35	±0.2	Coughlin (1954)
SnO_2	—138.8	±0.1	Coughlin (1954)
SrO	—140.5	±0.5	Flidlider *et al.* (1966)
SrO_2	—153	±4	Brewer (1953)
Ta_2O_5	—488.8	±0.5	Lewis *et al.* (1961)
$TbO_{1.5}$ (cubic)	—222.9	±0.9	Fitzgibbon and Holley (1968)
$TbO_{1.71}$	—227.6	±0.9	Fitzgibbon and Holley (1968)
TbO_2	—232.2	±0.7	Fitzgibbon and Holley (1968)
TcO_2	—103.7	±3	Cobble *et al.* (1953)
TcO_3	—129	±3	Cobble *et al.* (1953)
Tc_2O_7	—266	±3	Cobble *et al.* (1953)
TeO_2	—77.1	±0.7	Wagman *et al.* (1968)
ThO_2	—293.2	±0.4	Huber *et al.* (1952)
TiO	—124.2	±1	JANAF (1967)
Ti_2O_3	—363.5	±2	JANAF (1967)
Ti_3O_5 (β)	—585	±3	JANAF (1967)
TiO_2	—225.8	±1	JANAF (1967)
Tl_2O	—40.4	±1.4	Cubicciotti (1969)
Tl_2O_3	—94.3	±0.8	Cubicciotti (1968)
Tm_2O_3	—451.4	±1.4	Huber *et al.* (1960a)
UO_2	—259.0	±0.6	Rand and Kubaschewski (1963)
VO	—103.2	±0.3	Mah and Kelley (1961)
V_2O_3	—291.3	±0.4	Mah and Kelley (1961)
VO_2	—170.6	±0.2	Mah and Kelley (1961)
V_2O_5	—370.6	±0.5	Mah and Kelley (1961)
WO_2	—140.9	±0.2	Mah (1959)

TABLE VIII (continued)

Oxide	$\Delta H^\circ_{298}f$ (kcal/mole)		Ref.
WO_3	−201.5	±0.2	Mah (1959)
Y_2O_3	−455.5	±0.5	Huber *et al.* (1957a)
Yb_2O_3	−433.7	±0.5	Huber *et al.* (1956a)
ZnO	−83.24	±0.1	Wagman *et al.* (1968)
ZrO_2	−263.0	±0.5	Huber *et al.* (1964b), Kornilov *et al.* (1967)

TABLE IX. Free Energy Functions of Gaseous Diatomic Oxides

T(°K)	AgO		AlO		AsO		AuO		BO		BaO	
	$-\frac{(F^\circ-H^\circ_0)^a}{T}$ (elec)	$-\frac{(F^\circ-H^\circ_{298})^a}{T}$	$-\frac{(F^\circ-H^\circ_0)^a}{T}$ (elec)	$-\frac{(F^\circ-H^\circ_{298})^a}{T}$	$-\frac{(F^\circ-H^\circ_0)^a}{T}$ (elec)	$-\frac{(F^\circ-H^\circ_{298})^a}{T}$	$-\frac{(F^\circ-H^\circ_0)^a}{T}$ (elec)	$-\frac{(F^\circ-H^\circ_{298})^a}{T}$	$-\frac{(F^\circ-H^\circ_0)^a}{T}$ (elec)	$-\frac{(F^\circ-H^\circ_{298})^a}{T}$	$-\frac{(F^\circ-H^\circ_0)^a}{T}$ (elec)	$-\frac{(F^\circ-H^\circ_{298})^a}{T}$
298	3.56	59.58	1.38	52.16	1.39	55.04	3.56	61.66	1.38	48.62	0.	56.24
400	3.56	59.91	1.38	52.45	1.43	55.36	3.56	61.99	1.38	48.90	0.	56.56
500	3.56	60.54	1.38	53.02	1.48	55.96	3.56	62.63	1.38	49.42	0.	57.16
600	3.56	61.24	1.38	53.67	1.54	56.67	3.56	63.34	1.38	50.01	0.	57.84
700	3.56	61.96	1.38	54.33	1.61	57.39	3.56	64.06	1.38	50.61	0.	58.53
800	3.56	62.65	1.38	54.97	1.67	58.09	3.56	64.76	1.38	51.19	.00	59.21
900	3.56	63.31	1.38	55.59	1.73	58.77	3.56	65.43	1.38	51.76	.00	59.86
1000	3.56	63.94	1.38	56.18	1.79	59.42	3.56	66.06	1.38	52.30	.00	60.47
1100	3.56	64.54	1.38	56.75	1.84	60.03	3.57	66.66	1.38	52.81	.00	61.05
1200	3.57	65.10	1.38	57.28	1.89	60.61	3.57	67.23	1.38	53.30	.00	61.61
1300	3.57	65.64	1.38	57.79	1.93	61.17	3.58	67.77	1.38	53.77	.00	62.13
1400	3.57	66.15	1.38	58.28	1.97	61.69	3.58	68.29	1.38	54.22	.00	62.63
1500	3.58	66.63	1.38	58.74	2.01	62.19	3.59	68.78	1.38	54.65	.00	63.10
1600	3.58	67.09	1.38	59.18	2.04	62.66	3.61	69.25	1.38	55.06	.00	63.55
1700	3.59	67.54	1.38	59.60	2.07	63.11	3.62	69.71	1.38	55.45	.00	63.99
1800	3.59	67.96	1.38	60.01	2.10	63.55	3.63	70.14	1.38	55.83	.00	64.40
1900	3.60	68.37	1.38	60.39	2.13	63.96	3.65	70.56	1.38	56.19	.00	64.79
2000	3.61	68.76	1.39	60.77	2.15	64.35	3.67	70.96	1.38	56.54	.00	65.17
2100	3.62	69.13	1.39	61.13	2.18	64.73	3.69	71.35	1.38	56.88	.00	65.54
2200	3.62	69.49	1.39	61.47	2.20	65.10	3.71	71.73	1.38	57.20	.00	65.89
2300	3.63	69.84	1.39	61.81	2.22	65.45	3.73	72.09	1.38	57.52	.00	66.22
2400	3.64	70.18	1.40	62.13	2.24	65.79	3.76	72.44	1.38	57.82	.00	66.55
2500	3.65	70.51	1.40	62.45	2.25	66.12	3.78	72.79	1.38	58.12	.00	66.86
2600	3.66	70.82	1.41	62.75	2.27	66.43	3.81	73.12	1.38	58.40	.00	67.17
2700	3.67	71.13	1.42	63.05	2.28	66.73	3.83	73.44	1.38	58.68	.00	67.46
2800	3.68	71.43	1.42	63.34	2.30	67.03	3.86	73.76	1.38	58.95	.00	67.75
2900	3.69	71.72	1.43	63.62	2.31	67.32	3.89	74.06	1.38	59.21	.00	68.02
3000	3.70	72.00	1.44	63.89	2.33	67.59	3.92	74.36	1.38	59.46	.00	68.29
$H^\circ_{298} - H^\circ_0$ [b] =	2222		2100		2122		2256		2074		2153	
$H^\circ_{2000} - H^\circ_0$ [b] =	17544		16760		17621		18083		15800		17009	

T(°K)	BeO $-(F^\circ-H^\circ_0)^a/T$ (elec)	BeO $-(F^\circ-H^\circ_{298})^a/T$	BiO $-(F^\circ-H^\circ_0)^a/T$ (elec)	BiO $-(F^\circ-H^\circ_{298})^a/T$	BrO $-(F^\circ-H^\circ_0)^a/T$ (elec)	BrO $-(F^\circ-H^\circ_{298})^a/T$	CO $-(F^\circ-H^\circ_0)^a/T$ (elec)	CO $-(F^\circ-H^\circ_{298})^a/T$	CaO $-(F^\circ-H^\circ_0)^a/T$ (elec)	CaO $-(F^\circ-H^\circ_{298})^a/T$	CdO $-(F^\circ-H^\circ_0)^a/T$ (elec)	CdO $-(F^\circ-H^\circ_{298})^a/T$
298	.09	47.74	1.38	58.80	1.38	55.39	0.	47.21	3.87	56.35	0.	55.67
400	.30	48.11	1.38	59.11	1.38	55.70	0.	47.49	3.87	56.65	0.	55.99
500	.58	48.85	1.38	59.71	1.38	56.30	0.	48.00	3.87	57.25	0.	56.61
600	.86	49.70	1.38	60.39	1.38	56.98	0.	48.59	3.87	57.93	0.	57.31
700	1.13	50.56	1.38	61.08	1.38	57.67	0.	49.18	3.87	58.61	0.	58.01
800	1.37	51.38	1.38	61.76	1.38	58.34	0.	49.76	3.87	59.29	0.	58.70
900	1.58	52.15	1.38	62.40	1.38	58.99	0.	50.31	3.88	59.93	0.	59.36
1000	1.76	52.87	1.38	63.02	1.39	59.61	0.	50.84	3.88	60.55	0.	59.98
1100	1.92	53.55	1.38	63.60	1.39	60.19	0.	51.35	3.89	61.14	0.	60.57
1200	2.05	54.19	1.38	64.15	1.40	60.75	0.	51.83	3.90	61.70	0.	61.13
1300	2.17	54.78	1.38	64.67	1.41	61.29	0.	52.29	3.91	62.23	0.	61.66
1400	2.28	55.35	1.38	65.17	1.42	61.79	0.	52.73	3.93	62.74	0.	62.16
1500	2.37	55.88	1.38	65.64	1.43	62.28	0.	53.15	3.94	63.22	0.	62.64
1600	2.46	56.38	1.38	66.10	1.45	62.74	0.	53.55	3.95	63.69	0.	63.10
1700	2.53	56.85	1.38	66.53	1.46	63.19	0.	53.94	3.97	64.13	0.	63.53
1800	2.60	57.31	1.38	66.94	1.48	63.62	0.	54.31	3.99	64.56	0.	63.95
1900	2.66	57.74	1.38	67.34	1.50	64.03	0.	54.67	4.00	64.97	0.	64.35
2000	2.72	58.15	1.38	67.72	1.51	64.42	0.	55.01	4.02	65.36	0.	64.73
2100	2.77	58.54	1.39	68.08	1.53	64.80	0.	55.34	4.04	65.74	0.	65.09
2200	2.82	58.92	1.39	68.44	1.55	65.17	0.	55.66	4.05	66.11	0.	65.45
2300	2.87	59.29	1.39	68.78	1.57	65.52	0.	55.97	4.07	66.46	0.	65.79
2400	2.91	59.64	1.39	69.10	1.58	65.87	0.	56.27	4.09	66.80	0.	66.11
2500	2.95	59.98	1.40	69.42	1.60	66.20	0.	56.56	4.10	67.13	0.	66.43
2600	2.99	60.30	1.40	69.73	1.62	66.52	0.	56.84	4.12	67.45	0.	66.74
2700	3.02	60.62	1.41	70.03	1.64	66.83	0.	57.11	4.13	67.76	0.	67.03
2800	3.06	60.92	1.41	70.32	1.66	67.13	0.	57.38	4.15	68.06	0.	67.32
2900	3.09	61.22	1.42	70.60	1.67	67.43	0.	57.64	4.17	68.35	0.	67.60
3000	3.12	61.50	1.42	70.87	1.69	67.71	0.	57.89	4.18	68.64	0.	67.86
$H^\circ_{298}-H^\circ_0$ [b] =	2209		2149		2145		2074		2139		2193	
$H^\circ_{2000}-H^\circ_0$ [b] =	18349		17061		17663		15578		17603		17150	

[a] cal/deg mole [b] cal/mole

TABLE IX (continued)

$T(°K)$	CeO $-(F°-H_0°)^a/T$ (elec)	CeO $-(F°-H_{298}°)^a/T$	ClO $-(F°-H_0°)^a/T$ (elec)	ClO $-(F°-H_{298}°)^a/T$	CoO $-(F°-H_0°)^a/T$ (elec)	CoO $-(F°-H_{298}°)^a/T$	CrO $-(F°-H_0°)^a/T$ (elec)	CrO $-(F°-H_{298}°)^a/T$	CsO $-(F°-H_0°)^a/T$ (elec)	CsO $-(F°-H_{298}°)^a/T$	CuO $-(F°-H_0°)^a/T$ (elec)	CuO $-(F°-H_{298}°)^a/T$
298	3.22	59.20	1.41	52.97	4.60	57.75	3.91	58.21	2.89	60.23	3.56	57.21
400	3.29	59.54	1.46	53.29	4.66	58.08	4.39	58.55	2.97	60.58	3.56	57.53
500	3.39	60.21	1.53	53.92	4.73	58.71	4.71	59.19	3.04	61.23	3.56	58.14
600	3.53	61.00	1.60	54.65	4.82	59.45	4.95	59.91	3.10	61.97	3.57	58.84
700	3.69	61.82	1.68	55.38	4.91	60.20	5.13	60.64	3.15	62.71	3.58	59.54
800	3.86	62.64	1.75	56.10	4.99	60.94	5.27	61.34	3.19	63.43	3.59	60.24
900	4.02	63.43	1.81	56.79	5.08	61.65	5.38	62.01	3.22	64.11	3.61	60.90
1000	4.18	64.19	1.87	57.45	5.16	62.33	5.48	62.64	3.24	64.76	3.63	61.54
1100	4.34	64.92	1.92	58.07	5.23	62.97	5.55	63.24	3.27	65.37	3.65	62.14
1200	4.48	65.60	1.97	58.65	5.30	63.58	5.62	63.81	3.29	65.95	3.67	62.72
1300	4.62	66.25	2.01	59.21	5.36	64.15	5.67	64.34	3.31	66.49	3.69	63.27
1400	4.75	66.87	2.05	59.73	5.42	64.70	5.72	64.85	3.32	67.01	3.71	63.79
1500	4.86	67.45	2.09	60.23	5.48	65.22	5.76	65.33	3.34	67.50	3.73	64.29
1600	4.98	68.01	2.12	60.71	5.53	65.71	5.80	65.79	3.35	67.97	3.76	64.77
1700	5.08	68.54	2.15	61.16	5.57	66.18	5.84	66.23	3.36	68.41	3.78	65.22
1800	5.18	69.04	2.18	61.60	5.61	66.63	5.87	66.65	3.37	68.84	3.80	65.65
1900	5.27	69.52	2.20	62.01	5.65	67.06	5.89	67.05	3.38	69.24	3.82	66.07
2000	5.35	69.98	2.22	62.41	5.69	67.47	5.92	67.44	3.39	69.63	3.84	66.47
2100	5.43	70.42	2.24	62.79	5.72	67.87	5.94	67.80	3.39	70.00	3.86	66.85
2200	5.51	70.84	2.26	63.15	5.76	68.24	5.96	68.16	3.40	70.36	3.88	67.22
2300	5.58	71.24	2.28	63.50	5.79	68.61	5.98	68.50	3.41	70.71	3.89	67.58
2400	5.65	71.63	2.30	63.84	5.81	68.96	5.99	68.83	3.41	71.04	3.91	67.92
2500	5.71	72.01	2.31	64.17	5.84	69.29	6.01	69.15	3.42	71.36	3.93	68.25
2600	5.77	72.37	2.33	64.48	5.87	69.62	6.02	69.45	3.42	71.67	3.94	68.57
2700	5.83	72.72	2.34	64.79	5.89	69.93	6.04	69.75	3.43	71.97	3.96	68.88
2800	5.88	73.06	2.36	65.08	5.91	70.24	6.05	70.04	3.43	72.26	3.97	69.18
2900	5.94	73.38	2.37	65.37	5.93	70.53	6.06	70.32	3.44	72.54	3.99	69.47
3000	5.99	73.70	2.38	65.64	5.95	70.82	6.07	70.59	3.44	72.82	4.00	69.76
$H_{298}° - H_0°$ [b] =	2153		2149		2151		2623		2296		2167	
$H_{2000}° - H_0°$ [b] =	20115		17662		18222		17673		17539		17835	

T(°K)	FO $-(F^\circ-H_0^\circ)^a/T$ (elec)	FO $-(F^\circ-H_{298}^\circ)^a/T$	FeO $-(F^\circ-H_0^\circ)^a/T$ (elec)	FeO $-(F^\circ-H_{298}^\circ)^a/T$	GaO $-(F^\circ-H_0^\circ)^a/T$ (elec)	GaO $-(F^\circ-H_{298}^\circ)^a/T$	GdO $-(F^\circ-H_0^\circ)^a/T$ (elec)	GdO $-(F^\circ-H_{298}^\circ)^a/T$	GeO $-(F^\circ-H_0^\circ)^a/T$ (elec)	GeO $-(F^\circ-H_{298}^\circ)^a/T$	OH $-(F^\circ-H_0^\circ)^a/T$ (elec)	OH $-(F^\circ-H_{298}^\circ)^a/T$
298	1.64	51.13	5.35	59.19	1.38	55.16	3.91	61.08	0.	53.48	2.20	43.87
400	1.79	51.45	5.43	59.51	1.38	55.47	4.23	61.44	0.	53.77	2.32	44.15
500	1.92	52.06	5.51	60.12	1.38	56.06	4.50	62.14	0.	54.34	2.40	44.68
600	2.02	52.75	5.57	60.82	1.38	56.73	4.73	62.94	0.	54.98	2.45	45.26
700	2.10	53.45	5.63	61.53	1.38	57.41	4.93	63.74	0.	55.64	2.49	45.85
800	2.16	54.13	5.69	62.22	1.38	58.08	5.11	64.52	0.	56.29	2.52	46.42
900	2.21	54.77	5.73	62.89	1.38	58.72	5.26	65.26	0.	56.91	2.55	46.96
1000	2.26	55.39	5.78	63.52	1.39	59.34	5.40	65.97	0.	57.50	2.57	47.47
1100	2.30	55.97	5.81	64.12	1.40	59.92	5.53	66.64	0.	58.06	2.58	47.96
1200	2.33	56.53	5.85	64.69	1.41	60.48	5.64	67.27	0.	58.59	2.60	48.42
1300	2.36	57.05	5.88	65.23	1.42	61.01	5.74	67.86	0.	59.10	2.61	48.86
1400	2.38	57.55	5.90	65.74	1.44	61.53	5.83	68.43	0.	59.58	2.62	49.28
1500	2.41	58.03	5.93	66.23	1.46	62.02	5.91	68.96	0.	60.05	2.63	49.68
1600	2.43	58.48	5.95	66.70	1.48	62.49	5.99	69.47	0.	60.49	2.63	50.06
1700	2.44	58.91	5.97	67.14	1.51	62.94	6.05	69.95	0.	60.91	2.64	50.43
1800	2.46	59.32	5.99	67.56	1.53	63.38	6.12	70.42	0.	61.31	2.65	50.78
1900	2.47	59.72	6.01	67.96	1.56	63.80	6.18	70.86	0.	61.70	2.65	51.12
2000	2.49	60.10	6.02	68.35	1.59	64.20	6.23	71.28	0.	62.07	2.66	51.45
2100	2.50	60.46	6.04	68.72	1.62	64.59	6.28	71.68	0.	62.42	2.66	51.76
2200	2.51	60.81	6.05	69.08	1.65	64.97	6.33	72.07	0.	62.77	2.67	52.07
2300	2.52	61.15	6.06	69.43	1.68	65.34	6.38	72.44	0.	63.10	2.67	52.36
2400	2.53	61.47	6.08	69.76	1.71	65.69	6.42	72.80	0.	63.42	2.67	52.64
2500	2.54	61.79	6.09	70.08	1.74	66.03	6.46	73.15	0.	63.73	2.68	52.92
2600	2.55	62.09	6.10	70.39	1.77	66.37	6.50	73.48	0.	64.03	2.68	53.19
2700	2.55	62.38	6.11	70.69	1.80	66.69	6.53	73.81	0.	64.32	2.68	53.45
2800	2.56	62.67	6.12	70.98	1.83	67.00	6.57	74.12	0.	64.60	2.68	53.70
2900	2.57	62.94	6.12	71.26	1.86	67.31	6.60	74.42	0.	64.87	2.69	53.94
3000	2.57	63.21	6.13	71.53	1.89	67.60	6.63	74.71	0.	65.13	2.69	54.18
$H_{298}^\circ - H_0^\circ$ [b] =	2238		2190		2132		2417		2099		2208	
$H_{2000}^\circ - H_0^\circ$ [b] =	17089		17387		18059		18922		16650		15020	

[a] cal/deg mole [b] cal/mole

TABLE IX (continued)

T(°K)	HfO		HgO		IO		InO		IrO		KO	
	$-(F^\circ-H_0^\circ)^a/T$ (elec)	$-(F^\circ-H_{298}^\circ)^a/T$	$-(F^\circ-H_0^\circ)^a/T$ (elec)	$-(F^\circ-H_{298}^\circ)^a/T$	$-(F^\circ-H_0^\circ)^a/T$ (elec)	$-(F^\circ-H_{298}^\circ)^a/T$	$-(F^\circ-H_0^\circ)^a/T$ (elec)	$-(F^\circ-H_{298}^\circ)^a/T$	$-(F^\circ-H_0^\circ)^a/T$ (elec)	$-(F^\circ-H_{298}^\circ)^a/T$	$-(F^\circ-H_0^\circ)^a/T$ (elec)	$-(F^\circ-H_{298}^\circ)^a/T$
298	3.20	59.53	0.	57.72	1.38	57.24	1.38	56.83	4.58	61.47	2.89	56.18
400	3.21	59.84	0.	58.04	1.38	57.55	1.38	57.14	4.58	61.77	2.97	56.51
500	3.24	60.42	0.	58.67	1.38	58.15	1.38	57.74	4.58	62.36	3.04	57.14
600	3.28	61.11	0.	59.38	1.38	58.83	1.38	58.42	4.58	63.03	3.10	57.86
700	3.33	61.82	0.	60.09	1.38	59.52	1.38	59.11	4.58	63.71	3.15	58.58
800	3.39	62.53	0.	60.78	1.38	60.20	1.38	59.78	4.58	64.37	3.19	59.28
900	3.46	63.22	0.	61.45	1.38	60.84	1.38	60.43	4.58	65.01	3.22	59.95
1000	3.54	63.88	0.	62.07	1.38	61.46	1.38	61.04	4.59	65.62	3.24	60.58
1100	3.62	64.52	0.	62.67	1.38	62.04	1.38	61.62	4.60	66.21	3.27	61.18
1200	3.69	65.14	0.	63.23	1.38	62.59	1.38	62.17	4.61	66.77	3.29	61.75
1300	3.77	65.72	0.	63.76	1.38	63.12	1.38	62.69	4.62	67.30	3.31	62.28
1400	3.85	66.29	0.	64.27	1.38	63.62	1.38	63.19	4.64	67.81	3.32	62.79
1500	3.93	66.82	0.	64.75	1.38	64.09	1.38	63.66	4.66	68.29	3.34	63.28
1600	4.00	67.34	0.	65.21	1.38	64.54	1.38	64.11	4.68	68.76	3.35	63.74
1700	4.07	67.83	0.	65.65	1.38	64.97	1.38	64.55	4.70	69.21	3.36	64.18
1800	4.14	68.30	0.	66.06	1.38	65.39	1.38	64.96	4.73	69.65	3.37	64.60
1900	4.21	68.76	0.	66.46	1.38	65.78	1.38	65.35	4.75	70.06	3.38	65.00
2000	4.27	69.19	0.	66.85	1.39	66.17	1.38	65.73	4.78	70.47	3.39	65.38
2100	4.33	69.61	0.	67.21	1.39	66.53	1.38	66.09	4.81	70.86	3.39	65.75
2200	4.39	70.01	0.	67.57	1.39	66.88	1.38	66.44	4.84	71.23	3.40	66.11
2300	4.45	70.40	0.	67.91	1.39	67.23	1.38	66.78	4.87	71.60	3.41	66.45
2400	4.51	70.78	0.	68.24	1.40	67.55	1.38	67.10	4.90	71.95	3.41	66.78
2500	4.56	71.14	0.	68.55	1.40	67.87	1.38	67.42	4.93	72.30	3.42	67.10
2600	4.61	71.49	0.	68.86	1.41	68.18	1.38	67.72	4.96	72.63	3.42	67.41
2700	4.66	71.83	0.	69.16	1.41	68.48	1.38	68.02	4.99	72.95	3.43	67.70
2800	4.71	72.16	0.	69.44	1.42	68.77	1.38	68.30	5.02	73.27	3.43	67.99
2900	4.76	72.48	0.	69.72	1.42	69.05	1.38	68.58	5.05	73.58	3.44	68.27
3000	4.80	72.79	0.	69.99	1.43	69.33	1.38	68.84	5.09	73.87	3.44	68.54
$H_{298}^\circ - H_0^{\circ\,b}$ =	2104		2224		2151		2145		2126		2229	
$H_{2000}^\circ - H_0^{\circ\,b}$ =	19192		17238		17091		16973		17978		17318	

T(°K)	LaO $-(F^\circ-H^\circ_0)/T$ (elec)[a]	LaO $-(F^\circ-H^\circ_{298})/T$[a]	LiO $-(F^\circ-H^\circ_0)/T$ (elec)[a]	LiO $-(F^\circ-H^\circ_{298})/T$[a]	MgO $-(F^\circ-H^\circ_0)/T$ (elec)[a]	MgO $-(F^\circ-H^\circ_{298})/T$[a]	MnO $-(F^\circ-H^\circ_0)/T$ (elec)[a]	MnO $-(F^\circ-H^\circ_{298})/T$[a]	MoO $-(F^\circ-H^\circ_0)/T$ (elec)[a]	MoO $-(F^\circ-H^\circ_{298})/T$[a]	NO $-(F^\circ-H^\circ_0)/T$ (elec)[a]	NO $-(F^\circ-H^\circ_{298})/T$[a]
298	2.76	58.65	2.89	50.76	3.87	54.80	3.56	56.45	2.29	58.73	2.25	50.37
400	2.76	58.96	2.97	51.08	3.87	55.11	3.56	56.75	2.87	59.11	2.36	50.65
500	2.78	59.56	3.04	51.71	3.87	55.70	3.56	57.33	3.31	59.83	2.43	51.18
600	2.82	60.26	3.10	52.41	3.87	56.37	3.56	58.00	3.66	60.64	2.48	51.77
700	2.86	60.98	3.15	53.12	3.87	57.05	3.56	58.67	3.94	61.45	2.52	52.38
800	2.91	61.69	3.19	53.81	3.87	57.71	3.56	59.33	4.18	62.23	2.55	52.97
900	2.97	62.38	3.22	54.47	3.87	58.35	3.56	59.96	4.37	62.97	2.57	53.54
1000	3.03	63.05	3.24	55.10	3.87	58.96	3.56	60.56	4.54	63.66	2.59	54.08
1100	3.09	63.68	3.27	55.70	3.87	59.53	3.56	61.14	4.68	64.32	2.60	54.60
1200	3.15	64.28	3.29	56.26	3.88	60.08	3.56	61.68	4.80	64.93	2.61	55.09
1300	3.21	64.86	3.31	56.79	3.88	60.60	3.56	62.19	4.90	65.51	2.62	55.56
1400	3.26	65.40	3.32	57.30	3.88	61.10	3.56	62.69	5.00	66.06	2.63	56.01
1500	3.31	65.92	3.34	57.78	3.89	61.58	3.56	63.15	5.08	66.57	2.64	56.44
1600	3.36	66.41	3.35	58.24	3.89	62.03	3.56	63.60	5.15	67.06	2.65	56.85
1700	3.40	66.89	3.36	58.67	3.90	62.46	3.56	64.02	5.22	67.53	2.65	57.25
1800	3.45	67.34	3.37	59.09	3.90	62.88	3.56	64.43	5.28	67.97	2.66	57.63
1900	3.49	67.77	3.38	59.49	3.91	63.27	3.56	64.82	5.33	68.40	2.66	57.99
2000	3.52	68.18	3.39	59.87	3.91	63.66	3.56	65.20	5.38	68.80	2.67	58.34
2100	3.56	68.58	3.39	60.24	3.92	64.02	3.56	65.56	5.42	69.19	2.67	58.68
2200	3.59	68.96	3.40	60.60	3.93	64.38	3.56	65.90	5.46	69.56	2.68	59.01
2300	3.63	69.33	3.41	60.94	3.93	64.72	3.56	66.24	5.50	69.92	2.68	59.32
2400	3.66	69.68	3.41	61.27	3.94	65.05	3.56	66.56	5.54	70.27	2.68	59.62
2500	3.68	70.02	3.42	61.58	3.95	65.37	3.56	66.87	5.57	70.60	2.69	59.92
2600	3.71	70.35	3.42	61.89	3.95	65.68	3.56	67.17	5.60	70.92	2.69	60.20
2700	3.74	70.67	3.43	62.19	3.96	65.98	3.56	67.47	5.63	71.23	2.69	60.48
2800	3.76	70.97	3.43	62.47	3.97	66.27	3.56	67.75	5.65	71.53	2.69	60.75
2900	3.78	71.27	3.44	62.75	3.97	66.55	3.56	68.02	5.68	71.82	2.70	61.01
3000	3.80	71.56	3.44	63.02	3.98	66.82	3.56	68.29	5.70	72.10	2.70	61.26
$H^\circ_{298} - H^\circ_0$ [b] =	2125		2211		2128		2118		2673		2200	
$H^\circ_{2000} - H^\circ_0$ [b] =	18315		17229		17121		16815		18539		15957	

[a] cal/deg mole [b] cal/mole

TABLE IX (continued)

T(°K)	NaO		NbO		NiO		OsO		PO		PbO	
	$-(F^\circ-H^\circ_0)^a/T$ (elec)	$-(F^\circ-H^\circ_{298})^a/T$	$-(F^\circ-H^\circ_0)^a/T$ (elec)	$-(F^\circ-H^\circ_{298})^a/T$	$-(F^\circ-H^\circ_0)^a/T$ (elec)	$-(F^\circ-H^\circ_{298})^a/T$	$-(F^\circ-H^\circ_0)^a/T$ (elec)	$-(F^\circ-H^\circ_{298})^a/T$	$-(F^\circ-H^\circ_0)^a/T$ (elec)	$-(F^\circ-H^\circ_{298})^a/T$	$-(F^\circ-H^\circ_0)^a/T$ (elec)	$-(F^\circ-H^\circ_{298})^a/T$
298	2.89	53.98	3.00	57.94	4.37	57.63	3.87	60.73	1.96	53.21	0.	57.34
400	2.97	54.31	3.22	58.30	4.38	57.95	3.87	61.03	2.11	53.51	0.	57.65
500	3.04	54.93	3.45	59.00	4.40	58.58	3.87	61.62	2.22	54.08	0.	58.24
600	3.10	55.63	3.66	59.80	4.43	59.30	3.87	62.29	2.29	54.72	0.	58.92
700	3.15	56.34	3.86	60.61	4.47	60.03	3.87	62.97	2.35	55.38	0.	59.61
800	3.19	57.03	4.04	61.41	4.51	60.76	3.87	63.63	2.39	56.02	0.	60.28
900	3.22	57.69	4.20	62.16	4.56	61.46	3.87	64.27	2.43	56.63	0.	60.92
1000	3.24	58.32	4.35	62.88	4.61	62.12	3.87	64.87	2.46	57.22	0.	61.53
1100	3.27	58.91	4.49	63.56	4.66	62.76	3.87	65.45	2.48	57.77	0.	62.11
1200	3.29	59.47	4.61	64.20	4.71	63.36	3.87	65.99	2.51	58.30	0.	62.66
1300	3.31	60.00	4.72	64.81	4.76	63.94	3.87	66.51	2.52	58.80	0.	63.18
1400	3.32	60.51	4.81	65.38	4.80	64.48	3.87	67.00	2.54	59.28	0.	63.68
1500	3.34	60.99	4.90	65.92	4.85	65.00	3.87	67.47	2.55	59.74	0.	64.15
1600	3.35	61.45	4.99	66.43	4.89	65.50	3.87	67.92	2.56	60.18	0.	64.60
1700	3.36	61.88	5.06	66.92	4.93	65.97	3.87	68.35	2.58	60.59	0.	65.03
1800	3.37	62.30	5.13	67.39	4.97	66.42	3.87	68.76	2.58	60.99	0.	65.44
1900	3.38	62.70	5.19	67.83	5.00	66.85	3.87	69.15	2.59	61.38	0.	65.83
2000	3.39	63.08	5.25	68.26	5.04	67.27	3.87	69.52	2.60	61.75	0.	66.21
2100	3.39	63.45	5.30	68.66	5.07	67.67	3.87	69.89	2.61	62.10	0.	66.58
2200	3.40	63.81	5.35	69.05	5.10	68.05	3.87	70.23	2.61	62.44	0.	66.92
2300	3.41	64.15	5.40	69.43	5.13	68.41	3.87	70.57	2.62	62.77	0.	67.26
2400	3.41	64.47	5.44	69.79	5.15	68.77	3.87	70.89	2.63	63.09	0.	67.59
2500	3.42	64.79	5.48	70.13	5.18	69.11	3.87	71.21	2.63	63.39	0.	67.90
2600	3.42	65.10	5.52	70.47	5.20	69.44	3.87	71.51	2.64	63.69	0.	68.20
2700	3.43	65.39	5.56	70.79	5.23	69.75	3.87	71.80	2.64	63.98	0.	68.50
2800	3.43	65.68	5.59	71.10	5.25	70.06	3.87	72.09	2.64	64.26	0.	68.78
2900	3.44	65.96	5.62	71.40	5.27	70.36	3.87	72.36	2.65	64.53	0.	69.06
3000	3.44	66.23	5.65	71.69	5.29	70.65	3.88	72.63	2.65	64.79	0.	69.32
$H^\circ_{298}-H^\circ_0$ [b] =	2208		2282		2173		2126		2246		2141	
$H^\circ_{2000}-H^\circ_0$ [b] =	17213		18881		18385		16873		16687		16950	

T(°K)	PdO $-(F^\circ-H^\circ_0)^a/T$ (elec)	PdO $-(F^\circ-H^\circ_{298})^a/T$	PtO $-(F^\circ-H^\circ_0)^a/T$ (elec)	PtO $-(F^\circ-H^\circ_{298})^a/T$	RbO $-(F^\circ-H^\circ_0)^a/T$ (elec)	RbO $-(F^\circ-H^\circ_{298})^a/T$	ReO $-(F^\circ-H^\circ_0)^a/T$ (elec)	ReO $-(F^\circ-H^\circ_{298})^a/T$	RhO $-(F^\circ-H^\circ_0)^a/T$ (elec)	RhO $-(F^\circ-H^\circ_{298})^a/T$	RuO $-(F^\circ-H^\circ_0)^a/T$ (elec)	RuO $-(F^\circ-H^\circ_{298})^a/T$
298	4.37	59.38	4.37	61.32	2.89	58.71	1.38	58.16	4.58	59.50	4.37	59.06
400	4.37	59.68	4.37	61.63	2.97	59.05	1.39	58.47	4.58	59.80	4.39	59.37
500	4.37	60.27	4.37	62.22	3.04	59.69	1.43	59.09	4.58	60.39	4.43	59.98
600	4.37	60.93	4.37	62.89	3.10	60.42	1.48	59.82	4.59	61.06	4.48	60.69
700	4.37	61.61	4.37	63.57	3.15	61.16	1.55	60.57	4.60	61.74	4.54	61.41
800	4.37	62.27	4.38	64.23	3.19	61.87	1.64	61.32	4.61	62.42	4.60	62.13
900	4.38	62.91	4.38	64.88	3.22	62.54	1.74	62.05	4.63	63.07	4.67	62.83
1000	4.38	63.52	4.39	65.49	3.24	63.19	1.84	62.76	4.66	63.70	4.73	63.49
1100	4.39	64.10	4.41	66.08	3.27	63.79	1.95	63.45	4.68	64.30	4.80	64.13
1200	4.40	64.66	4.42	66.64	3.29	64.36	2.06	64.11	4.71	64.88	4.86	64.73
1300	4.42	65.19	4.44	67.18	3.31	64.91	2.17	64.74	4.75	65.43	4.92	65.31
1400	4.43	65.70	4.47	67.70	3.32	65.42	2.28	65.34	4.78	65.95	4.98	65.85
1500	4.45	66.18	4.49	68.19	3.34	65.91	2.40	65.93	4.82	66.46	5.04	66.38
1600	4.46	66.65	4.52	68.67	3.35	66.37	2.51	66.49	4.85	66.94	5.09	66.87
1700	4.48	67.09	4.55	69.13	3.36	66.82	2.62	67.03	4.89	67.40	5.14	67.34
1800	4.50	67.52	4.58	69.57	3.37	67.24	2.73	67.54	4.93	67.85	5.18	67.80
1900	4.53	67.93	4.62	70.00	3.38	67.64	2.84	68.04	4.96	68.28	5.22	68.23
2000	4.55	68.33	4.65	70.41	3.39	68.03	2.94	68.53	5.00	68.69	5.26	68.64
2100	4.57	68.71	4.69	70.81	3.39	68.40	3.04	68.99	5.04	69.08	5.30	69.04
2200	4.59	69.08	4.73	71.19	3.40	68.76	3.14	69.44	5.07	69.47	5.34	69.42
2300	4.62	69.44	4.77	71.57	3.41	69.11	3.24	69.87	5.10	69.83	5.37	69.79
2400	4.64	69.79	4.81	71.93	3.41	69.44	3.34	70.29	5.14	70.19	5.40	70.14
2500	4.66	70.12	4.85	72.29	3.42	69.76	3.43	70.70	5.17	70.54	5.43	70.48
2600	4.69	70.45	4.90	72.63	3.42	70.07	3.52	71.09	5.20	70.87	5.46	70.81
2700	4.71	70.76	4.94	72.97	3.43	70.37	3.61	71.47	5.23	71.19	5.49	71.13
2800	4.73	71.07	4.98	73.29	3.43	70.66	3.70	71.84	5.26	71.51	5.51	71.44
2900	4.75	71.37	5.02	73.61	3.44	70.94	3.78	72.20	5.29	71.81	5.54	71.74
3000	4.78	71.65	5.07	73.92	3.44	71.21	3.87	72.55	5.32	72.11	5.56	72.02
$H^\circ_{298} - H^\circ_0$ [b] =	2121		2126		2265		2136		2121		2124	
$H^\circ_{2000} - H^\circ_0$ [b] =	17725		18348		17452		21056		18270		18350	

[a] cal/deg mole [b] cal/mole

TABLE IX (continued)

T(°K)	SO $-(F^\circ-H_0^\circ)^a/T$ (elec)	SO $-(F^\circ-H_{298}^\circ)^a/T$	SbO $-(F^\circ-H_0^\circ)^a/T$ (elec)	SbO $-(F^\circ-H_{298}^\circ)^a/T$	ScO $-(F^\circ-H_0^\circ)^a/T$ (elec)	ScO $-(F^\circ-H_{298}^\circ)^a/T$	SeO $-(F^\circ-H_0^\circ)^a/T$ (elec)	SeO $-(F^\circ-H_{298}^\circ)^a/T$	SiO $-(F^\circ-H_0^\circ)^a/T$ (elec)	SiO $-(F^\circ-H_{298}^\circ)^a/T$	SnO $-(F^\circ-H_0^\circ)^a/T$ (elec)	SnO $-(F^\circ-H_{298}^\circ)^a/T$
298	2.18	53.02	1.38	56.93	3.66	56.62	1.25	55.83	0.	50.54	0.	55.44
400	2.18	53.30	1.38	57.23	3.85	56.92	1.45	56.14	0.	50.82	0.	55.75
500	2.18	53.85	1.38	57.82	3.98	57.51	1.58	56.73	0.	51.37	0.	56.33
600	2.18	54.48	1.39	58.49	4.07	58.18	1.68	57.39	0.	51.99	0.	57.00
700	2.18	55.13	1.40	59.17	4.13	58.86	1.74	58.07	0.	52.62	0.	57.67
800	2.18	55.76	1.41	59.85	4.18	59.52	1.80	58.73	0.	53.25	0.	58.33
900	2.18	56.36	1.43	60.50	4.22	60.15	1.84	59.37	0.	53.85	0.	58.96
1000	2.18	56.94	1.45	61.12	4.26	60.75	1.87	59.97	0.	54.42	0.	59.57
1100	2.18	57.49	1.48	61.72	4.28	61.32	1.90	60.55	0.	54.96	0.	60.14
1200	2.18	58.02	1.50	62.29	4.31	61.86	1.92	61.09	0.	55.48	0.	60.68
1300	2.18	58.52	1.53	62.84	4.33	62.38	1.94	61.61	0.	55.98	0.	61.20
1400	2.18	58.99	1.56	63.36	4.34	62.87	1.96	62.10	0.	56.45	0.	61.69
1500	2.18	59.45	1.59	63.86	4.36	63.34	1.97	62.57	0.	56.90	0.	62.16
1600	2.18	59.88	1.62	64.33	4.37	63.78	1.98	63.01	0.	57.33	0.	62.61
1700	2.18	60.30	1.65	64.79	4.38	64.21	2.00	63.44	0.	57.75	0.	63.03
1800	2.18	60.70	1.68	65.22	4.39	64.61	2.01	63.85	0.	58.14	0.	63.44
1900	2.18	61.08	1.70	65.64	4.40	65.00	2.01	64.24	0.	58.52	0.	63.83
2000	2.18	61.45	1.73	66.05	4.41	65.38	2.02	64.61	0.	58.88	0.	64.21
2100	2.18	61.80	1.76	66.43	4.42	65.74	2.03	64.97	0.	59.24	0.	64.57
2200	2.18	62.14	1.78	66.80	4.43	66.08	2.04	65.32	0.	59.57	0.	64.92
2300	2.18	62.47	1.81	67.16	4.43	66.42	2.04	65.66	0.	59.90	0.	65.25
2400	2.18	62.78	1.83	67.51	4.44	66.74	2.05	65.98	0.	60.21	0.	65.57
2500	2.18	63.09	1.85	67.84	4.44	67.05	2.05	66.29	0.	60.52	0.	65.88
2600	2.18	63.39	1.87	68.17	4.45	67.35	2.06	66.59	0.	60.81	0.	66.19
2700	2.18	63.67	1.89	68.48	4.45	67.64	2.06	66.88	0.	61.10	0.	66.48
2800	2.18	63.95	1.91	68.78	4.46	67.92	2.07	67.17	0.	61.38	0.	66.76
2900	2.18	64.22	1.93	69.08	4.46	68.20	2.07	67.44	0.	61.64	0.	67.03
3000	2.18	64.48	1.95	69.36	4.46	68.46	2.08	67.71	0.	61.91	0.	67.30
$H_{298}^\circ - H_0^\circ$ [b] =	2087		2121		2307		2336		2083		2120	
$H_{2000}^\circ - H_0^\circ$ [b] =	16479		17896		16984		17044		16383		16833	

T(°K)	SrO $-(F^\circ-H^\circ_0)^a/T$ (elec)	SrO $-(F^\circ-H^\circ_{298})^a/T$	TaO $-(F^\circ-H^\circ_0)^a/T$ (elec)	TaO $-(F^\circ-H^\circ_{298})^a/T$	TeO $-(F^\circ-H^\circ_0)^a/T$ (elec)	TeO $-(F^\circ-H^\circ_{298})^a/T$	ThO $-(F^\circ-H^\circ_0)^a/T$ (elec)	ThO $-(F^\circ-H^\circ_{298})^a/T$	TiO $-(F^\circ-H^\circ_0)^a/T$ (elec)	TiO $-(F^\circ-H^\circ_{298})^a/T$	TlO $-(F^\circ-H^\circ_0)^a/T$ (elec)	TlO $-(F^\circ-H^\circ_{298})^a/T$
298	3.87	58.86	2.75	58.98	.00	55.66	3.27	60.91	4.38	57.77	1.38	58.75
400	3.89	59.18	2.75	59.27	.00	55.96	3.40	61.26	4.69	58.10	1.38	59.07
500	3.92	59.80	2.76	59.83	.00	56.55	3.56	61.95	4.90	58.72	1.38	59.67
600	3.95	60.52	2.76	60.47	.01	57.22	3.72	62.74	5.06	59.41	1.38	60.35
700	3.98	61.25	2.76	61.13	.01	57.91	3.88	63.55	5.18	60.11	1.38	61.05
800	4.02	61.96	2.77	61.78	.03	58.58	4.03	64.33	5.28	60.79	1.38	61.72
900	4.06	62.64	2.78	62.41	.05	59.24	4.17	65.09	5.36	61.44	1.38	62.37
1000	4.09	63.29	2.80	63.02	.07	59.87	4.30	65.80	5.42	62.05	1.38	62.99
1100	4.13	63.91	2.83	63.60	.11	60.48	4.42	66.49	5.47	62.64	1.38	63.57
1200	4.16	64.50	2.86	64.17	.14	61.06	4.54	67.14	5.51	63.19	1.38	64.12
1300	4.19	65.05	2.90	64.71	.18	61.62	4.65	67.76	5.55	63.71	1.38	64.65
1400	4.22	65.58	2.95	65.24	.22	62.15	4.76	68.35	5.59	64.21	1.38	65.15
1500	4.25	66.08	3.00	65.75	.27	62.66	4.86	68.91	5.62	64.69	1.38	65.62
1600	4.27	66.56	3.05	66.24	.31	63.15	4.97	69.46	5.64	65.14	1.38	66.07
1700	4.30	67.02	3.11	66.71	.35	63.63	5.07	69.98	5.66	65.57	1.38	66.50
1800	4.32	67.45	3.17	67.18	.40	64.08	5.17	70.48	5.68	65.98	1.38	66.92
1900	4.34	67.87	3.23	67.62	.44	64.51	5.27	70.97	5.70	66.38	1.38	67.31
2000	4.37	68.27	3.29	68.06	.49	64.93	5.36	71.43	5.72	66.76	1.38	67.69
2100	4.39	68.66	3.36	68.48	.53	65.34	5.46	71.88	5.74	67.12	1.38	68.05
2200	4.41	69.03	3.42	68.89	.57	65.73	5.55	72.32	5.75	67.47	1.38	68.41
2300	4.42	69.38	3.49	69.28	.61	66.10	5.64	72.74	5.76	67.81	1.38	68.74
2400	4.44	69.73	3.56	69.67	.65	66.46	5.73	73.15	5.78	68.13	1.38	69.07
2500	4.46	70.06	3.62	70.04	.68	66.81	5.82	73.55	5.79	68.45	1.38	69.38
2600	4.47	70.38	3.69	70.41	.72	67.15	5.90	73.93	5.80	68.75	1.38	69.69
2700	4.49	70.69	3.75	70.76	.75	67.48	5.98	74.30	5.81	69.05	1.38	69.98
2800	4.50	70.99	3.82	71.10	.79	67.79	6.07	74.66	5.82	69.34	1.38	70.27
2900	4.52	71.28	3.88	71.44	.82	68.10	6.14	75.02	5.83	69.61	1.38	70.54
3000	4.53	71.56	3.94	71.77	.85	68.40	6.22	75.36	5.84	69.88	1.38	70.81
$H^\circ_{298}-H^\circ_0$ [b] =	2169		2095		2125		2196		2420		2155	
$H^\circ_{2000}-H^\circ_0$ [b] =	17857		19178		18566		20552		17280		17015	

[a] cal/deg mole [b] cal/mole

TABLE IX (continued)

T(°K)	UO		VO		WO		YO		ZnO		ZrO	
	$-\frac{(F^\circ-H_0^\circ)}{T}$[a] (elec)	$-\frac{(F^\circ-H_{298}^\circ)}{T}$[a]	$-\frac{(F^\circ-H_0^\circ)}{T}$[a] (elec)	$-\frac{(F^\circ-H_{298}^\circ)}{T}$[a]	$-\frac{(F^\circ-H_0^\circ)}{T}$[a] (elec)	$-\frac{(F^\circ-H_{298}^\circ)}{T}$[a]	$-\frac{(F^\circ-H_0^\circ)}{T}$[a] (elec)	$-\frac{(F^\circ-H_{298}^\circ)}{T}$[a]	$-\frac{(F^\circ-H_0^\circ)}{T}$[a] (elec)	$-\frac{(F^\circ-H_{298}^\circ)}{T}$[a]	$-\frac{(F^\circ-H_0^\circ)}{T}$[a] (elec)	$-\frac{(F^\circ-H_{298}^\circ)}{T}$[a]
298	3.20	60.85	4.40	58.17	2.21	58.78	2.84	57.64	0.	53.68	3.30	57.97
400	3.20	61.15	4.79	58.51	2.26	59.10	2.96	57.99	0.	53.99	3.44	58.32
500	3.20	61.72	5.07	59.15	2.35	59.73	3.09	58.65	0.	54.59	3.60	58.99
600	3.20	62.37	5.28	59.86	2.45	60.46	3.22	59.41	0.	55.28	3.76	59.76
700	3.21	63.04	5.44	60.58	2.57	61.22	3.33	60.17	0.	55.97	3.91	60.55
800	3.21	63.70	5.57	61.28	2.69	61.97	3.43	60.91	0.	56.64	4.05	61.32
900	3.23	64.34	5.67	61.94	2.81	62.69	3.52	61.62	0.	57.29	4.18	62.05
1000	3.25	64.95	5.76	62.57	2.93	63.40	3.60	62.29	0.	57.90	4.30	62.75
1100	3.27	65.54	5.83	63.16	3.05	64.07	3.67	62.92	0.	58.48	4.40	63.41
1200	3.30	66.11	5.89	63.72	3.17	64.72	3.72	63.51	0.	59.03	4.50	64.03
1300	3.33	66.65	5.94	64.26	3.29	65.34	3.78	64.07	0.	59.56	4.58	64.62
1400	3.36	67.17	5.98	64.76	3.41	65.94	3.82	64.60	0.	60.05	4.66	65.17
1500	3.41	67.68	6.02	65.24	3.52	66.51	3.87	65.11	0.	60.53	4.73	65.70
1600	3.45	68.16	6.06	65.70	3.63	67.06	3.90	65.59	0.	60.98	4.80	66.20
1700	3.50	68.63	6.09	66.14	3.74	67.58	3.94	66.04	0.	61.41	4.86	66.68
1800	3.55	69.09	6.12	66.55	3.85	68.09	3.97	66.48	0.	61.82	4.92	67.14
1900	3.60	69.53	6.14	66.95	3.96	68.58	4.00	66.89	0.	62.22	4.97	67.57
2000	3.66	69.96	6.17	67.33	4.06	69.05	4.02	67.29	0.	62.60	5.02	67.99
2100	3.71	70.37	6.19	67.70	4.16	69.50	4.05	67.67	0.	62.96	5.06	68.39
2200	3.77	70.77	6.21	68.06	4.26	69.94	4.07	68.04	0.	63.31	5.10	68.77
2300	3.83	71.17	6.22	68.40	4.35	70.37	4.09	68.39	0.	63.65	5.14	69.14
2400	3.89	71.55	6.24	68.72	4.44	70.77	4.11	68.73	0.	63.97	5.18	69.50
2500	3.95	71.92	6.26	69.04	4.53	71.17	4.13	69.06	0.	64.29	5.22	69.84
2600	4.01	72.28	6.27	69.35	4.62	71.56	4.14	69.38	0.	64.59	5.25	70.17
2700	4.07	72.63	6.28	69.65	4.70	71.93	4.16	69.69	0.	64.88	5.29	70.50
2800	4.13	72.97	6.30	69.93	4.79	72.29	4.18	69.98	0.	65.17	5.32	70.81
2900	4.19	73.31	6.31	70.21	4.87	72.64	4.19	70.27	0.	65.44	5.35	71.11
3000	4.25	73.63	6.32	70.49	4.94	72.98	4.20	70.55	0.	65.71	5.38	71.40
$H_{298}^\circ - H_0^\circ$ [b] =	2105		2507		2126		2205		2150		2202	
$H_{2000}^\circ - H_0^\circ$ [b] =	18949		17501		20611		17796		16991		18531	

[a] cal/deg mole [b] cal/mole

III. Sources and Evaluation of Data

1. AgO

A linear Birge–Sponer extrapolation of the ground-state vibrational constants (Uhler, 1953) based upon measurements on four vibrational levels gives 57 kcal/mole for the dissociation energy of this molecule. The increase in the vapor pressure of silver in a stream of oxygen (Fox and Esdaile, 1963) sets an upper limit of $\sim 10^{-6}$ at 1150°K to the equilibrium constant for the reaction

$$Ag(s) + \tfrac{1}{2}\, O_2(g) = AgO(g)$$

and a corresponding upper limit to the dissociation energy of AgO(g) at 0°K of 68 kcal/mole.

$$D_0^\circ = 50 \pm 20 \quad \text{kcal}$$

2. AlO

The partial pressures of the various species in equilibrium with Al_2O_3 (s or *l*) in tungsten and molybdenum Knudsen cells, and also in molybdenum cells with metallic uranium added, have been measured with a mass spectrometer (Drowart *et al.*, 1960). Recalculation of data for the reaction

$$AlO(g) = Al(g) + O(g)$$

in tungsten and molybdenum cells yields dissociation energies that agree with the originally reported values within 0.2 kcal/mole. Thus the reported average dissociation energy of 115 ± 5 kcal at 0°K based upon measurements in all three types of cells is unchanged. Data obtained in the tungsten cell yield $\Delta H^\circ_{298} = 252$ kcal for the reaction

$$\tfrac{1}{2}Al_2O_3(s) = AlO(g) + O(g)$$

using the following values for the free energy function of $Al_2O_3(s)$: at 2000°K, 39.72, and at 2300°K, 43.30 cal/mole deg (JANAF, 1964). The latter enthalpy leads to $D_0^\circ(AlO) = 117$ kcal/mole when combined with the heat of formation of $Al_2O_3(\alpha)$, -400.5 kcal/mole at 298°K (JANAF, 1964; Wagman *et al.*, 1968), the enthalpy of vaporization of aluminum, 78.7 kcal/g-atom at 298°K (Hultgren *et al.*, 1968), and the dissociation energy of oxygen. The thermochemical data are in agreement with an approximate spectroscopic value of 116 kcal/mole based upon a linear Birge–Sponer extrapolation (Becart and Declerck, 1960). Tyte (1967) has studied the absorption spectrum of AlO produced at 4000°K in a shock tube. He ascribes the continuum seen at 2700 to 2800 Å to absorption from the ground state of AlO and obtains

$D_0 = 104.7$ kcal/mole from the long-wavelength edge at 2730 Å. The dissociation energy based upon mass spectrometric studies corresponds to a dissociation limit 250 Å farther toward the violet. Under the conditions of the shock experiment Al_2O and AlOH are present. In addition, by comparison with the red system of the isoelectronic CN molecule, a $^2\Pi$ state of AlO is expected to fall within 9000 cm^{-1} of the ground state, and this state would be populated at the shock temperatures. As there is considerable doubt that the observed absorption is due to the ground state of AlO, no weight has been given to the shock-tube result.

$$D_0^\circ = 115 \pm 5 \quad \text{kcal}$$

3. ArO

Although ArO has been observed spectroscopically (Herman *et al.*, 1951; Cooper and Lichtenstein, 1958), all evidence indicates this molecule to be very much less stable than XeO.

$$D_0^\circ < 1 \quad \text{kcal}$$

4. AsO

An upper limit of D_0 of 114.8 kcal/mole is set by the predissociation observed in the $v=0$ level of the B state (Meyer, 1965; Collomon and Morgan, 1965). The upper limit is 17% below the dissociation energy obtained from linear Birge–Sponer extrapolations of the vibrational levels of the X $^2\Pi_{1/2}$ and A $^2\Sigma^+$ states (Collomon and Morgan, 1965).

$$D_0^\circ = 114 \pm 3 \quad \text{kcal}$$

5. BO

Blackburn *et al.* (1966) measured BO^+ and $B_2O_2^+$ ion currents as a function of temperature during a mass spectrometric study of vapor species in the Al–B–O system under reducing (excess aluminum) conditions. They report $\Delta H^\circ_{298} = 56.6 \pm 1.8$ kcal/mole for

$$\tfrac{1}{2}\, B_2O_2(g) = BO(g)$$

from a least-squares fit of the data. Taking $\Delta H_f^\circ(B_2O_2, g) = -108.7 \pm 2$ kcal/mole (Wagman *et al.*, 1968) and ΔH_f° (B, g) $= 136.5 \pm 3$ (Hultgren *et al.*, 1964) leads to ΔH_f°(BO, g) $= 2.2$ and D°_{298}(BO) $= 194$ kcal/mole. Coppens and associates (1968) have studied several isomolecular oxygen exchange reactions involving BO. Third-law treatment of their data using the free energy functions and dissociation enthalpies of this paper yields the results shown in the tabulation.

Reaction	ΔH°_{298} (kcal)	D°_{298}(BO), (kcal)
$B+LaO=BO+La$	−8.0	196
$B+UO=BO+U$	−5.8	188
$B+UO_2=BO+UO$	−15.8	189

The mass spectrometric results are in very good agreement with the value $D_0=8.3\pm0.3$ eV ($D^\circ_{298}=192$ kcal/mole) obtained from the radial dependence of boron atom concentration across an arc (De Galen, 1965). The spectroscopic data are less precise, but, on the whole, consistent with the thermochemical results. Linear Birge–Sponer extrapolations of the vibrational levels of the X $^2\Sigma^+$, A $^2\Pi$, B $^2\Sigma^+$, C $^2\Pi$ states, assuming reasonable atomic products, yield ground-state dissociation energies at 0°K of 214, 171, 189, and 173 kcal/mole, respectively, (Lagerqvist *et al.*, 1958; Mal'tsev *et al.*, 1960). $D_0=179$ kcal/mole has been obtained by fitting an empirical potential function containing D_0 as a parameter to Rydberg–Klein–Rees–Vanderslice potential energy curves (Singh and Rai, 1965).

$$D_0^\circ = 191\pm5 \quad \text{kcal}$$

6. BaO

Numerous mass spectrometic studies have established that BaO(g) is the only major species upon vaporization of BaO(s) under neutral conditions (Inghram *et al.*, 1955; Shchukarev and Semenov, 1957; Newbury *et al.*, 1968; additional references are listed by Inghram and Drowart, 1960). Third-law treatment of the various vapor–pressure measurements on BaO yield the following enthalpies at 298°K for the reaction

$$BaO(s)=BaO(g)$$

Hermann (1937), 102.2; Claasen and Veenemans (1933), 102.6; Blewett *et al.* (1939), 103.4; Nikonov and Otmakhova (1961), 99.7; Shchukarev and Semenov (1957), 101.8; Inghram *et al.* (1955), 104.8; and Newbury *et al.* (1968), 102.6 kcal. Free energy functions for BaO(s) are based upon data tabulated by Kelley (1960) and Kelley and King (1961). The weighted average of these results, $\Delta H^\circ_{298}=103$ kcal, gives the dissociation energy of barium monoxide as 131 kcal/mole at 298°K and 130.5 ± 6 kcal/mole at 0°K. Along with the dissociation energy of oxygen the data needed for this calculation are the heat of formation of BaO(s) at 298°K, −132.1 kcal/mole (Flidlider *et al.*, 1966), and the heat of sublimation of barium, 42.5 kcal/g-atom (Lewis *et al.*, 1961). The heat of formation of BaO(s) is not unambiguously established: use of the values determined by Mah (1963) or Grebenshchikov and Pasechnova (1966) would raise the dissociation energy by 7 kcal. $D_0^\circ = 130.5$ kcal/mole is in reasonable agreement with some of the results from flame studies: James (1954), $D_0^\circ = 134.5$ kcal/mole; Veits and Gurvich (1956), 138.1; Gurvich and Ryabova (1965), 135.4; and Lagerqvist and Huldt (1954), 128 ± 8 kcal/mole. However, recent mass spectrometric investigation (Stafford and Berkowitz, 1964) indicates that the concentration of hydroxide species in these flames has been underestimated by several orders of magnitude, implying that the flame studies only set an upper limit to the monoxide dissociation energy. The flame work has been reviewed by Ryabova and Gurvich (1965), Kalff *et al.* (1965), and Schofield (1967). A mass spectrometric study of the reaction

$$\mathrm{Ba(g) + SO(g) = BaO(g) + S(g)}$$

by Colin *et al.* (1964) also indicates $D_0^\circ = 130.4 \pm 6$. Our tentative conclusions about the heat of vaporization and the dissociation energy of BaO differ markedly from those reached by Schofield (1967), primarily because Schofield chose a different value for ΔH_f°(BaO, s) and included extensive electronic degeneracy in the free energy functions for BaO(g). A large electronic contribution to the entropy of BaO(g) would explain the deviation between the second- and third-law heats of Newbury *et al.* (1968) but is not in accord with molecular-beam electric (Wharton *et al.*, 1962) and magnetic (Brooks and Kaufman, 1965) resonance observations.

$$D_0^\circ = 131 \pm 6 \quad \text{kcal}$$

7. BeO

Two groups (Chupka *et al.*, 1959; Theard and Hildenbrand, 1964) have examined the vapor over BeO(s) in a tungsten Knudsen cell with a mass spectrometer. The dissociation energy of gaseous beryllium monoxide has been recalculated from these data in the same two ways followed by Chupka *et al.* (1959). First the equilibrium constant K_1 for the reaction

$$\mathrm{BeO(s) = Be(g) + O(g)} \qquad (1)$$

is calculated from the heat of formation of BeO(s), -143.1 kcal/mole (Cosgrove and Snyder, 1953), the heat of vaporization of beryllium at 298°K, 77.5 kcal/g-atom (Hultgren *et al.*, 1967), and $-(F^\circ - H^\circ_{298})/T$ for BeO(s) equal to 13.67 cal/deg mole at 2000° and 15.08 at 2300°K (JANAF, 1963). The equilibrium constant K_2 for the dissociation reaction

$$\mathrm{BeO(g) = Be(g) + O(g)} \qquad (2)$$

is then evaluated from the reported partial pressures via the relation $K_2 = K_1^{1/2}(p_{Be}/p_{BeO})$. This procedure is equivalent to considering the reaction to be

$$BeO(g) = \tfrac{1}{2}\, Be(g) + \tfrac{1}{2}\, O(g) + \tfrac{1}{2}\, BeO(s) \tag{3}$$

with $K_3 = p_{Be}/p_{BeO}$. Secondly, dissociation energies are computed from data for the equilibrium.

$$BeO(g) + O(g) = Be(g) + O_2(g) \tag{4}$$

Table X compares results using the present free energy functions, functions

TABLE X. EVALUATION OF BeO(g) EQUILIBRIA WITH DIFFERENT BeO(g) FREE ENERGY FUNCTIONS

	Present BeO(g) functions	Functions based upon observed electronic levels	Functions based upon calculated levels[a]
Electronic contribution to $-(F^\circ - H_0^\circ)/T$ at 2500°K (cal/deg)	2.95	0.02	0.25
$-(F^\circ - H^\circ_{298})/T$ at 2500°K (cal/deg)	59.98	56.99	57.23
$-\Delta H^\circ_{298}(3)$ (kcal)	41.4[b], 43.6[c]	35.4[b], 36.5[c]	35.7[b], 37.0[c]
$D^\circ_{298}(3)$ (kcal)	98.7[b], 96.5[c]	104.7[b], 103.6[c]	104.4[b], 103.1[c]
$-\Delta H^\circ_{298}(4)$ (kcal)	18.1[b], 22.5[c]	11.6[b], 15.5[c]	11.9[b], 15.9[c]
$D^\circ_{298}(4)$ (kcal)	101.0[b], 96.6[c]	107.5[b], 103.6[c]	107.2[b], 103.2[c]

[a] Calculated by Verhaegen and Richards (1966).
[b] Data of Chupka *et al.* (1959).
[c] Data of Theard and Hildenbrand (1964).

based upon observed electronic levels only (Herzberg, 1950), and functions based upon the electronic energy levels computed by Verhaegen and Richards (1966). The present functions lead to a BeO dissociation energy at 298.15° of 98 kcal/mole, ~7 kcal/mole lower than that computed using alternative choices for the electronic partition function. However, examination of the table indicates that alternative free energy functions do not particularly improve agreement between investigators or between reactions (3) and (4).

$$D_0^\circ = 97 \pm 7 \quad \text{kcal}$$

8. BiO

Predissociations in the C and B states set an approximate upper limit to D_0 of 31,000 cm^{-1}. From this Barrow *et al.* (1967) conclude that the dissociation energy is about 30,000 cm^{-1}, which agrees with a rough linear Birge–Sponer extrapolation of the ground-state levels (Scari, 1956).

$$D_0^\circ = 86 \pm 3 \quad \text{kcal}$$

9. BrO

The dissociation energy is based upon a graphical Birge–Sponer extrapolation (Durie and Ramsey, 1958).

$$D_0^\circ = 55.3 \pm 0.6 \quad \text{kcal}$$

10. CO

The value quoted (Wagman *et al.*, 1968) is based upon observed predissociations (Douglas and Moller, 1955) assuming dissociation to ground-state atoms, $O(^3P_2)+C(^3P_0)$. Assuming the atomic ^3P sublevels to be unknown would lower D_0 by 0.38 kcal/mole.

$$D_0^\circ = 256.16 \pm 0.77 \quad \text{kcal}$$

11. CaO

The numbers in parentheses below were calculated assuming a $^1\Sigma$ ground state with no other states close enough to contribute to the free energy function at 2400°K. The other results are based upon the free energy functions in Table IX. The basis of the present functions has been discussed in Section II.

A mass spectrometric investigation of the reaction

$$Ca(g)+SO(g)=CaO(g)+S(g)$$

yields $D_0^\circ(CaO)=84.5\ (93.7)\pm 6$ kcal/mole (Colin *et al.*, 1964) after correction to the present SO dissociation energy. Studies by Drowart *et al.* (1964a) of the equilibria

$$CaO(g)+O(g)=Ca(g)+O_2(g)$$
$$CaO(g)+MoO_2(g)=Ca(g)+MoO_3(g)$$
$$CaO(g)+WO_2(g)=Ca(g)+WO_3(g)$$

give D_0° equal to 80.7 (90.1), 83.6 (93.4), and 84.6 (94.1) kcal/mole respectively, taking WO_2/WO_3 and MoO_2/MoO_3 equilibria and enthalpies from a previous paper (De Maria *et al.*, 1960). Drowart and associates (1964a) and Schofield (1967) review earlier determinations of the dissociation energy.

The selected values are in accord with the observation of Ackermann and Thorn (1961) that $D_0°(\text{SrO}) - D_0°(\text{CaO}) \approx 10$ kcal/mole.

$$D_0° = 83 \pm 7 \quad \text{kcal}$$

12. CdO

Failure to see evidence of any cadmium oxide vapor species in transport studies of the dissociation of CdO(s) using oxygen-containing carrier gas sets an upper limit to the dissociation energy of CdO(g) (Glemser and Stoecker, 1963; Gilbert and Kitchener, 1956). The data indicate $K \geq 5$ at 1300°K for the reaction

$$\text{CdO(g)} = \text{Cd(g)} + \tfrac{1}{2}\,\text{O}_2\text{(g)}$$

which shows the dissociation energy at 298°K to be less than, or on the order of, 67 kcal/mole using free energy functions in Table IX and Stull and Sinke (1956).

$$D_0° \leq 66 \quad \text{kcal}$$

13. CeO

Independent mass spectrometric studies of the isomolecular exchange reaction

$$\text{Ce(g)} + \text{LaO(g)} = \text{CeO(g)} + \text{La(g)}$$

indicate $D_0(\text{LaO}) - D_0(\text{CeO}) = +1$ (Coppens *et al.*, 1967) or 0 (Ames *et al.*, 1967; White *et al.*, 1962; Walsh *et al.*, 1961) by the second-law method and -2 (Coppens, 1966) or $+1$ kcal/mole (Ames *et al.*, 1967; Walsh *et al.*, 1961) by the third-law method.

$$D_0° = 187 \pm 6 \quad \text{kcal}$$

14. ClO

The absorption spectrum has been observed to the dissociation limit (Durie and Ramsey, 1958).

$$D_0° = 63.33 \pm 0.03 \quad \text{kcal}$$

15. CoO

A mass spectrometric investigation (Grimley *et al.*, 1966) of equilibrium vapor species in the Co–O system has shown the vapor phase to consist of Co, O_2, and small amounts of CoO over the temperature range 1550–1750°K. Equilibrium data for the reaction

$$\text{CoO(g)} = \text{Co(g)} + \tfrac{1}{2}\,\text{O}_2\text{(g)}$$

yield $\Delta H^\circ_{298}=27.6$ kcal using the present CoO(g) free energy functions and Co(g) functions tabulated by Stull and Sinke (1956). This corresponds to $D^\circ_{298}=87.2$ kcal/mole. Data for the reaction

$$CoO(s)=CoO(g)$$

yield $\Delta H^\circ_{298}=130.3$, using CoO(s) free energy functions of 23.21 cal/deg at 1500°K and 25.23 at 1800°K (King, 1957; King and Christensen, 1958; Grimley *et al.*, 1966). Combination with the standard heat of formation of CoO(s), -57.1 kcal/mole (Boyle *et al.*, 1954), of Co(g), 102.4 kcal/mole (Hultgren *et al.*, 1966), and of O(g) results in $D^\circ_{298}(CoO)=88.8$ kcal/mole.

$$D_0^\circ = 87 \pm 5 \quad \text{kcal}$$

16. CrO

Grimley *et al.* (1961a) have measured with a mass spectrometer the partial pressures of the vapor species in equilibrium with $Cr_2O_3(s)$ under neutral and oxidizing conditions. The reported pressures of Cr(g) and O(g) are higher than expected from the heat of formation of $Cr_2O_3(s)$ and the heat of sublimation of chromium. Attempting to avoid this uncertainty the dissociation energy has been calculated assuming the reaction taking place to be

$$\tfrac{1}{2}\,Cr_2O_3(s)=CrO(g)+\tfrac{1}{4}\,O_2(g)$$

Free energy functions for Cr_2O_3 were calculated from tabulated heat contents (Kelley, 1960) and $S^\circ_{298}=19.4$ cal/deg mole (Anderson, 1937). The enthalpy of the above vaporization reaction at 298.15°K is calculated to be 181 kcal/mole from the data reported under " neutral " conditions and 184 kcal/mole from the observations with added oxygen. The 181 kcal/mole value was combined with the enthalpy of formation of chromium (III) oxide, -272.7 ± 0.4 kcal/mole (Mah, 1954; Novokhatskii and Lenev, 1966). the enthalpy of sublimation of chromium, and the dissociation energy of oxygen to calculate the dissociation energy at 298.15°K, 110 ± 10 kcal/mole. The enthalpy of sublimation of chromium was taken to be 95 ± 1 kcal (Hultgren *et al.*, 1966).

$$D_0^\circ = 109 \pm 10 \quad \text{kcal}$$

17. CsO

Gusarov *et al.* (1967) estimate $D_0=70\pm6$ by comparison with data for gaseous lithium oxides and alkali fluorides, in the same manner as for KO. Their estimate has been lowered to be consistent with the value of $D_0(LiO)$ presented in the present paper.

$$D_0^\circ = 66 \pm 8 \quad \text{kcal}$$

18. CuO

Vapor pressures of CuO reported (Mack *et al.*, 1923) for the range 600–950°C indicate $\Delta H^\circ_{298} \approx 78–90$ kcal for the reaction

$$CuO(s) = CuO(g)$$

after treatment with the present CuO(g) free energy functions and CuO(s) functions of 14.24, 16.03, and 17.66 cal/deg mole at 800, 1000, and 1200°K, respectively (Kelley and King, 1961; Mah *et al.*, 1967). Combination of the average value, 84 kcal, with the heat of formation of CuO(s), −37.2 kcal (Mah *et al.*, 1967), the heat of sublimation of copper, 80.7±0.3 kcal (Hultgren *et al.*, 1968), and the dissociation energy of oxygen yields a dissociation energy for CuO(g) at 298.15°K of 93±20 kcal/mole. This value is in agreement with an estimate of 89 kcal based upon flame photometry (Hinnov and Kohn, 1957). On the other hand, a preliminary mass spectrometric report (Grimley and Burns, 1961) of the equilibrium

$$CuO(g) = Cu(g) + \tfrac{1}{2}\, O_2(g)$$

indicates $\Delta H^\circ_{298} = 4.2$ and $D^\circ_{298}(CuO) = 64$ kcal/mole. The average value, $D^\circ_{298} = 82 \pm 15$, has been chosen.

$$D_0^\circ = 81 \pm 15 \quad \text{kcal}$$

19. DyO

Knudsen-effusion data (Ames *et al.*, 1967) for the reaction

$$Dy_2O_3(s) = 1.85\ DyO(g) + 0.15\ Dy(g) + 1.15\ O(g)$$

indicate $D_0(DyO) = 145$ kcal/mole after estimating DyO(g) free energy functions consistent with the functions in Table IX: $-(F^\circ - H_0^\circ)/T \simeq 72.4$ cal/deg mole at 2600°K. For $Dy_2O_3(s)$, $\Delta H^\circ_{298}f = -446$ kcal/mole (Huber *et al.*, 1956a); for Dy(g), $\Delta H^\circ_{298} = 71 \pm 1$ kcal/g-atom (Habermann and Daane, 1964; Savage *et al.*, 1959).

$$D_0^\circ = 145 \pm 10 \quad \text{kcal}$$

20. ErO

For ErO(g) we estimate $-(F^\circ - H_0^\circ)/T \simeq 72.0$ cal/deg mole at 2600°K. Utilizing this free energy function—along with the heat of formation at 25°C of $Er_2O_3(s)$, −453.6±0.5 kcal/mole (Huber *et al.*, 1956b), and the standard enthalpy of vaporization of erbium, 75.8±1 kcal/g-atom (Hultgren *et al.*, 1966)—we compute D_0 (ErO)=146 kcal from results presented by Ames *et al.* (1967) for the reaction

$$Er_2O_3(s) = 1.70\ ErO(g) + 0.30\ Er(g) + 1.30\ O(g)$$

$$D_0^\circ = 146 \pm 10 \quad kcal$$

21. EuO

Ames *et al.* (1967) have investigated the reaction

$$Eu_2O_3(s) = 2\ EuO(g) + O(g)$$

and also Eu(g)+SmO(g)=EuO(g)+Sm(g). Their effusion results for the vaporization of europium sesquioxide yield D_0(EuO)=131 kcal using estimated monoxide free energy functions of $-(F^\circ - H_0^\circ)/T \simeq 70.7$ cal/deg mole at 2200°K. The thermochemical cycle that leads to the dissociation energy uses the enthalpy of formation of monoclinic Eu_2O_3, $\Delta H^\circ_{298} = 394 \pm 1$ kcal/mole (Huber *et al.*, 1964a), and the enthalpy of vaporization of metallic europium, $\Delta H^\circ_{298} = 42.4 \pm 0.2$ kcal/g-atom. The latter value is a revision of the value selected by Hultgren *et al.* (1966) to agree with the Eu entropy of Gerstein *et al.* (1967). Third-law treatment of data for the oxygen exchange with samarium (Ames *et al.*, 1967) yields D_0(SmO)$-D_0$(EuO)=12 kcal or D_0(EuO)=121. Second-law treatment of data for this reaction (White *et al.*, 1962) yields D_0(SmO)$-D_0$(EuO)=7 kcal/mole.

$$D_0^\circ = 129 \pm 10 \quad kcal$$

22. FO

Data pertaining to the dissociation energy of FO have been reviewed by Arkell and associates (1965) and in the JANAF (1966b) tables. Electron impact of F_2O (Dibeler *et al.*, 1957) leads to D_0(FO)≈28 kcal using recent values for the electron affinity of fluorine (Berry and Rieman, 1963) and for the heats of formation of F_2O(g), F(g), and O(g) (Wagman *et al.*, 1968). However, theoretical estimates range from 45 to 56 kcal. Assuming equal F—O bond strengths in FO and F_2O gives 50 kcal. The vibrational frequency of FO observed by Arkell *et al.* (1965) in an argon matrix led these authors to favor the high value, ~56 kcal, originally proposed by Glockler (1948). Accepting their conclusions, we take

$$D_0^\circ = 55 \pm 10 \quad kcal$$

23. FeO

Mass spectrometric observations (Washburn, 1963) indicate $P_{FeO}/P_{Fe} \approx 0.2$ at 1600°C over the congruently vaporizing composition $FeO_{1.116}(l)$. This ratio, coupled with the O_2 pressure of 1.66×10^{-6} atm (Darken and Gurry, 1946), gives the equilibrium constant at 1873°K for the reaction

$$FeO(g) = Fe(g) + \tfrac{1}{2} O_2(g)$$

A third-law calculation yields the dissociation energy of FeO(g) at 298.15°K to be 96 kcal/mole, corresponding to $D_0^\circ = 95 \pm 5$. The result is in agreement with the upper limit of 93 ± 5 calculated from the data of Brewer and Mastick (1951), the value of 99 ± 12 kcal/mole reported from flame studies (Lagerqvist and Huldt, 1953), and rough spectroscopic extrapolations (Gaydon, 1953; Herzberg, 1950). The enthalpy of vaporization of iron is 99.3 ± 0.3 kcal/g-atom (Hultgren *et al.*, 1967).

$$D_0^\circ = 95 \pm 5 \quad \text{kcal}$$

24. GaO

Linear Birge–Sponer extrapolations of lower $^2\Sigma$ state vibrational constants yield dissociation energies for this state of 65.4 (Howell, 1945), 66.4 (Raziunas *et al.*, 1963), and 69.3 kcal/mole (Gurvich *et al.*, 1965). All evidence indicates that this state is most probably the ground state of GaO. By analogy with his results for InO, Howell (1945) expected the linear Birge–Sponer extrapolation to be $\sim 10\%$ high. On the other hand a linear Birge–Sponer extrapolation of upper-state vibrational constants suggests that the dissociation energy of GaO at 0°K is 80 kcal/mole if this state dissociates to Ga^+ and O^- (Gurvich *et al.*, 1965). The average of the lower-state vibrational extrapolations has been selected. Flame results (Gurvich and Veits, 1958) are unreliable. The enthalpy of vaporization of solid gallium given by Munir and Searcy (1964) and listed in Table V is 0.8 kcal higher than the value determined by Macur *et al.* (1966).

$$D_0^\circ = 67 \pm 15 \quad \text{kcal}$$

25. GdO

The vaporization of gadolinium sesquioxide has been investigated in a number of laboratories (Ames *et al.*, 1967; White *et al.*, 1962; Messier, 1967; Alcock and Peleg, 1967; Shchukarev and Semenov, 1961; Panish, 1961). Mass spectrometric analysis (Panish, 1961; Ames *et al.*, 1967; White *et al.*, 1962; Shchukarev and Semenov, 1961) shows the predominant reaction to be

$$Gd_2O_3(s) = 2\ GdO(g) + O(g)$$

Third-law treatment of the effusion data of Messier (1967), which are supported by the torsion-Langmuir results of Alcock and Peleg (1967), yields $\Delta H_{298}=582$ kcal for this reaction. Free energy functions for Gd_2O_3(s) needed for this computation, $-(F^\circ-H^\circ_{298})/T=66.24$ cal/deg mole at 2000°K and 72.3 at 2500°K, are based upon high temperature heat contents (Pankratz *et al.*, 1962) extrapolated to 2500°K and $S^\circ_{298}=36.0$ cal/deg mole (Justice and Westrum, 1963b). Combination of the third-law enthalpy with the heats of formation of Gd_2O_3(s), -433.9 kcal/mole (Huber and Holley, 1955), and Gd(g), 95.0 ± 0.5 kcal/g-atom (Hultgren *et al.*, 1967; Hoenig *et al.*, 1967), leads to D°_{298}(GdO)$=160$ kcal/mole. This value is 6 kcal lower than the result, $D^\circ_{298}=166$ kcal, computed from the data of White and co-workers (1962; Ames *et al.*, 1967) because the effusion rates of White and associates are about a factor of 4 greater than those of Messier (1967) and Alcock and Peleg (1967). An intermediate value has been selected.

$$D_0{}^\circ = 161\pm6 \quad \text{kcal}$$

26. GeO

Knudsen (Jolly and Latimer, 1952) and mass spectrometric (Drowart *et al.*, 1965b) data for the reaction

$$\tfrac{1}{2}\,\mathrm{Ge(c)}+\tfrac{1}{2}\,\mathrm{GeO_2(hex)}=\mathrm{GeO(g)}$$

have been treated by the third-law method using Stull and Sinke (1956) free energy functions for Ge(s), and GeO_2(hex) free energy functions obtained from high temperature heat content measurements (Kelley and Christensen, 1961) and $S^\circ_{298}=13.21$ cal/deg mole (Kelley and King, 1961). The Knudsen measurements, which are a lower limit as they are uncorrected for polymerization of the germanium (II) oxide vapor, yield $\Delta H^\circ_{298}=55.2\pm2$ kcal/mole for the above reaction; the mass spectrometric results give 57.8 ± 1 kcal/mole. Combination of the latter value with the standard enthalpy of formation of GeO_2(hex), -132.2 ± 1.2 kcal/mole (Bills and Cotton, 1964), the enthalpy of sublimation of germanium, 89.5 ± 0.5 kcal/mole (Hultgren *et al.*, 1965), and the dissociation energy of oxygen indicates the dissociation energy of germanium monoxide at 298.15°K to be 157.4 ± 3 kcal/mole. Transpiration studies (Belton, 1968) of the equilibrium

$$\mathrm{Ge(c)}+\mathrm{H_2O(g)}=\mathrm{GeO(g)}+\mathrm{H_2(g)}$$

yield $\Delta H^\circ_{298}=48.0\pm0.1$ by the third-law method, which corresponds to D°_{298}(GeO)$=158.9$ when combined with standard thermochemical data for H_2O (Wagman *et al.*, 1968). The thermochemical results are in satisfactory agreement with the spectroscopic dissociation energy at 0°K, 155.8 ± 5.3 kcal/mole (Barrow and Rowlinson, 1954).

$$D_0{}^\circ = 157\pm3 \quad \text{kcal}$$

27. HO

Barrow (1956) has studied the B $^2\Sigma^+$—A $^2\Sigma^+$ band system of OH and OD and carried out various short extrapolations of the observed vibrational levels to the dissociation limit. All the values obtained lie within the range $D_0^\circ = 101.58 \pm 0.3$ kcal/mole. Barrow recommends a most probable value of 101.36 ± 0.3 kcal/mole, which corresponds to a heat of formation of OH at 298.15°K of 9.31 kcal/mole. The spectroscopic dissociation energy is in excellent agreement with the thermochemical value, 100.8 ± 1.2 kcal/mole, computed from $\Delta H_0^\circ = -153.4 \pm 2.6$ kcal/mole for $4OH(g) = 2H_2O(g) + O_2(g)$ (Dwyer and Oldenburg, 1944) and formation enthalpies of H(g), H_2O(g), and O(g) tabulated by Wagman and co-workers (1968).

$$D_0^\circ = 101.4 \pm 0.3 \quad \text{kcal}$$

28. HeO

Self-consistent field calculations using approximate wave functions formed by linear combination of Slater-type atomic orbitals indicate that there are no bound states of HeO (Masse and Masse-Baerlocher, 1967). Only continuum spectra are expected upon interaction of He and O atoms. The calculations indicate that HeO^+ and HeO^{2+} form stable bound states and have $^2\Pi$ and $^3\Sigma^-$ ground states, respectively.

$$D_0^\circ = 0 \quad \text{kcal}$$

29. HfO

The vaporization of hafnia has been studied by Knudsen effusion and mass spectrometric techniques (Panish and Reif, 1964) and also by the torsion–Langmuir method (Alcock and Peleg,1967). Assuming the vaporization reaction to be

$$HfO_2(s) = HfO(g) + O(g)$$

with $K = P^2_{HfO}(16/194.5)^{1/2}$ the Knudsen data yield $\Delta H^\circ_{298} = 359 \pm 7$ kcal for this reaction, while the torsion results give 352 kcal/mole. For these calculations free energy functions for HfO_2(s) (Lewis *et al.*, 1961) were extrapolated to the value $-(F^\circ - H^\circ_{298})/T = 36.5$ cal/deg mole at 2500°K. A dissociation energy of 185 kcal/mole at 298°K is calculated upon combining the average of these enthalpy changes with the heat of formation of HfO_2(s), -273.6 ± 0.3 kcal/mole (Huber and Holley, 1968a), the heat of sublimation of hafnium, 148 ± 1 kcal/g-atom (Hultgren *et al.*, 1966), and the dissociation energy of oxygen. The thermochemical data are in reasonable agreement with rough linear Birge–Sponer extrapolations (Gaydon, 1968) of C and D system vibrational levels (Gatterer *et al.*, 1957) which lead to HfO dissociation energies at 298°K of 169 and 183 kcal/mole, respectively.

$$D_0^\circ = 184 \pm 10 \quad \text{kcal}$$

30. HoO

Free energy functions for gaseous holmium monoxide were estimated in order to treat available equilibrium data on a basis consistent with other monoxide data in this paper: at 2000°K, $-(F^\circ - H_0^\circ)/T \simeq 69.7$; at 2600°K, $\sim$72.2 cal/deg mole. These free energy functions lead to $D_0^\circ(\text{HoO}) = 148$ kcal/mole, based upon effusion results (Ames *et al.*, 1967) for the reaction

$$Ho_2O_3(s) = 1.69\ HoO(g) + 0.31\ Ho(g) + 1.31\ O(g)$$

The following heats of formation at 25°C were used for this computation: Ho_2O_3, -449.6 ± 1.2 (Huber *et al.*, 1957b); Ho(g), 71.9 ± 0.3 (Hultgren *et al.*, 1966); O(g), 59.6 kcal/mole. The computed dissociation energy agrees with the value $D_0(\text{HoO}) = 147$ obtained from a third-law treatment of mass spectrometric data (Ames *et al.*, 1967) for the reaction

$$Ho(g) + TbO(g) = HoO(g) + Tb(g)$$

which indicate $D_0(\text{TbO}) - D_0(\text{HoO}) = 17$ kcal/mole.

$$D_0^\circ = 148 \pm 10 \quad \text{kcal}$$

31. IO

Coleman *et al.* (1948) report the dissociation energy at 0°K to be 44 ± 5 kcal/mole from a graphical Birge–Sponer extrapolation of 12 ground-state vibrational levels. Durie and Ramsey (1958) have carried out a linear Birge–Sponer extrapolation of four upper-state vibrational levels, which yields 45 ± 7 kcal/mole. However, they prefer the value 42 ± 5 obtained upon decreasing the linear extrapolation by 10% in analogy with ClO and BrO. Phillips and Sugden (1961) determined the dissociation energy to be 57 ± 6 kcal/mole by flame photometry. The range, 46 ± 7 kcal/mole, appears likely to include the correct value.

$$D_0^\circ = 46 \pm 7 \quad \text{kcal}$$

32. InO

The InO^+ intensity observed in a mass spectrometric investigation of the vaporization of $In_2O_3(s)$ indicates the dissociation energy of indium monoxide to be 76 kcal/mole at 0°K using the present free energy functions (Burns *et al.*, 1963). This value represents an upper limit, since part of the InO^+ intensity could be due to fragmentation. Conflicting values of 26 and 103 kcal/mole have been obtained from the visible spectrum (Howell, 1945) and from flame equilibria (Gurvich and Veits, 1958), respectively.

$$D_0^\circ \leq 76 \quad \text{kcal}$$

33. IrO

A mass spectrometric study (Norman *et al.*, 1965a) of the vapor species in equilibrium with iridium metal and oxygen gas indicates the partial pressure of IrO(g) to be less than or on the order of one fiftieth of the partial pressure of IrO_2(g) at the highest temperatures and oxygen pressures studied. The data yield $K \leq 10^{-7.7}$ at 2023°K for the reaction

$$\text{Ir(s)} + \tfrac{1}{2}\,\text{O}_2\text{(g)} = \text{IrO(g)}$$

and $\Delta H^\circ_{298} \geq 126$ kcal using tabulated Ir(s) free energy functions (Hultgren *et al.*, 1963). Taking the heat of sublimation of iridium to be 160 ± 1 kcal/g-atom (Panish and Reif, 1961; Hampson and Walker, 1961; Paule and Margrave, 1963; Hultgren *et al.*, 1963) this result shows the dissociation energy of iridium monoxide to be less than or equal to 94 kcal/mole at 298°K.

$$D_0^\circ \leq 93 \quad \text{kcal}$$

34. KO

Gusarov *et al.* (1967) estimate the dimerization energy of KO to be ~81 kcal by comparison with experimental data for LiO, Li_2O_2, and the alkali fluorides. From this they deduce $D_0(\text{KO}) \approx 60 \pm 6$. This value has been lowered by 4 kcal to be consistent with the chosen dissociation energy of LiO.

$$D_0^\circ = 56 \pm 8 \quad \text{kcal}$$

35. KrO

Spectroscopic observations (Cooper *et al.*, 1961) indicate that this molecule probably has a dissociation energy less than, or on the order of, 1 kcal/mole.

$$D_0^\circ \leq 1 \quad \text{kcal}$$

36. LaO

The dissociation energy is based primarily upon Knudsen studies (White *et al.*, 1962; Goldstein *et al.*, 1961; Walsh *et al.*, 1960) of the vaporization of lanthanum sesquioxide via the equilibrium

$$\text{La}_2\text{O}_3\text{(s)} = 2\ \text{LaO(g)} + \text{O(g)}$$

Free energy functions for lanthanum sesquioxide were calculated from high temperature heat contents (King *et al.*, 1961) and $S^\circ_{298}=30.43$ cal/deg mole (Justice and Westrum, 1963a) to obtain the values $-(F^\circ-H^\circ_{298})/T=61.68$ at 2000°K and $-(F^\circ-H^\circ_{298})/T=67.3$ cal/deg mole (extrapolated) at 2500°K. The following enthalpies at 298.15°K were used to calculate the dissociation energy: vaporization of La_2O_3(s) by the above reaction, 437 kcal/mole, calculated by the same procedure used for YO; formation of La_2O_3(s), -428.6 kcal/mole (Huber and Holley, 1953a; Fitzgibbon *et al.*, 1965); sublimation of lanthanum metal, 103 ± 1 kcal/g-atom (Hultgren *et al.*, 1967).

The dissociation energy of $D^\circ_{298}=188\pm8$ kcal/mole calculated from these data is in satisfactory agreement with that obtained from other equilibria. A second-law treatment of mass spectrometric data for the same vaporization reaction (Goldstein *et al.*, 1961) yields $\Delta H^\circ_{298}=434$ and D°_{298}(LaO)$=190$ kcal/mole. Mass spectrometric results for the reaction

$$\tfrac{1}{3}\ \mathrm{La(g)}+\tfrac{1}{3}\ \mathrm{La_2O_3(s)}=\mathrm{LaO(g)}$$

yield $\Delta H^\circ_{298}=87$ and $D^\circ_{298}=184$ kcal/mole after a third-law calculation (Chupka *et al.*, 1956). Clausius–Clapeyron data for this equilibrium give $\Delta H^\circ_{298}=89$, $D^\circ_{298}=182$ kcal/mole (Chupka *et al.*, 1956) and $\Delta H^\circ_{298}=76$, $D^\circ_{298}=195$ kcal/mole (White *et al.*, 1962). Mass spectrometric studies of the isomolecular exchange reaction

$$\mathrm{Y(g)+LaO(g)=YO(g)+La(g)}$$

indicate D°_{298}(LaO)$=185$ (White *et al.*, 1963), or 178 kcal/mole (Smoes *et al.*, 1965) by the second law of thermodynamics and $D^\circ_{298}=183$ kcal/mole (Smoes *et al.*, 1965; Ames *et al.*, 1967) by the third-law method when combined with the dissociation energy of YO. Third-law treatment of mass spectrometric data (Coppens *et al.*, 1967) for the reaction

$$\mathrm{La(g)+SiO(g)=LaO(g)+Si(g)}$$

yields $\Delta H^\circ_{298}=3.8$ or D°_{298}(LaO)$=189$ kcal/mole.

$$D_0^\circ=187\pm5\quad \mathrm{kcal}$$

37. LiO

The vapor over Li_2O(s) has been analyzed mass spectrometrically in two laboratories (Berkowitz *et al.*, 1959; White *et al.*, 1963). Both groups observed a small amount of LiO^+, the primary species being Li^+, O_2^+ and Li_2O^+. Evidence reported in these papers coupled with the observation that LiO^+ arises from a molecule having a dipole moment (Büchler *et al.*, 1963) indicates that the LiO^+ peak is produced by simple ionization of LiO(g).

Berkowitz *et al.* (1959) present equilibrium constants for the reaction

$$LiO(g) = Li(g) + \tfrac{1}{2} O_2(g)$$

which can be combined with free energy functions in this paper and Stull and Sinke (1956) to calculate $\Delta H^\circ_{298} = 18$ kcal/mole, corresponding to a LiO(g) dissociation energy at 298.15°K of 77 ± 7 kcal/mole.

White and co-workers (1963) present data that may be treated in terms of the reaction

$$Li_2O(s) = Li(g) + LiO(g)$$

avoiding the introduction of thermal functions and enthalpies for $Li_2O(g)$. The slopes of log IT against $1/T$ curves indicate $\Delta H^\circ_{1500} = 197.7$ kcal/mole for this reaction. This corresponds to $\Delta H^\circ_{298} = 204.6 \pm 6$ kcal/mole taking $H^\circ_{1500} - H^\circ_{298}$ for LiO(g) to be 10.96 kcal/mole using the same molecular constants used to calculate the free energy functions and extrapolating tabulated data (Kelley, 1960) to obtain $H^\circ_{1500} - H^\circ_{298} = 23.40$ kcal/mole for $Li_2O(s)$. A third-law calculation with $P_{LiO} = 3.4 \times 10^{-8}$ atm and $P_{Li} = 0.47 \times 3.03 \times 10^{-6}$ atm at 1500°K yields $\Delta H^\circ_{298} = 199.6 \pm 3$ kcal/mole, taking the free energy function of $Li_2O(s)$ at 1500°K as 22.84 cal/deg mole (Lewis *et al.*, 1961; Kelley, 1960; Kelley and King, 1961). Using $P_{Li} = 1.3 \times 10^{-6}$ atm fixed by the heat of formation of $Li_2O(s)$ would raise this result by 0.3 kcal/mole. Combination of these results with the heat of formation of $Li_2O(s)$, -143.0 ± 0.5 kcal/mole (JANAF, 1964), the heat of sublimation of lithium, 38.6 ± 0.4 kcal (Hultgren *et al.*, 1963), and the dissociation energy of oxygen yields the dissociation energy of LiO(g) at 298.15°K: 80.2 ± 4 (third law) and 75.2 ± 7 kcal/mole (second law). The heat of formation of $Li_2O(s)$ is based upon heat of solution measurements (Kolesov *et al.*, 1959) and the heat of formation of LiOH (Gunn, 1967).

$$D_0^\circ = 77 \pm 6 \quad \text{kcal}$$

38. LuO

Effusion data (Ames *et al.*, 1967) for the reaction

$$Lu_2O_3(s) = 1.94\ LuO(g) + 0.06\ Lu(g) + 1.06\ O(g)$$

indicate $D_0(LuO) = 156$ kcal/mole upon estimating LuO(g) free energy functions consistent with those in Table IX: $-(F^\circ - H_0^\circ)/T \simeq 52.5$ cal/deg mole at 2200°K and 53.6 at 2600°K. The heat of formation at 25°C of $Lu_2O_3(s)$ is -449 ± 2 kcal/mole (Huber *et al.*, 1960b), and that of Lu(g) is 102.2 ± 0.4 (Hultgren *et al.*, 1966). Mass spectrometric data for the isomolecular exchange reaction

$$Lu(g) + LaO(g) = LuO(g) + La(g)$$

indicate $D_0(\text{LaO}) - D_0(\text{LuO}) = 24$ kcal/mole by the second-law method (Ames *et al.*, 1967; White *et al.*, 1962) and 26 using the above free energy functions (Ames *et al.*, 1967). The third-law result corresponds to $D_0(\text{LuO}) =$ 161 kcal/mole.

$$D_0^\circ = 158 \pm 8 \quad \text{kcal}$$

39. MgO

The numbers in parentheses below were calculated assuming a $^1\Sigma$ ground state and a small electronic contribution to the free energy function due to the observed $^1\Pi$ state at 3503 cm^{-1}. The other results are based upon the MgO(g) free energy functions in Table IX. The basis of the present MgO(g) functions has been discussed in Section II.

Drowart *et al.* (1964a) have looked at the reactions

$$\text{MgO(g)} + \text{O(g)} = \text{Mg(g)} + \text{O}_2\text{(g)}$$
$$\text{MgO(g)} + \text{WO}_2\text{(g)} = \text{Mg(g)} + \text{WO}_3\text{(g)}$$

in a mass spectrometer. In combination with $p_{\text{WO}_3}/p_{\text{WO}_2} \cdot p_{\text{O}}$ given by De Maria *et al.* (1960), as well as $\Delta H_0^\circ = 147.3$ kcal/mole for $\text{WO}_3 = \text{WO}_2 + \text{O}$ from these authors, the oxygen exchange data yield dissociation energies at 0°K of 78.2 (85.9) and 77.3 (85.0) kcal/mole, respectively. Recalculation of the results of an earlier mass spectrometric study (Porter *et al.*, 1955) sets an upper limit of 88 (95) kcal/mole to the dissociation energy. Transpiration studies of the vaporization of magnesium oxide in oxygen via the reaction

$$\text{MgO(s)} = \text{MgO(g)}$$

give $D_0^\circ = 85$ (92) kcal/mole (Alexander *et al.*, 1963) and $D_0^\circ = 72$ (80) kcal/mole (Altman, 1963). However, interpretation of the transpiration results of Alexander *et al.* (1963) may be somewhat ambiguous because of the presence of hydroxide in the vapor (Alexander, 1968). All data for solid magnesium oxide are from Pankratz and Kelley (1963b). The enthalpy of sublimation of magnesium is 35.0 ± 0.3 kcal/g-atom (Hultgren *et al.*, 1966). Flame studies on MgO have recently been reviewed by Schofield (1967). The dissociation enthalpy chosen is based primarily upon the mass spectrometric results for the MgO–O reaction

$$D_0^\circ = 78 \pm 7 \quad \text{kcal}$$

40. MnO

The dissociation energy is based primarily upon flame photometric studies of MnO emission bands. The value $D_0^\circ = 96 \pm 3$ kcal/mole has been reported from measurements of the intensity of these bands as a function of temperature in flames of known H_2:H_2O ratio (Padley and Sugden, 1959).

An upper limit of 92±9 kcal has been obtained from spectrographic measurements of the Mn(g) concentration in acetylene–air flames assuming MnO(g) to be the only other species containing manganese (Huldt and Lagerqvist, 1951). Linear Birge–Sponer extrapolations based upon reported ground-state vibrational constants yield 104 kcal [9 levels, Das Sarma (1959)], 102 kcal [8 levels, Sen Gupta, (1934)], and 111 kcal [6 levels, Joshi (1962)]. Application of the Gaydon (1953) correction lowers the first of these to 94±10 kcal/mole. The reported value is well below the upper limit set by vaporization studies (Brewer and Mastick, 1951).

$$D_0^\circ = 95 \pm 8 \quad \text{kcal}$$

41. MoO

De Maria *et al.* (1960) measured with a mass spectrometer partial pressures of MoO and the other species effusing from molybdenum Knudsen cells containing Al_2O_3. Third-law evaluation of their data yields $\Delta H^\circ_{298} = 43$ kcal for the reaction

$$Mo(s) + O(g) = MoO(g)$$

corresponding to a MoO(g) dissociation energy at 298.15°K of 115 kcal/mole. The standard enthalpy of vaporization of molybdenum is taken to be 157.3±0.5 (Hultgren *et al.*, 1965). Pressures reported by De Maria and co-workers (1960) along with those in Table II of these authors' paper on Al_2O_3 (Drowart *et al.*, 1960) yield equilibrium constants for the reaction

$$Mo(s) + \tfrac{1}{3}\,Al_2O_3(s) = MoO(g) + \tfrac{2}{3}\,Al(g)$$

Using free energy functions for Al_2O_3(s) noted in the paragraph on AlO (JANAF, 1964) leads to $\Delta H^\circ_{298} = 288$ kcal for the latter reaction. The corresponding dissociation energy of the gaseous monoxide is 115 kcal/mole utilizing heats of formation in Tables V and VIII. The dissociation energy obtained from these data is about 12 kcal/mole less than is calculated from data for the reaction $MoO_2(g) = MoO(g) + O(g)$ taking $\Delta H^\circ_{298} = 282$ kcal for $MoO_2(g) = Mo(g) + 2\ O(g)$ (Brewer and Rosenblatt, 1961).

$$D_0^\circ = 114 \pm 12 \quad \text{kcal}$$

42. NO

Frisch and Margrave (1965) carried out a calorimetric study of the reaction $NO(g) + CO(g) = \tfrac{1}{2}\ N_2(g) + CO_2(g)$ and obtain a heat of formation of NO(g) at 298.15°K of 21.556±0.06 kcal/mole corresponding to 21.43±0.06 at 0°K. Combination with one half the dissociation energy at 0°K of N_2, 112.53±0.06 kcal/g-atom (Douglas, 1952; Wagman *et al.*, 1968), and of

O_2, 58.98±0.02 kcal/g-atom (Brix and Herzberg, 1954; Wagman *et al.*, 1968) leads to the thermochemically based result, $D_0(NO) = 150.08 \pm 0.14$ kcal/mole. This is essentially the dissociation energy computed from the tabulation of Wagman and associates (1968), as they list a value for ΔHf° (NO) only 0.01 kcal/mole higher. However, Callear and Smith (1964) have observed predissociation in the C $^2\Pi$ ($v=0$) state of NO. They conclude that all rotational levels of C $^2\Pi$ ($v=0$) predissociate and that the energy of the lowest rotational level at 6.493 eV places an upper limit on the dissociation energy of nitric oxide, $D_0(NO) \leq 149.73$ kcal/mole, which is independent of calorimetric data. We concur with Gaydon (1968) that the range $D_0 = 149.9 \pm 0.2$ seems likely to include the correct value.

$$D_0^\circ = 149.9 \pm 0.2 \quad \text{kcal}$$

43. NaO

Bawn and Evans (1937) observed the reaction of sodium atoms with NO_2 to have a very small activation energy and concluded that the enthalpy of the reaction

$$Na(g) + NO_2(g) = NaO(g) + NO(g)$$

must be negative or close to zero. Combining this observation with the heats of formation of NO_2 (Wagman *et al.*, 1968) and NO and the dissociation energy of O_2 indicates that the dissociation energy of NaO at 298.15°K is greater than or equal to 73 kcal/mole. Since estimates based upon comparison with neighboring elements in the periodic table or upon the average bond strength in NaO_2 (McEwan and Phillips, 1966) suggest $D_{298} \simeq 60$–63 kcal/mole, the limit set by Bawn and Evans should not be greatly below the correct value.

$$D_0^\circ = 72 \pm 12 \quad \text{kcal}$$

44. NbO

Shchukarev *et al.* (1966) have investigated the volatization of NbO(s, *l*) with a mass spectrometer. A third-law treatment of their results yields $\Delta H^\circ_{298} = 141$ kcal for the reaction NbO(s) = NbO(g) using the present free energy functions for NbO(g) and $-(F^\circ - H^\circ_{298})/T = 28.6$ cal/deg mole at 2000°K for NbO(s). The latter function is based upon $S^\circ_{298} = 12.0$ cal/deg mole and an extrapolation of high temperature heat content data (Gel'd and Kusenko, 1960). After combination with the enthalpy of formation of NbO(s), -98 ± 2 kcal/mole (Schäfer and Liedmeier, 1964; Morozova and Stolyarova, 1960), and with $\Delta H^\circ_{298} = 172.4 \pm 1$ for the sublimation of elemental niobium (Hultgren *et al.*, 1966; Scheer and Fine, 1965) these data indicate D°_{298}(NbO) = 189 kcal/mole. A very rough spectroscopic value of 182 kcal is obtained from a linear Birge–Sponer extrapolation of lower-state vibrational constants (Uhler, 1954) assuming dissociation to normal atoms.

$$D_0^\circ = 187 \pm 10 \quad \text{kcal}$$

45. NdO

White and his collaborators (1962; Walsh *et al.*, 1960; Goldstein, *et al.*, 1961) studied the reaction

$$Nd_2O_3(c) = 2\ NdO(g) + O(g)$$

in a tungsten effusion cell. Their data lead to D_0(NdO) = 167 kcal/mole. For this calculation we estimated NdO(g) free energy functions consistent with those presented in this paper: $-(F^\circ - H_0^\circ)/T \simeq 54.4$ cal/deg mole at 2400°K. The following heats of formation were used: $\Delta H^\circ_{298}(Nd_2O_3) = -432.1$ (Fitzgibbon *et al.*, 1968); ΔH°_{298}(Nd, g) = 77 ± 1 kcal/mole (Habermann and Daane, 1964; White *et al.*, 1961; Spedding and Daane, 1954; Johnson *et al.*, 1956).

$$D_0^\circ = 167 \pm 8 \quad \text{kcal}$$

46. NeO

This molecule has not been observed. Comparison with HeO and ArO suggests that if stable states exist the dissociation energy will be less than 1 kcal/mole.

$$D_0^\circ < 1 \quad \text{kcal}$$

47. NiO

Third-law treatment of NiO(g) pressures observed with a mass spectrometer (Grimley *et al.*, 1961b) yields $\Delta H^\circ_{298} = 131$ kcal for the reaction

$$NiO(s) = NiO(g)$$

taking $-(F^\circ - H^\circ_{298})/T$ for NiO(s) to equal 19.51 cal/deg mole at 1500°K and 21.51 cal/deg mole at 1800°K (Kelley, 1960; Kelley and King, 1961). This corresponds to a dissociation energy at 25°C for NiO(g) of 88.6 kcal/mole, using the following heats of formation: Ni(g), 102.8 ± 0.5 (Hultgren *et al.*, 1966); NiO(s), −57.3 ± 0.1 (Lewis *et al.*, 1961; Coughlin, 1954; Boyle *et al.*, 1954; Hahn and Muan, 1961); O(g), 59.6 kcal/mole. The dissociation energy is consistent with the upper limit of ~97 kcal/mole set by the results of Brewer and Mastick (1951) and Huldt and Lagerqvist (1954).

$$D_0^\circ = 88 \pm 5 \quad \text{kcal}$$

48. O_2

The dissociation energy has been obtained from spectroscopic measurements near the convergence limit (Brix and Herzberg, 1954).

$$D_0^\circ = 117.97 \pm 0.04 \quad \text{kcal}$$

49. OsO

Failure to observe OsO in a mass spectrometric study of the osmium–oxygen system (Grimley *et al.*, 1960) suggests that the equilibrium constant for the reaction

$$OsO(g) + O_2(g) = OsO_3(g)$$

is greater than 10^8 at 1700°K. Combining this limit with the free energy of formation of $OsO_3(g)$ (Alcock, 1961) along with free energy functions for OsO(g), Os(s) (Stull and Sinke, 1956), and $O_2(g)$ sets a lower limit of 106 kcal/mole to the standard enthalpy of formation of OsO(g). Taking the heat of sublimation of osmium to be 188 kcal/g-atom (Panish and Reif) 1962; Carrera *et al.*, 1964) sets an upper limit of 142 kcal/mole to the dissociation energy of OsO at 298°K.

$$D_0^\circ < 141 \quad \text{kcal}$$

50. PO

Extrapolations of the rapidly converging B state vibrational levels (Dressler, 1955; Cordes and Warkehr, 1965; Meinel and Krauss, 1966) yield dissociation energies of the B state between 22,100 and 24,000 cm^{-1}. The lowest of these values is based upon a cubic equation for the vibrational levels (Meinel and Krauss, 1966) and corresponds to a convergence limit above $v=0$ of the ground state at 52,907 cm^{-1}. If the B $^2\Sigma^+$ state dissociates to normal oxygen and excited $^2D_{3/2}$ phosphorous atoms (Meinel and Krauss, 1966; Cordes and Warkehr, 1965) the dissociation energy of PO is 41,545 cm^{-1} or 118.8 kcal/mole. The $^2\Sigma^+$ state which correlates with ground-state atoms is expected to be repulsive, by analogy with NO state assignments (Gilmore, 1965; Dressler and Miescher, 1965). On the other hand, if the B state correlates with ground state atoms (Santharam and Rao, 1963) the dissociation energy should be 32.5 kcal/mole higher than the value listed, on the order of 151 kcal/mole. The present dissociation energy is supported by the predissociation observed in the D′ $^2\Pi_r$ (or D $^2\Pi_r$) state about 49,650 cm^{-1}, 142 kcal, above $v''=0$, $J''=\frac{1}{2}$ of the X $^2\Pi_{1/2}$ ground state (Couet *et al.*, 1967, 1968).

$$D_0^\circ = 118.8 \pm 3 \quad \text{kcal}$$

51. PbO

Third-law calculations based upon a mass spectrometric study (Drowart *et al.*, 1965a) of the reaction

$$PbO(\beta) = PbO(g)$$

yield $\Delta H^\circ_{298} = 68.7 \pm 1$ kcal. Free energy functions for PbO(s) are based upon tabulated heat contents (Kelley, 1960) and $S^\circ_{298} = 16.42$ cal/deg mole (Kostryukov and Morozova, 1960): at 1000°K, $-(F^\circ - H^\circ_{298})/T = 22.60$. The mass spectrometric enthalpy is notably higher than the lower limits set by third-law treatments of total vapor pressure data, uncorrected for polymerization and disproportionation: Nesmeyanov *et al.* (1960a), Knudsen, 62.5; Nesmeyanov *et al.* (1960b), transpiration, 63.4; Feiser (1929), transpiration, 60.9; Knacke and Prescher (1964), Knudsen, 62.4; Richardson and Webb (1955), Knudsen, 64.4; Richards (1956), transpiration, 65.4; and Hörbe and Knacke (1959), 62.2 kcal. The mass spectrometric data yield $D^\circ_{298}(PbO) = 89.4 \pm 2$ kcal/mole when combined with $\Delta H^\circ_{298}f(PbO, \beta) = -51.94$ (Wagman *et al.*, 1968), $\Delta H^\circ_{298}f(Pb, g) = 46.62 \pm 0.3$ kcal/mole (Hultgren *et al.*, 1965), and the dissociation energy of oxygen.

$$D_0^\circ = 88.4 \pm 2 \quad \text{kcal}$$

52. PdO

Mass spectrometric observations (Norman *et al.*, 1964, 1965b) indicate that at 1900°K, $K = 10^{-1.9}$ for

$$Pd(g) + \tfrac{1}{2}\, O_2(g) = PdO(g)$$

which corresponds to $D^\circ_{298} = 56.3$ kcal/mole using the present free energy functions. This value is about 8 kcal/mole lower than the upper limit set by the increase in volatility of Pd in a stream of O_2 (Alcock and Hooper, 1960).

$$D_0^\circ = 55 \pm 7 \quad \text{kcal}$$

53. PrO

Mass spectrometric study (Walsh *et al.*, 1961; White *et al.*, 1962) of the isomolecular exchange reaction

$$Pr(g) + LaO(g) = PrO(g) + La(g)$$

indicates $D_0(LaO) - D_0(PrO) = 15$ kcal/mole by the second-law method (Ames *et al.*, 1967) and 7 kcal/mole using an estimated PrO(g) free energy function of $-(F^\circ - H_0^\circ)/T \simeq 53.0$ cal/deg mole at 2000°K. Similar data for the reaction

$$Pr(g) + NdO(g) = PrO(g) + Nd(g)$$

yield $D_0(PrO) - D_0(NdO) = -7$ (second law) or -10 kcal/mole (third law).

The third-law results correspond to $D_0(\mathrm{PrO}) = 180$ and 177 kcal/mole, respectively.

$$D_0^\circ = 179 \pm 8 \quad \text{kcal}$$

54. PtO

Norman *et al.* (1967) have carried out a mass spectrometric investigation of the reaction of oxygen with metallic platinum. Their data for the reaction

$$\mathrm{Pt(s)} + \tfrac{1}{2}\,\mathrm{O_2(g)} = \mathrm{PtO(g)}$$

indicate $\Delta H^\circ_{298} = 104$ kcal by the second-law method and 111 kcal by the third-law method. The thermal functions for Pt(s) used for these computations were obtained from Hultgren *et al.* (1963). The third-law result corresponds to a dissociation energy of platinum monoxide at 298.15°K of 83 kcal, in exact agreement with the value calculated directly from the reported partial pressures of Pt, O, and PtO at 2018°K. The enthalpy of sublimation of platinum is 135.2 kcal/g-atom (Dreger and Margrave, 1960; Hampson and Walker, 1961).

$$D_0^\circ = 82 \pm 8 \quad \text{kcal}$$

55. PuO

Knudsen (Phipps *et al.*, 1950; Pardue and Keller, 1964; Ackermann *et al.*, 1966) and transpiration (Mulford and Lamar, 1961) vapor-pressure studies of PuO_2(s) indicate plutonium monoxide to be an important vapor species in the plutonium–oxygen system. Standard free energies of formation presented by Ackermann *et al.* (1966) indicate $\Delta F^\circ = 170{,}750 - 27.07\,T$ for the dissociation of PuO in the temperature range, 1600–2150°K. An approximate dissociation energy has been obtained from this equation using tabulated data for Pu(g) (Hultgren *et al.*, 1965) and an estimated free energy function for PuO(g) of 68.8 cal/deg mole at 1800°K.

$$D_0^\circ = 162 \pm 15 \quad \text{kcal}$$

56. RbO

The dissociation energy has been estimated by comparison with neighboring elements in the periodic table.

$$D_0^\circ = (60) \pm 20 \quad \text{kcal}$$

57. ReO

This molecule has never been observed. ReO^+ has been seen only as an electron-impact fragmentation product. Mass spectrometric investigations of the vapor over condensed ReO_3 and ReO_2 (Studier, 1962; Semenov and Ovchinnikov, 1965) and kinetic studies of the oxidation of rhenium (Phillips, 1963; Hamamura and Tomita, 1967) indicate Re_2O_7(g) and ReO_3(g) to be the most stable gaseous oxides of rhenium. The average Re—O bond energy in these molecules is 148 kcal/mole (Semenov and Ovchinnikov, 1965).

58. RhO

The reaction of oxygen with rhodium has been investigated with a mass spectrometer (Norman *et al.*, 1964). The authors give a second-law heat of formation for RhO(g) at 2000°K that corresponds to the value 93 ± 10 kcal/mole at 298.15°K. These authors also give intensity ratios at 2000°K that can be used in an approximate third-law calculation. For this calculation we took $P_{\mathrm{Rh}} = 1.7 \times 10^{-7}$ atm from the known vapor pressure of rhodium (Panish and Reif, 1961; Hampson and Walker, 1961; Dreger and Margrave, 1961) and then calculated $P_{\mathrm{O}_2} = 2.8 \times 10^{-4}$ atm and $P_{\mathrm{RhO}} = 8.6 \times 10^{-10}$ atm from the observed intensity ratios and the ionization cross sections of Mann (1967). The calculated heat of formation of RhO(g) is 117 ± 15 kcal/mole. The reason for the large discrepancy between the second- and third-law results is not apparent. An intermediate value, 103 kcal/mole, was used along with the heat of sublimation of rhodium, 133 ± 1 kcal/g-atom (Panish and Reif, 1961; Hampson and Walker, 1961; Dreger and Margrave, 1961; Strassmair and Stark, 1967), to calculate the dissociation energy $D^\circ_{298} = 90 \pm 15$ kcal/mole.

$$D_0^\circ = 89 \pm 15 \quad \text{kcal}$$

59. RuO

The ruthenium–oxygen system has been investigated by mass spectrometric Knudsen cell methods (Norman *et al.*, 1968). Norman and co-workers (1968) present a second-law heat of formation for RuO(g) at 2000°K that corresponds to the value 88 ± 5 kcal/mole at 298.15°K. These authors also carried out an approximate pressure calibration that leads to $\Delta F^\circ_{2000} = 61$ kcal/mole for the formation of RuO(g). This latter result yields a third-law standard heat of formation for RuO(g) of 116 kcal/mole using free energy functions in Table IX and $-(F^\circ - H^\circ_{298})/T = 11.95$ cal/deg mole for Ru(s) at 2000°K (Kelley, 1960; Kelley and King, 1961). The reason for the large discrepancy between the second- and third-law values, which also occurs in these authors' (Norman *et al.*, 1964) study of RhO, is not apparent, although some suggestions have been discussed (Norman *et al.*, 1968). An intermediate formation enthalpy of 100 kcal/mole, somewhat weighted toward the second-law result favored by Norman *et al.* (1968), leads to $D^\circ_{298}(\mathrm{RuO}) = 115 \pm 15$

kcal/mole. The enthalpy of vaporization of ruthenium was taken as 155.5 ± 1.5 kcal/g-atom (Panish and Reif, 1962; Paule and Margrave, 1963; Carrera *et al.*, 1964). A rough linear Birge–Sponer extrapolation based upon six observed vibrational levels gives $D_0^\circ = 41$ kcal/mole (Raziunas *et al.*, 1965b).

$$D_0^\circ = 114 \pm 15 \quad \text{kcal}$$

60. SO

The selected value, based upon predissociations observed by Martin (1932) and the vibrational numbering of Norrish and Oldershaw (1959), is in agreement with the values, 123.7 ± 0.3 (Abadie and Herman, 1963) and 127.0 ± 3.7 kcal (McGrath and McGarvey, 1964), derived from short extrapolations of the vibrational levels of the excited state observed by McGrath and McGarvey (1962). The thermochemical data are in accord with the spectroscopic value. $D_0^\circ = 123.58 \pm 0.03$ kcal/mole is consistent with formation enthalpies tabulated by Wagman and co-workers (1968).

$$D_0^\circ = 123.58 \pm 0.03 \quad \text{kcal}$$

61. SbO

Rough linear Birge–Sponer extrapolations of the ground-state vibrational constants measured in several transitions indicate the dissociation energy of this molecule to be 88 (Sen Gupta, 1939), 92 (Sen Gupta, 1943), or 104 kcal/mole (Laksham, 1960).

$$D_0^\circ = 88 \pm 20 \quad \text{kcal}$$

62. ScO

The volatization of scandium sesquioxide has been studied by Semenov (1965) and by Ames *et al.* (1967). Recalculation of Semenov's mass spectrometric data yields $\Delta H_{298}^\circ = 256$ kcal for the reaction

$$\tfrac{1}{2}\, Sc_2O_3(s) = ScO(g) + \tfrac{1}{2}\, O(g)$$

using free energy functions for $Sc_2O_3(s)$ of 53.2 and 54.2 at 2500 and 2600°K extrapolated from high temperature heat content data (Pankratz and Kelley, 1963a) and $S_{298}^\circ = 18.4$ cal/deg mole (Weller and King, 1963). Combination with the heat of formation of scandium sesquioxide, -456 kcal/mole (Huber *et al.*, 1963), and the heat of vaporization of scandium metal, 90.3 kcal/g-atom (Hultgren *et al.*, 1966), leads to $D_{298}^\circ(ScO) = 152$ kcal/mole. Recomputation of effusion results for the reaction (Ames *et al.*, 1967).

$$Sc_2O_3(s) = 1.97\, ScO(g) + 0.03\, Sc(g) + 1.03\, O(g)$$

leads to $D_{298}^\circ(ScO) = 155$ kcal/mole.

Mass spectrometric studies (Smoes *et al.*, 1965) of the isomolecular exchange reaction

$$YO(g)+Sc(g)=ScO(g)+Y(g)$$

yield $D_0(YO)-D_0(ScO)=10.8$ (second law) or 9.9 kcal/mole (third law). The third-law result corresponds to $D_0(ScO)=152$ kcal/mole. Clausius–Clapeyron enthalpies for the exchange reaction

$$LaO(g)+Sc(g)=ScO(g)+Y(g)$$

indicate $D_0(LaO)-D_0(ScO)=30.9$ (Smoes *et al.*, 1965) or 27.7 kcal/mole (Ames *et al.*, 1967). The same data treated with the present monoxide free energy functions yield 30.5 (Smoes *et al.*, 1965) or 29 (Ames *et al.*, 1967), corresponding to $D_0(ScO)=156$ or 158 kcal/mole, respectively. Third-law treatment of equilibrium constants (Coppens *et al.*, 1967) for the reaction

$$GeO(g)+Sc(g)=ScO(g)+Ge(g)$$

indicates $D_0(GeO)-D_0(ScO)=+2.4$ and $D_0(ScO)=154$ kcal/mole. The thermochemical dissociation energy is consistent with that derived from spectroscopic data: a linear Birge–Sponer extrapolation yields 171 kcal/mole, and from this Gaydon (1953) estimated $D_0=138\pm23$ kcal/mole.

$$D_0{}^\circ = 154\pm5 \quad \text{kcal}$$

63. SeO

The dissociation energy is based upon the predissociation observed by Barrow and Deutsch (1963), as well as their interpretation of a linear Birge–Sponer extrapolation.

$$D_0{}^\circ = 100\pm15 \quad \text{kcal}$$

64. SiO

Heats of formation of SiO(g), Si(g), and O(g) tabulated by Wagman and associates (1968) lead to $D_0{}^\circ(SiO)=190.9$ and $D^\circ_{298}=192.3$ kcal/mole. The tabulated formation enthalpy of SiO(g) agrees closely with that recommended by Wise and co-workers (1963) in a careful review of data affected by the heat of formation of α-silica. The chosen enthalpy of formation of SiO(g) is 1 kcal more negative—and the dissociation energy 1 kcal higher—than the value obtained by Ancey-Moret *et al.* (1964) utilizing a thermodynamic cycle independent of the heat of formation of $SiO_2(\alpha)$. The thermochemical data are in satisfactory agreement with the spectroscopic result, $D_0{}^\circ=185\pm7$ kcal/mole (Barrow and Rowlinson, 1954).

$$D_0{}^\circ = 190.9\pm2 \quad \text{kcal}$$

65. SmO

According to White and associates (1962; Ames *et al.*, 1967) samarium sesquioxide vaporizes via the reaction

$$Sm_2O_3(c) = 1.80\ SmO(g) + 0.20\ Sm(g) + 1.20\ O(g)$$

Using $-(F° - H_0°)/T \approx 71.5$ cal/deg mole at 2400°K, along with $\Delta H^\circ_{298} = -433.9$ kcal/mole for the formation of Sm_2O_3 (Huber *et al.*, 1955) and $\Delta H^\circ_{298} = 49.4$ (Hultgren *et al.*, 1966) for the vaporization of samarium metal, their Knudsen results indicate $D_0(SmO) = 132$ kcal/mole. Mass spectrometric data (White *et al.*, 1962; Ames *et al.*, 1967), for the oxygen exchange reaction

$$Sm(g) + YO(g) = SmO(g) + Y(g)$$

yield $D_0(YO) - D_0(SmO) = 27$ kcal by both the second- and third-law methods. From this one computes $D_0(SmO) = 134$ kcal/mole.

$$D_0° = 133 \pm 8 \quad \text{kcal}$$

66. SnO

Mass spectrometric (Colin *et al.*, 1965) and Knudsen effusion (Hoenig and Searcy, 1966) studies of the reaction

$$SnO_2(s) = SnO(g) + \tfrac{1}{2}\ O_2(g)$$

yield $\Delta H^\circ_{298} = 142.6$ and 143.4 kcal, respectively, by the third-law method. For $SnO_2(s)$, $-(F° - H^\circ_{298})/T = 20.67$ cal/deg mole at 1000°K and 26.20 at 1500°K (Kelley, 1960; Kelley and King, 1961). The average of the above two enthalpies corresponds to a dissociation energy of SnO(g) at 298.15°K of 127.6 kcal/mole using the following heats of formation: $SnO_2(s)$, -138.8 kcal/mole (Coughlin, 1954) and Sn(g), 72.2 ± 0.5 kcal/g-atom (Hultgren *et al.*, 1963). Colin *et al.* (1965) mass spectroscopic data for the reaction

$$\tfrac{1}{2}\ Sn(c) + \tfrac{1}{2}\ SnO_2(s) = SnO(g)$$

yield $\Delta H^\circ_{298} = 73.7$ kcal using free energy functions for Sn(c) tabulated by Hultgren and associates (1963). Transport (Platteeuw, 1953; Platteeuw and Meyer, 1956) and Knudsen (Vesselovskii, 1943) results, uncorrected for polymerization of the vapor, set lower limits to the enthalpy of this reaction of 64.7 and 68.6 kcal, respectively. The mass spectrometric data correspond to $D^\circ_{298}(SnO) = 127.5$ kcal/mole. A spectroscopic upper limit to $D_0°$ of 130.9 kcal/mole is obtained from a very short graphical Birge–Sponer extrapolation of the E state vibrational levels (Barrow and Rowlinson, 1954; Eisler and

Barrow, 1949). If the E state dissociates to Sn($^3P_1$) and O($^3P_1$) the spectroscopic limit yields $D_0^\circ = 125.5$ kcal/mole, in excellent agreement with the thermochemical value, 126.6.

$$D_0^\circ = 126 \pm 2 \quad \text{kcal}$$

67. SrO

The numbers in parentheses below were calculated assuming a $^1\Sigma$ ground state with no other states low enough to contribute to the free energy function at 2300°K. The other results are based upon the free energy functions in Table IX, discussed in Section II of this chapter.

Colin *et al.* (1964) have studied the equilibrium

$$\mathrm{Sr(g)+SO(g)=S(g)+SrO(g)}$$

in a mass spectrometer. Their data yield D_0°(SrO)$=92.3$ (101.2)± 6 kcal/mole after correction to be consistent with the dissociation energy of SO in Table VI. In the same laboratory Drowart *et al.* (1964a) have investigated the reactions

$$\mathrm{SrO(g)+O(g)=Sr(g)+O_2(g)}$$
$$\mathrm{SrO(g)+WO_2(g)=Sr(g)+WO_3(g)}$$
$$\mathrm{SrO(g)+MoO_2(g)=Sr(g)+MoO_3(g)}$$

Third-law calculations give $D_0^\circ = 92.6$ (102.5), 91.8 (102.1), and 92.8 (102.0) kcal/mole, respectively, from these data. The equilibrium constants $K = p_{\mathrm{WO_2}}\, p_{\mathrm{O}}/p_{\mathrm{WO_3}}$ and $K = p_{\mathrm{MoO_2}}\, p_{\mathrm{O}}/p_{\mathrm{MoO_3}}$, as well as $\Delta H_0^\circ = 147.3$ kcal/mole for $\mathrm{WO_3 = WO_2 + O}$ and $\Delta H_0^\circ = 149.6$ for $\mathrm{MoO_3 = MoO_2 + 0}$, were taken from the work of De Maria *et al.* (1960). Other, less reliable, determinations of the dissociation energy of SrO have been reviewed by Drowart and co-workers (1964a) and Schofield (1967)

$$D_0^\circ = 92 \pm 6 \quad \text{kcal}$$

68. TaO

Two mass spectrometric investigations of the vaporization of Ta–Ta_2O_5 mixtures in tantalum Knudsen cells agree that TaO(g) and TaO_2(g) are the predominant vapor species under these conditions. However, although both investigations show the pressures of TaO and TaO_2 to be roughly equal around 2200°K, they disagree by about a factor of 1000 concerning the magnitude of those pressures: Krikorian and Carpenter (1965) report TaO pressures on the order of $10^{-8.6}$ atm at 2200°K, while Inghram *et al.* (1957) report pressures on the order of $10^{-5.5}$ atm at 2200°K. This results in a 11–13 kcal discrepancy (depending upon ionization cross sections) in the enthalpy of the reaction

$$\tfrac{1}{2}\,\mathrm{Ta(s)}+\tfrac{1}{2}\,\mathrm{TaO_2(g)}=\mathrm{TaO(g)}$$

for which third-law results range from 74 kcal [data of Inghram *et al.* (1957)] to 87.5 kcal [data of Krikorian and Carpenter (1965)]. Some possible reasons for the discrepancy have been discussed by Krikorian and Carpenter (1965). Until further data are available there appears to be no basis for revision of their conclusion that the enthalpy of the above reaction is $\approx$84 kcal and that the dissociation energy of tantalum monoxide is about 182 kcal/mole. A linear Birge–Sponer extrapolation of ground-state vibrational constants (Cheetham and Barrow, 1967) yields $D_0=214$ kcal/mole.

$$D_0^\circ = 182\pm15 \quad \mathrm{kcal}$$

69. TbO

Mass spectrometric investigations of isomolecular oxygen exchange reactions of the type $\mathrm{Tb(g)+MO(g)=TbO(g)+M(g)}$ (White *et al.*, 1962; Ames *et al.*, 1967) indicate $D_0(\mathrm{LaO})-D_0(\mathrm{TbO})=21$, $D_0(\mathrm{GdO})-D_0(\mathrm{TbO})=1$, and $D_0(\mathrm{PrO})-D_0(\mathrm{TbO})=12$ kcal/mole using estimated TbO free energy functions, based upon 0°K, of $\approx$69.9 cal/deg mole at 2000°K. Clausius–Clapeyron enthalpies for these three reactions indicate $D_0(\mathrm{MO})-D_0(\mathrm{TbO})=$ 18, 0, and 4 kcal/mole, respectively. The third-law results correspond to $D_0(\mathrm{TbO})=166$, 160, and 167 kcal/mole, respectively.

$$D_0^\circ = 164\pm8 \quad \mathrm{kcal}$$

70. TeO

An upper limit to the dissociation energy is set by a predissociation in the A $0^+\rightarrow$X 0^+ system between 31,300 and 31,595 cm^{-1} (Shin-Piaw, 1938; Chandler *et al.*, 1965). The higher vibrational levels of the A $0^+(^3\Sigma^-_{0^+})$ state converge rapidly to a limit at about 32,542 cm^{-1} above $v''=0$ of the ground state (Shin-Piaw, 1938; Chandler *et al.*, 1965). A higher excited state, tentatively assigned to be $^3\Sigma^+$, converges to a limit at 38,239 cm^{-1} (Haranath *et al.*, 1959). Although it is generally agreed that the ground state dissociates into normal atoms, $\mathrm{Te(^3P_2)+O(^3P)}$, the dissociation products and potential energy curves of the excited states are somewhat controversial. Haranath *et al.*, (1959) assume that A 0^+ extrapolates to $\mathrm{Te(^3P_0}$, 4707 cm$^{-1})+\mathrm{O(^3P)}$ and that the higher excited state ($^3\Sigma^+$?) extrapolates to $\mathrm{Te(^1D_1}$, 10,559 cm^{-1}) $+\mathrm{O(^3P)}$. These products lead to $D_0=27{,}835$ (79.6 kcal/mole) and 27,680 cm^{-1}, respectively. Chandler *et al.* (1965) assign the extrapolated convergences to potential maxima and take the observed predissociation as corresponding to the level of $Te(^3P_2)+O(^3P)$, which yields $D_0=90\pm3$ kcal. Other interpretations of the spectrum yield $D_0=62.85$ (Gaydon, 1953; Shin-Piaw, 1938). We have chosen the value based upon the simplest interpretation with uncertainty limits including the 90 kcal value.

$$D_0{}^\circ = 80^{+10}_{-1} \quad \text{kcal/mole}$$

71. ThO

Equilibrium constants of $10^{-4.49}$ at 2369°K (Ackermann *et al.*, 1963) and $10^{-4.95}$ at 2300°K (Darnell and McCollum, 1961) have been reported for the reaction

$$\tfrac{1}{2}\,\mathrm{Th(s)}+\tfrac{1}{2}\,\mathrm{ThO_2(s)}=\mathrm{ThO(g)}$$

Both sets of data yield $\Delta H^\circ_{298}=152$ kcal for this reaction after a third-law calculation. Free energy functions for $ThO_2(s)$ were extrapolated above 2000°K as reported previously (Brewer and Rosenblatt, 1961). Combination with the enthalpy of formation of thorium (IV) oxide, -293.2 kcal/mole (Huber *et al.*, 1952), the enthalpy of sublimation of thorium, 137.5 kcal/g-atom (Hultgren *et al.*, 1967), and the dissociation energy of oxygen leads to $D^\circ_{298}=192$ kcal/mole. The Knudsen results are supported by weight-loss (Wolff and Alcock, 1962) and torsion-Langmuir (Alcock and Peleg, 1967) data.

$$D_0{}^\circ = 191\pm10 \quad \text{kcal}$$

72. TiO

Wahlbeck and Gilles (1967) have studied the vaporization of $Ti_3O_5(\beta)$ in a tungsten crucible by Knudsen-effusion and mass spectrometric methods. They conclude that the primary reaction is

$$\tfrac{1}{3}\,\mathrm{Ti_3O_5}(\beta)=\mathrm{TiO(g)}+\tfrac{2}{3}\,\mathrm{O(g)}$$

Their data lead to $\Delta H^\circ_{298}=242\pm1$ kcal using the present free energy functions for TiO(g), $Ti_3O_5(\beta)$ functions from the JANAF (1967) tables, and O(g) functions from the JANAF (1962) tables. The second-law result is $\Delta H^\circ_{298}=$ 237 kcal. These enthalpies differ from those quoted in Wahlbeck and Gilles' paper (1967), 239 and 238 kcal, respectively, largely because different TiO(g) free energy functions were used. The third-law enthalpy, 242 kcal, yields D°_{298}(TiO)$=165\pm5$ kcal/mole from a thermodynamic cycle utilizing the heat of formation of $Ti_3O_5(\beta)$, -585 kcal/mole (JANAF, 1967), the heat of sublimation of titanium, 112.3 kcal/g-atom (Hultgren *et al.*, 1966), and the dissociation energy of oxygen. However, an appreciably lower dissociation energy is calculated from mass spectrometric observations by Berkowitz *et al.* (1957b) of the volatilization of TiO(s) in molybdenum and thoria Knudsen cells. Third-law treatment of their data, taking $-(F^\circ - H^\circ_{298})/T =$ 21.62 cal/deg mole at 2000°K for TiO(s) (Kelley, 1960; Kelley and King, 1961), yields an enthalpy at 298.15°K of 141 kcal for the reaction

$$\mathrm{TiO(s) = TiO(g)}$$

The second-law enthalpy at 298.15°K is 140 kcal. The third-law result leads to $D^\circ_{298}(\mathrm{TiO}) = 155 \pm 5$, taking the enthalpy of formation of TiO(s) to be -124.2 kcal/mole (Lewis *et al.*, 1961). The latter dissociation energy can be combined with the dissociation energy of TiO_2(g) (Brewer and Rosenblatt, 1961) to calculate $\Delta H^\circ_{298} = 5 \pm 15$ for the reaction $\mathrm{Ti(g) + TiO_2(g) = 2\ TiO(g)}$, in good agreement with the value $\Delta H^\circ_{298} = 8$ kcal calculated from the pressure ratios observed in the mass spectrometer (Berkowitz *et al.*, 1957b). A recent communication (Gilles *et al.*, 1969) indicates that mass spectrometric measurements on isomolecular reactions involving ScO, YO, and TiO (Drowart *et al.*, 1969) favor the lower value for the dissociation energy of TiO. The discrepancy between the dissociation energy of Wahlbeck and Gilles (1967) and that obtained from other experiments has been discussed (Gilles, 1967, *et al.*, 1969; Drowart *et al.*, 1969). The entropies and free energy functions of the high temperature forms of Ti_3O_5(c) and TiO(c) are also quite uncertain, because of both a lack of heat capacity measurements below 50°K and uncertainties regarding the heats associated with high-temperature phase transitions. In addition, the TiO(c) thermal data contain uncertainties associated with nonstoichiometry, magnetic ordering, and vacancies. Pending further clarification of the thermodynamics of the Ti–O system we have chosen an intermediate value for the monoxide dissociation energy, somewhat weighted towards the mass spectrometer results, $D^\circ_{298} = 158 \pm 8$ kcal/mole.

TABLE XI. Comparison of Free Energy Functions at 2000°K and Derived Dissociation Enthalpies for TiO Gas

Basis of electronic contribution	$-(F^\circ - H^\circ_0)/T$ (electronic) (cal/deg mole)	$-(F^\circ - H^\circ_{298})/T$ (total) (cal/deg mole)	D°_{298} [a] (kcal/mole)	D°_{298} [b] (kcal/mole)
Ti^{2+} levels	5.72	66.76	164.8	154.7
Observed TiO states	3.99	64.88	168.5	158.5
Carlson and Nesbet (1964) calculations	3.58	64.45	169.3	159.4

[a] Data of Wahlbeck-Gilles (1967).
[b] Data of Berkowitz *et al.* (1957b).

The absolute magnitude of the dissociation energy, but not the difference between conflicting values, depends upon the free energy functions for TiO(g). Table XI summarizes the electronic contribution to $-(F^\circ - H_0^\circ)/T$ at 2000°K, $-(F^\circ - H_{298}^\circ)/T$ total at 2000°K, and calculated dissociation energies using the present free energy functions, functions based upon experimentally observed levels, and functions based upon the quantum mechanical calculations of Carlson and Nesbet (1964). The results in the table—and possibly the discrepancy between second- and third-law enthalpies introduced into the data of Wahlbeck and Gilles by the present functions—indicate that for TiO, where good spectroscopic data and calculations are available, the recipe used in the present chapter overestimates the electronic contribution to the partition function.

$$D_0^\circ = 157 \pm 8 \quad \text{kcal}$$

73. TlO

Failure to observe TlO^+ in a mass spectrometric examination of the vapor over Tl_2O_3 while GaO^+ and InO^+ were seen over the corresponding oxides (Shchukarev *et al.*, 1962), the observation that thallium atoms can account for 70 to 90% of the total thallium in flames (Gurvich and Veits, 1958), failure to observe any bands in a thallium arc in air under conditions where GaO and InO were observed (Watson and Shambon, 1936), and Howell's (1945) prediction that all TlO states are repulsive all indicate that the dissociation energy of TlO(g) is very small.

$$D_0^\circ \ll 75 \quad \text{kcal}$$

74. TmO

Effusion (Ames *et al.*, 1967) and mass spectrometric (Panish, 1961) studies indicate that thulium sesquioxide forms both TmO(g) and Tm(g) upon vaporization. Taking the heat of formation of Tm_2O_3(s) at 298°K determined by Huber *et al.* (1960a), -451.4 ± 1.4 kcal, and the heat of vaporization of Tm evaluated by Hultgren and associates (1967), 55.5 ± 1 kcal, both types of data suggest that vaporization to the elements predominates. The effusion data indicate $D_0(\text{TmO}) \leq 127$ kcal/mole, in good agreement with the value $D_0(\text{TmO}) = 121$ kcal obtained from an estimated TmO(g) heat of formation of -7 kcal/mole. The estimate is based upon the correlation noted by White and associates (1962; Ames *et al.*, 1967). For consistency with the free energy functions in Table IX, $-(F^\circ - H_0^\circ)/T \simeq 71.0$ cal/deg mole at 2400°K for TmO(g).

$$D_0^\circ = 121 \pm 15 \quad \text{kcal}$$

75. UO

Third-law treatment of the partial pressures over mixtures of uranium dioxide and alumina measured with a mass spectrometer (De Maria *et al.*, 1960) gives $\Delta H^\circ_{298} = 173$ kcal for the reaction

$$UO_2(g) = UO(g) + O(g)$$

This leads to $D^\circ_{298}(UO) = 182$ kcal/mole assuming $D^\circ_{298}(UO_2) = 355$ kcal/mole (Brewer and Rosenblatt, 1961). The dissociation energy of $UO_2(g)$ is based upon $\Delta H^\circ_{298} = 149$ kcal/mole for $UO_2(s) = UO_2(g)$ (Ackerman *et al.*, 1956; Ohse, 1966) and a heat of sublimation of metallic uranium of 126 kcal/g-atom (De Maria *et al.*, 1960; Drowart *et al.*, 1964b, 1965c, 1967; Storms, 1965, 1966; Alexander *et al.*, 1969). Taking the heat of sublimation of uranium dioxide from the mass spectrometric study (De Maria *et al.*, 1960), which corresponds to considering the reaction occurring to be $UO_2(s) = UO(g) + O(g)$, would lower $D^\circ_{298}(UO)$ to 174 kcal/mole. Third-law treatment of pressures reported over mixtures of excess uranium and Al_2O_3 (De Maria *et al.*, 1960) yields $\Delta H^\circ_{298} = 8$ kcal for the equilibrium..

$$2\ UO(g) = UO_2(g) + O(g)$$

corresponding to $D_{298}(UO) = 182$ kcal/mole. For these calculations, free energy functions for $UO_2(g)$ are from Brewer and Rosenblatt (1961) and those for U(g) from Stull and Sinke (1956). The enthalpy of formation of $UO_2(s)$ is -259.0 kcal/mole (Rand and Kubaschewski, 1963). The above data are supported by recent work (Drowart *et al.*, 1967) on the exchange reaction

$$UO(g) + Si(g) = SiO(g) + U(g)$$

which indicates $D_0(UO) - D_0(SiO) = -8.9$ kcal/mole or $D^\circ_{298}(UO) = 183$, after correction to be consistent with the free energy functions and enthalpies in this chapter.

$$D_0^\circ = 181 \pm 8 \quad \text{kcal}$$

76. VO

Berkowitz *et al.* (1957a) measured the vapor pressure of VO(g) over VO(s) in a tungsten Knudsen cell. As reported earlier (Brewer and Rosenblatt, 1961) third-law analysis of their data yields $\Delta H^\circ_{298} = 133.9 \pm 5$ kcal/mole for the reaction

$$VO(s) = VO(g)$$

Taking $H^\circ_{1800} - H^\circ_{298}$ of VO(g) to be 13.2 kcal and the heat content of VO(s) given by Kelley (1960) along with the reported $\Delta H^\circ_{1800} = 124.9$ kcal/mole leads to the second-law value, $\Delta H^\circ_{298} = 133.3$ kcal/mole. The third-law value can be combined with a heat of formation of VO(s) of −103.2 (Mah and Kelley, 1961), the heat of sublimation of vanadium, 122.9±0.3 (Hultgren *et al.*, 1966), and the dissociation energy of oxygen to calculate the dissociation energy of gaseous vanadium monoxide at 298.15°, 152±10 kcal/mole. In addition, mass spectrometric equilibrium constants (Coppens *et al.*, 1967) for the isomolecular exchange reaction

$$VO(g)+Ge(g)=GeO(g)+V(g)$$

indicate $\Delta H^\circ_{298}=D^\circ_{298}(VO)-D^\circ_{298}(GeO)=-2.8$, or $D^\circ_{298}(VO)=155\pm4$ kcal/mole, using the present free energy functions and taking the equilibrium data to refer to the reverse of the reaction quoted (Drowart, 1968).

$$D_0^\circ = 153\pm5 \quad \text{kcal}$$

77. WO

The partial pressure of WO and other vapor species in the W–Al_2O_3 system have been measured with a mass spectrometer (De Maria *et al.*, 1960). Recalculation of the reported data leads to $\Delta H^\circ_{298}=47$ kcal for the reaction

$$W(s)+O(g)=WO(g)$$

using W(s) free energy functions of 14.39 cal/deg mole at 2000° and 15.72 at 2500°K (JANAF, 1966a) and O functions in the JANAF (1962) tables. The tungsten metal functions selected agree with those of Kirillin and associates (1963) in the temperature range of interest. Subtraction of the above result from the enthalpy of vaporization of tungsten, 203.1 kcal/g-atom (Hultgren *et al.*, 1965; Szwarc *et al.*, 1965; JANAF, 1966a), indicates the dissociation energy of WO(g) to be 156 kcal/mole at 298°K. Other results (Drowart *et al.*, 1960; De Maria *et al.*, 1960) lead to $D^\circ_{298}(WO)=158$ kcal/mole via the equilibrium

$$W(s)+\tfrac{1}{3}\,Al_2O_3(c)=WO(g)+\tfrac{2}{3}\,Al(g)$$

for which the data indicate $\Delta H^\circ_{298}=291$ kcal. Free energy functions and the heat of formation of $Al_2O_3(\alpha)$ are from the JANAF (1964, 1965, 1966a) tables. The dissociation energy can also be calculated from the reaction

$$WO_2(g)=WO(g)\pm O(g)$$

for which $\Delta H^\circ_{298}=152$ kcal, using the $WO_2(g)$ free energy functions and dissociation energy reported previously (Brewer and Rosenblatt, 1961). The result is $D^\circ_{298}(WO)=155$ kcal/mole, indicating very good internal consistency in these data.

$$D_0^\circ = 155\pm6 \quad \text{kcal}$$

78. XeO

Vibrational analysis (Cooper *et al.*, 1961) gives $\omega_e=372$ and $\omega_e x_e=12\text{cm}^{-1}$ for this molecule, which extrapolates to a dissociation limit of 8 kcal/mole.

$$D_0^\circ = 8 \quad \text{kcal/mole}$$

79. YO

Measurements of the absolute effusion rate of the vapor from solid yttrium sesquioxide in a tungsten Knudsen cell, as well as mass spectrometric analysis of that vapor, have been carried out in two laboratories (Ames *et al.*, 1967; White *et al.*, 1962; Walsh *et al.*, 1960; Ackermann *et al.*, 1964). Uncertainty concerning the relative cross sections of YO and Y makes it difficult to calculate the dissociation energy of YO(g) from the mass spectrometric results. The absolute effusion rates observed, however, indicate the overall reaction to be

$$Y_2O_3(s)=2\ YO(g)+O(g)$$

The pressure of Y(g) is one or two orders of magnitude lower than the pressure of YO(g). The absolute effusion rates of White and co-workers (1962; Ames *et al.*, 1967) yield $\Delta H^\circ_{298}=509$ kcal/mole for this reaction, while those of Ackermann *et al.* (1964) yield $\Delta H^\circ_{298}=515$ kcal/mole. The latter data are supported by recent effusion and torsion-Langmuir results (Alcock and Peleg, 1967). These enthalpies were calculated, utilizing the same procedure outlined by Ackermann *et al.* from the reported mass rates of effusion assuming only YO and O in the vapor. Free energy functions for $Y_2O_3(s)$ were obtained from $S^\circ_{298}=23.58$ (Kelley and King, 1961) and experimental high temperature heat contents and entropies (Pankratz *et al.*, 1962). The $Y_2O_3(s)$ data were extrapolated above 2000°K with a constant C_p of 31.5 cal/deg mole to obtain $-(F^\circ-H^\circ_{298})/T=59.63$ cal/deg mole at 2500° K. The weighted average of the two enthalpies, 513 kcal, was combined with the enthalpy of formation of yttrium sesquioxide, -455.45 kcal/mole (Huber *et al.*, 1957a), the enthalpy of sublimation of metallic yttrium, 101.5 kcal/g-atom (Hultgren *et al.*, 1967), and the dissociation energy of oxygen to obtain $\Delta H^\circ_{298}=162\pm5$ kcal/mole for the dissociation of YO(g).

$$D_0^\circ = 161 \pm 5 \quad \text{kcal}$$

80. YbO

Effusion and mass spectrometric studies (Ames *et al.*, 1967; Alcock and Peleg, 1967; Panish, 1961) indicate that ytterbium sesquioxide vaporizes predominantly by dissociation to the elements. Following the trend noted by White and co-workers (1962; Ames *et al.*, 1967) the heat of formation of gaseous ytterbium monoxide at 298°K can be estimated to be -2 ± 15 kcal/mole. This corresponds to $D_0(\text{YbO})=97$ kcal/mole, taking the heat of vaporization of Yb at 298°K to equal 36.4 ± 0.2 kcal/g-atom (Hultgren *et al.*, 1966). The estimated dissociation energy is 10–15 kcal/mole greater than the upper limit estimated by Ames *et al.*, (1967) from their effusion results. At 2400°K, $-(F^\circ-H_0^\circ)/T\simeq70.6$ cal/deg mole for YbO(g).

$$D_0^\circ = 97\pm15 \quad \text{kcal}$$

81. ZnO

A mass spectrometric search of the vapor over ZnO(s) showed no peaks that could be ascribed to zinc monoxide gaseous molecules (Anthrop and Searcy, 1964; Anthrop, 1963). The results indicate, at 1250°K, $P_{\text{ZnO}}<1.2\times10^{-10}$ atm when $P_{\text{Zn}}=1.6\times10^{-6}$ atm and $P_{\text{O}_2}=5.6\times10^{-7}$ atm. These data along with free energy functions in Table IX and Stull and Sinke (1956) set an upper limit of 66 kcal/mole to the dissociation energy of ZnO(g) at 298.15°K.

$$D_0^\circ \leq 65 \quad \text{kcal}$$

82. ZrO

The ZrO(g) pressures measured in a mass spectrometer (Chupka *et al.*, 1957) lead to $\Delta H^\circ_{298}=156$ kcal for the reaction

$$\tfrac{1}{2}\,\text{Zr(c)}+\tfrac{1}{2}\,\text{ZrO}_2\text{(s)}=\text{ZrO(g)}$$

using the present free energy functions for ZrO(g). Heat content and entropy data (Kelley, 1960; Kelley and King, 1961) for Zr(s) extrapolate to $-(F^\circ-H^\circ_{298})/T=19.0$ cal/deg mole at 2500°K. Tabulated ZrO_2(c) free energy functions (Lewis *et al.*, 1961) extrapolate to the value 33.3 cal/deg mole at 2500°K, assuming Cp $=17.8$ cal/deg mole. This result, along with the heat of formation of solid zirconium dioxide, -263.0 kcal/mole (Huber *et al.*, 1964b; Kornilov *et al.*, 1967), and the heat of sublimation of zirconium, 145.5 kcal/g-atom (Hultgren *et al.*, 1967), yields the dissociation energy of gaseous zirconium monoxide, 181 kcal/mole at 298.15°K.

$$D_0^\circ = 180\pm10 \quad \text{kcal}$$

REFERENCES

Abadie, D., and Herman, R. (1963). *Compt. Rend.* **257**, 2820.

Ackermann, R. J., and Thorn, R. J. (1961). *Progr. Ceram. Sci.* **1**, 39.

Ackermann, R. J., Gilles, P. W., and Thorn, R. J. (1956). *J. Chem. Phys.* **25**, 1089.

Ackermann, R. J., Rauh, E. G., Thorn, R. J., and Cannon, M. C. (1963). *J. Phys. Chem.* **67**, 762.

Ackermann, R. J., Rauh, E. G., and Thorn, R. J. (1964). *J. Chem. Phys.* **40**, 883.

Ackermann, R. J., Faircloth, R. F., and Rand, M. H. (1966). *J. Phys. Chem.* **70**, 3698.

Akerlind, L. (1962a). *Arkiv Fysik* **22**, 41.

Akerlind, L. (1962b). *Arkiv Fysik* **22**, 65.

Alcock, C. B. (1961). *Platinum Metals Rev.* **5**, 134.

Alcock, C. B., and Hooper, G. W. (1960). *Proc. Roy. Soc.* **A254**, 551.

Alcock, C. B., and Peleg, M. (1967). *Trans. Brit. Ceram. Soc.* **66**, 217.

Alexander, C. A. (1968). Private communication.

Alexander, C. A., Ogden, J. S., and Levy A. (1963). *J. Chem. Phys.* **39**, 3057.

Alexander, C. A., Ogden, J. S., and Pardue, W. M. (1969). *J. Nucl. Mater.* **31**, 13.

Altman, R. L. (1963). *J. Phys. Chem.* **67**, 366.

Amano, T., Hirota, E., and Morino, Y. (1967). *J. Phys. Soc. Japan* **22**, 399.

Ames, L. L., and Barrow, R. F. (1967). *Proc. Phys. Soc.* (*London*) **90**, 869.

Ames, L. L., Walsh, P. N., and White, D. (1967). *J. Phys. Chem.* **71**, 2707.

Ancey-Moret, M. F., Olette, M., and Richardson, F. D. (1964). *Mem. Sci. Rev. Met.* **61**, 169.

Anderson, C. T. (1937). *J. Am. Chem. Soc.* **59**, 488.

Anthrop, D. F. (1963). UCRL-10708. Lawrence Radiation Lab., Univ. of California, Berkeley, California.

Anthrop, D. F., and Searcy, A. W. (1964). *J. Phys. Chem.* **68**, 2335.

Arkell, A., Reinhard, R. R., and Larson, L. P. (1965). *J. Am. Chem. Soc.* **87**, 1016.

Baker, F. B., and Holley, Jr., C. E. (1968). *J. Chem. Eng. Data* **13**, 405.

Barrow, R. F. (1956). *Arkiv Fysik* **11**, 281.

Barrow, R. F., and Deutsch, E. W. (1963). *Proc. Phys. Soc.* (*London*) **82**, 548.

Barrow, R. F., and Rowlinson, H. C. (1954). *Proc. Roy. Soc.* **A224**, 374.

Barrow, R. F., Deutsch, J. L., and Travis, D. N. (1961). *Nature* **191**, 374.

Barrow, R. F., Gissane, W. J. M., and Richards, D. (1967). *Proc. Roy. Soc.* **A300**, 469.

Bawn, C. E. H., and Evans, A. G. (1937). *Trans. Faraday Soc.* **33**, 1571.

Becart, M., and Declerck, F. (1960). *Compt. Rend.* **251**, 2153.

Bell, W. E., and Tagami, M. (1963). *J. Phys. Chem.* **67**, 2432.

Bell, W. E., Tagami, M., and Inyard, R. E. (1966). *J. Phys. Chem.* **70**, 2048.

Belton, G. R. (1968). Private communication.

Berg, R. A., and Sinanogiu, O. (1960). *J. Chem. Phys.* **32**, 1082.

Berkowitz, J., Chupka, W. A., and Inghram, M. G. (1957a). *J. Chem. Phys.* **27**, 87.

Berkowitz, J., Chupka, W. A., and Inghram, M. G. (1957b). *J. Phys. Chem.* **61**, 1569.

Berkowitz, J., Chupka, W. A., Blue, G. D., and Margrave, J. L. (1959). *J. Phys. Chem.* **63**, 644.

Berry, R. S., and Rieman, C. W. (1963). *J. Chem. Phys.* **38**, 1540.

Bills, J. L., and Cotton, F. A. (1964). *J. Phys. Chem.* **68**, 802.

Blackburn, P. E. (1966). *J. Phys. Chem.* **70**, 311.
Blackburn, P. E., Büchler, A., and Stauffer, J. L. (1966). *J. Phys. Chem.* **70**, 2469.
Blewett, J. P., Liebhafsky, H. A., and Hennely, E. F. (1939). *J. Chem. Phys.* **7**, 478.
Boyle, B. J., King, E. G., and Conway, K. C. (1954). *J. Am. Chem. Soc.* **76**, 3835.
Brewer, L. (1953). *Chem. Rev.* **52**, 1.
Brewer, L. (1963). *Proc. Robert A. Welch Found. Conf. Chem. Res. 6th, Topics Mod. Inorg. Chem., Houston, Texas, November 1962*, pp. 47–92. Foundation Offices, Houston, Texas.
Brewer, L., and Chandrasekharaiah, M. S. (1960). UCRL-8713 (Revised). Lawrence Radiation Lab. Univ. of California, Berkeley, California.
Brewer, L., and Green, D. W. (1969). *High Temp. Sci.* **1**, 26.
Brewer, L., and Mastick, D. F. (1951). *J. Chem. Phys.* **19**, 834.
Brewer, L., and Rosenblatt, G. M. (1961). *Chem. Rev.* **61**, 257.
Brewer, L., Somayajulu, G. R., and Brackett, E. (1963). *Chem. Rev.* **63**, 111.
Brix, P., and Herzberg, G. (1954). *Can. J. Phys.* **32**, 110.
Brooks, R., and Kaufman, M. (1965). *J. Chem. Phys.* **43**, 3406.
Büchler, A., Stauffer, J. L., Klemperer, W., and Wharton, L. (1963). *J. Chem. Phys.* **39**, 2299.
Burns, R. P., De Maria, G., Drowart, J., and Inghram, M. G. (1963). *J. Chem. Phys.* **38**, 1035.
Callahan, W. R. (1963). *J. Opt. Sci. Am.* **53**, 695.
Callear, A. B., and Smith, I. W. M. (1964). *Discussions Faraday Soc.* **37**, 96.
Carlson, K. D., and Moser, C. M. (1963). *J. Phys. Chem.* **67**, 2644.
Carlson, K. D., and Moser, C. M. (1967). *J. Chem. Phys.* **46**, 35.
Carlson, K. D., and Nesbet, R. K. (1964). *J. Chem. Phys.* **41**, 1051.
Carrera, N. J., Walker, R. F., and Plante, E. R. (1964). *J. Res. Natl. Bur. Stds.* **A68**, 325.
Carrington, A., Dyer, P. N., and Levy, D. H. (1967). *J. Chem. Phys.* **47**, 1756.
Chandler, G. G., Hurst, H. J., and Barrow, R. F. (1965). *Proc. Phys. Soc.* (*London*) **86**, 105.
Charles, G. W. (1958). ORNL-2319. Oak Ridge Natl. Lab., Oak Ridge, Tennessee.
Cheetham, C. J., and Barrow, R. F. (1967). *Trans. Faraday Soc.* **63**, 1835.
Cheetham, C. J., Gissane, W. J. M., and Barrow, R. F. (1965). *Trans. Faraday Soc.* **61**, 1308.
Chupka, W. A., Inghram, M. G., and Porter, R. F. (1956). *J. Chem. Phys.* **24**, 792.
Chupka, W. A., Berkowitz, J., and Inghram, M. G. (1957). *J. Chem. Phys.* **26**, 1207.
Chupka, W. A., Berkowitz, J., and Giese, C. F. (1959). *J. Chem. Phys.* **30**, 827.
Claasen, A., and Veenemans, C. F. (1933). *Z. Physik* **80**, 342.
Cobble, J. S., Smith, W. T., and Boyd, G. E. (1953). *J. Am. Chem. Soc.* **75**, 5777, 5783.
Coleman, E. H., and Gaydon, A. G. (1947). *Discussions Faraday Soc.* **2**, 166.
Coleman, E. H., Gaydon, A. G., and Vaidya, V. M. (1948). *Nature* **162**, 108.
Colin, R., Goldfinger, P., and Jeunehomme, M. (1964). *Trans. Faraday Soc.* **60**, 306.
Colin, R., Drowart, J., and Verhaegen, G. (1965). *Trans. Faraday Soc.* **61**, 1364.
Collomon, J. H., and Morgan, J. E. (1965). *Proc. Phys. Soc.* (*London*) **86**, 1091.
Cooper, C. D., and Lichtenstein, M. (1958). *Phys. Rev.* **109**, 2026.
Cooper, C. D., Cobb, G. C., and Tolnas, E. L. (1961). *J. Mol. Spectry.* **7**, 223.
Coppens, P. (1966). Dissertation. Free Univ. of Brussels, Belgium.
Coppens, P., Smoes, S., and Drowart, J. (1967). *Trans. Faraday Soc.* **63**, 2140.
Coppens, P., Smoes, S., and Drowart, J. (1968). *Trans. Faraday Soc.* **64**, 630.
Cordes, H., and Warkehr, E. (1965). *Z. Physik. Chem.* (*Frankfurt*) **46**, 26.
Cosgrove, L. A., and Snyder, P. E. (1953). *J. Am. Chem. Soc.* **75**, 3102.

Couet, C., Coquart, B., Tuan Anh, N., and Guenebaut, H. (1967). *Compt. Rend.* **265B**, 479.
Couet, C., Coquart, B., Tuan Anh, N., and Guenebaut, H. (1968). *Compt. Rend.* **266B**, 1219.
Coughlin, J. P. (1954). *U. S. Bur. Mines Bull.* **542**.
Cubicciotti, D. (1969). *High Temp. Sci.* **1**, 11.
Daniels, J. M., and Dorain, P. B. (1966). *J. Chem. Phys.* **45**, 26.
Darken, L. S., and Gurry, R. W. (1946). *J. Am. Chem. Soc.* **68**, 798.
Darnell, A. J., and McCollum, W. A. (1961). NAA-SR-6498. Atomics Intern., Canoga Park, California [(1962). *Chem. Abstr.* **56**, 32f].
Das Sarma, J. M. (1959). *Z. Physik* **157**, 98.
De Galen, L. (1965). *Physica* **31**, 1286.
De Maria, G., Burns, R. P., Drowart, J., and Inghram, M. G. (1960). *J. Chem. Phys.* **32**, 1373.
Dhumwad, R. K., and Narasimham, N. A. (1966). *Proc. Indian Acad. Sci. Sect. A* **64**, 283.
Dibeler, V. H., Reese, R. M., and Franklin, J. L. (1957). *J. Chem. Phys.* **27**, 1296.
D'Orazio, L. A., and Wood, R. H. (1965). *J. Phys. Chem.* **69**, 2550.
Douglas, A. E. (1952). *Can. J. Phys.* **30**, 302.
Douglas, A. E., and Moller, C. K. (1955). *Can. J. Phys.* **33**, 125.
Dreger, L. H., and Margrave, J. L. (1960). *J. Phys. Chem.* **64**, 1323.
Dreger, L. H., and Margrave, J. L. (1961). *J. Phys. Chem.* **65**, 2106.
Dressler, K. (1955). *Helv. Phys. Acta* **28**, 563.
Dressler, K., and Miescher, E. (1965). *Astrophys. J.* **141**, 1266.
Drowart, J. (1968). Private communication.
Drowart, J., De Maria, G., Burns, R. P., and Inghram, M. G. (1960). *J. Chem. Phys.* **32**, 1366.
Drowart, J., Exsteen, G., and Verhaegen, G. (1964a). *Trans. Faraday Soc.* **60**, 1920.
Drowart, J., Pattoret, A., and Smoes, S. (1964b). *J. Nucl. Mater.* **12**, 319.
Drowart, J., Colin, R., and Exsteen, G. (1965a). *Trans. Faraday Soc.* **61**, 1376.
Drowart, J., Degreve, F., Verhaegen, G., and Colin, R. (1965b). *Trans. Faraday Soc.* **61**, 1072.
Drowart, J., Pattoret, A., and Smoes, S. (1965c). *J. Chem. Phys.* **42**, 2629.
Drowart, J., Pattoret, A., and Smoes, S. (1967). *Proc. Brit. Ceram. Soc.* **8**, 67.
Drowart, J., Coppens, P., and Smoes, S. (1969). *J. Chem. Phys.* **50**, 1046.
Durie, R. A., and Ramsey, D. A. (1958). *Can. J. Phys.* **36**, 35.
Durie, R. A., Legay, F., and Ramsey, D. A. (1960). *Can. J. Phys.* **38**, 444.
Dwyer, R. J., and Oldenberg, O. (1944). *J. Chem. Phys.* **12**, 351.
Edvinsson, E., Selin, L. E., and Aslund, N. (1965). *Arkiv Fysik* **30**, 283.
Eisler, B., and Barrow, R. F. (1949). *Proc. Phys. Soc.* (*London*) **A62**, 740.
Feiser, J. (1929). *Metall u. Erz* **26**, 269.
Fitzgibbon, G. C., and Holley, Jr., C. E. (1968). *J. Chem. Eng. Data* **13**, 63.
Fitzgibbon, G. C., Holley, Jr., C. E., and Wadsö, I. (1965). *J. Phys. Chem.* **69**, 2464.
Fitzgibbon, G. C., Pavone, D., and Holley, Jr., C. E. (1968). *J. Chem. Eng. Data* **13**, 547.
Flidlider, G. V., Kovtunenko, P. V., and Bundel, A. A. (1966). *Zh. Fiz. Khim.* **40**, 2168.
Fox, P. G., and Esdaile, R. J. (1963). *Acta Met.* **11**, 1363.
Frisch, M. A., and Margrave, J. L. (1965). *J. Phys. Chem.* **69**, 3863.
Gatterer, A., and Krishnamurty, S. G. (1952). *Nature* **109**, 543.

Gatterer, A., Junkes, J., Salpeter, E. W., and Rosen, B. (1957). " Molecular Spectra of Metallic Oxides." Vatican Press, Vatican City.
Gaydon, A. G. (1953). " Dissociation Energies and Spectra of Diatomic Molecules," 2nd ed. Chapman & Hall, London.
Gaydon, A. G. (1968). " Dissociation Energies and Spectra of Diatomic Molecules," 3rd ed. Chapman & Hall, London.
Gel'd, P. V., and Kusenko, F. G. (1960). *Izv. Akad. Nauk SSSR Otd. Tekhn. Nauk. Met. i Toplivo* p. 79.
Gerstein, B. C., Jelinek, F. J., Mullaly, J. R., Shickell, W. D., and Spedding, F. H. (1967). *J. Chem. Phys.* **47**, 5194.
Ghosh, C. (1932). *Z. Physik* **78**, 521.
Gilbert, I. G. F., and Kitchener, J. A. (1956). *J. Chem. Soc.* p. 3919.
Gilles, P. W. (1967). *J. Chem. Phys.* **46**, 4987.
Gilles, P. W. Hampson, P. J., and Wahlbeck, P. G. (1969) *J. Chem. Phys.* **50**, 1048.
Gilles, P. W., and Margrave, J. L. (1956). *J. Phys. Chem.* **60**, 1333.
Gilmore, F. R. (1965). *J. Quant. Spectry. Radiative Transfer* **5**, 369.
Glockler, G. (1948). *J. Chem. Phys.* **16**, 604.
Glemser, O., and Stoecker, U. (1963). *Ber. Bunsenges. Physik. Chem.* **67**, 505.
Goldberg, R. N., and Hepler, L. G. (1968). *Chem. Rev.* **68**, 229.
Goldstein, H. W., Walsh, P. N., and White, D. (1961). *J. Phys. Chem.* **65**, 1400.
Grebenshchikov, R. G., and Pasechnova, R. A. (1966). *Russ. J. Inorg. Chem. (English Transl.)* **11**, 264.
Grimley, R. T., and Burns, R. P. (1961). Presented at *Ann. Conf. Mass Spectrometry Allied Topics, 9th Chicago, June* 1961.
Grimley, R. T., Burns, R. P., and Inghram, M. G. (1960). *J. Chem. Phys.* **33**, 308.
Grimley, R. T., Burns, R. P., and Inghram, M. G. (1961a). *J. Chem. Phys.* **34**, 664.
Grimley, R. T., Burns, R. P., and Inghram, M. G. (1961b). *J. Chem. Phys.* **35**, 551.
Grimley, R. T., Burns, R. P., and Inghram, M. G. (1966). *J. Chem. Phys.* **45**, 4158.
Gunn, S. R. (1967). *J. Phys. Chem.* **71**, 1386.
Gurvich, G. V., and Ryabova, V. G. (1965). *Opt. i Spektroskopiya* **18**, 143.
Gurvich, G. V., and Veits, I. V. (1958). *Izv. Akad. Nauk SSSR, Ser. Fiz.* **22**, 673.
Gurvich, L. V., Novikov, M. M., and Ryabova, V. G. (1965). *Opt. i Spektroskopiya* **18**, 132.
Gusarov, A. V., Gorokhov, L. N., and Efimova, A. G. (1967). *Teplofiz. Vysokikh Temperatur Akad. Nauk SSSR* **5**, 584 [English Transl. (1967). *High Temp.* **5**, 524].
Haar, L., and Friedman, A. S. (1955). *J. Chem. Phys.* **23**, 869.
Habermann, C. E., and Daane, A. H. (1964). *J. Chem. Phys.* **41**, 2818.
Hahn, Jr., W. C., and Muan, A. (1961). *Phys. Chem. Solids* **19**, 338.
Hall, R. T., and Dowling, J. M. (1966). *J. Chem. Phys.* **45**, 1899.
Hamamura, T., and Tomita, G. (1967). *Bull. Chem. Soc. Japan* **40**, 1066.
Hampson, Jr., R. F., and Walker, R. F. (1961). *J. Res. Natl. Bur. Stds.* **A65**, 289.
Haranath, P. B. V., Rao, R. T., and Sivaramamurty, V. (1959). *Z. Physik* **155**, 507.
Herman, R., Weigner, C., and Herman, L. (1951). *Phys. Rev.* **82**, 751.
Hermann, G. (1937). *Z. Physik Chem* **35B**, 298.
Herzberg, G. (1950). " Molecular Spectra and Molecular Structure. I. Spectra of Diatomic Molecules," 2nd ed. Van Nostrand, Princeton, New Jersey.
Hildenbrand, D. L., and Hall, W. F. (1962). *J. Phys. Chem.* **66**, 754.
Hinnov, E., and Kohn, H. (1957). *J. Opt. Sci. Am.* **47**, 156.
Hoenig, C. L., and Searcy, A. W. (1966). *J. Am. Ceram. Soc.* **49**, 128.

Hoenig, C. L., Stout, N. D., and Nordine, P. C. (1967). *J. Am. Ceram. Soc.* **50**, 385.
Holley, Jr., C. E., and Huber, Jr., E. J. (1951). *J. Am. Chem. Soc.* **73**, 5577.
Hörbe, R., and Knacke, O. (1959). *Z. Erzbergbau Metallhuettenw.* **12**, 321.
Howell, H. G. (1945). *Proc. Phys. Soc.* (*London*) **57**, 32.
Huber, Jr., E. J., and Holley, Jr., C. E. (1953a). *J. Am. Chem. Soc.* **75**, 3594.
Huber, Jr., E. J., and Holley, Jr., C. E. (1953b). *J. Am. Chem. Soc.* **75**, 5645.
Huber, Jr., E. J., and Holley, Jr., C. E. (1955). *J. Am. Chem. Soc.* **77**, 1444.
Huber, Jr., E. J., and Holley, Jr., C. E. (1956). *J. Phys. Chem.* **60**, 498.
Huber, Jr., E. J., and Holley, Jr., C. E. (1968a). *J. Chem. Eng. Data* **13**, 252.
Huber, Jr., E. J., and Holley, Jr., C. E. (1968b). *J. Chem. Eng. Data* **13**, 545.
Huber, Jr., E. J., Holley, Jr., C. E., and Meierkord, E. H. (1952). *J. Am. Chem. Soc.* **74**, 3406.
Huber, Jr., E. J., Matthews, C. O., and Holley, Jr., C. E. (1955). *J. Am. Chem. Soc.* **77**, 6493.
Huber, Jr., E. J., Head, E. L., and Holley, Jr., C. E. (1956a). *J. Phys. Chem.* **60**, 1457.
Huber, Jr., E. J., Head, E. L., and Holley, Jr., C. E. (1956b). *J. Phys. Chem.* **60**, 1582.
Huber, Jr., E. J., Head, E. L., and Holley, Jr., C. E. (1957a). *J. Phys. Chem.* **61**, 497.
Huber, Jr., E. J., Head, E. L., and Holley, Jr., C. E. (1957b). *J. Phys. Chem.* **61**, 1021.
Huber, Jr., E. J., Head, E. L., and Holley, Jr., C. E. (1960a). *J. Phys. Chem.* **64**, 379.
Huber, Jr., E. J., Head, E. L., and Holley, Jr., C. E. (1960b). *J. Phys. Chem.* **64**, 1768.
Huber, Jr., E. J., Fitzgibbon, G. C., Head, E. L., and Holley, Jr., C. E. (1963). *J. Phys. Chem.* **67**, 1731.
Huber, Jr., E. J., Fitzgibbon, G. C., and Holley, Jr., C. E. (1964a). *J. Phys. Chem.* **68**, 2720.
Huber, Jr., E. J., Head, E. L., and Holley, Jr., C. E. (1964b). *J. Phys. Chem.* **68**, 3040.
Huldt, L., and Lagerqvist, A. (1951). *Arkiv Fysik* **3**, 525.
Huldt, L., and Lagerqvist, A. (1954). *Z. Naturforsch.* **9a**, 358.
Hultgren, R., Orr, R. L., Anderson, P. D., and Kelley, K. K. (1963). "Selected Values of Thermodynamic Properties of Metals and Alloys." Wiley, New York.
Hultgren, R., Orr, R. L., and Kelley, K. K. (1964). Supplement to selected values of thermodynamic properties of metals and alloys. Loose-leaf sheets issued by Univ. of California, Berkeley, California.
Hultgren, R., Orr, R. L., and Kelley, K. K. (1965). Supplement to selected values of thermodynamic properties of metals and alloys. Loose-leaf sheets issued by Univ. of California, Berkeley, California.
Hultgren, R., Orr, R. L., and Kelley, K. K. (1966). Supplement to selected values of thermodynamic properties of metals and alloys. Loose-leaf sheets issued by Univ. of California, Berkeley, California.
Hultgren, R., Orr, R. L., and Kelley, K. K. (1967). Supplement to selected values of thermodynamic properties of metals and alloys. Loose-leaf sheets issued by Univ. of California, Berkeley, California.
Hultgren, R., Orr, R. L., and Kelley, K. K. (1968). Supplement to selected values of thermodynamic properties of metals and alloys. Loose-leaf sheets issued by Univ. of California, Berkeley, California.
Hultin, M., and Lagerqvist, A. (1951). *Arkiv Fysik* **2**, 471.
Humphrey, G. L. (1954). *J. Am. Chem. Soc.* **76**, 978.
Huo, W. H., Freed, K. F., and Klemperer, W. (1967). *J. Chem. Phys.* **46**, 3556.
Iglesias, L. (1966). *Can. J. Phys.* **44**, 895.
Inghram, M. G., and Drowart, J. (1960). *Proc. Intern. Conf. High Temp. Technol. Asilomar, California*, 1959, p. 219. McGraw-Hill, New York.

Inghram, M. G., Chupka, W. A., and Porter, R. F. (1955). *J. Chem. Phys.* **23**, 2159.
Inghram, M. G., Chupka, W. A., and Berkowitz, J. (1957). *J. Chem. Phys.* **27**, 569.
James, C. G. (1954). Doctoral Dissertation. Cambridge Univ., Cambridge [quoted by V. A. Medvedev (1961). *Zh. Fiz. Khim.* **35**, 1481].
James, T. C., and Thibault, R. J. (1964). *J. Chem. Phys.* **41**, 2806.
JANAF (1962). "JANAF Thermochemical Tables," PB 168 370. Dow Chem. Co., Midland, Michigan, June 1962.
JANAF (1963). "JANAF Thermochemical Tables," PB 168 370. Dow Chem. Co., Midland, Michigan, September 1963.
JANAF (1964). "JANAF Thermochemical Tables," PB 168 370. Dow Chem. Co., Midland, Michigan, March 1964.
JANAF (1965). "JANAF Thermochemical Tables. First Addendum." PB 168 370-1. Dow Chem. Co., Midland, Michigan, December 1965.
JANAF (1966a). "JANAF Thermochemical Tables. Second Addendum." PB 168 370-2. Dow Chem. Co., Midland, Michigan, June 1966.
JANAF (1966b). "JANAF Thermochemical Tables. Second Addendum." PB 168 370-2. Dow Chem. Co., Midland, Michigan, December 1966.
JANAF (1967). "JANAF Thermochemical Tables. Second Addendum," PB 168 370-2. Dow Chem. Co., Midland, Michigan, March 1967.
Jevons, W. (1938). *Proc. Phys. Soc. (London)* **50**, 910.
Johnson, R. G., Hudson, D. E., Caldwell, W. C., Spedding, F. H., and Savage, W. R. (1956). *J. Chem. Phys.* **25**, 917.
Jolly, W. L., and Latimer, W. M. (1952). *J. Am. Chem. Soc.* **74**, 5757.
Joshi, K. C. (1962). *Spectrochim. Acta* **18**, 265.
Justice, B. H., and Westrum, Jr., E. F. (1963a). *J. Phys. Chem.* **67**, 339.
Justice, B. H., and Westrum, Jr., E. F. (1963b). *J. Phys. Chem.* **67**, 345.
Kalff, P. J., Hollander, Tj., and Alkemade, C. Th. J. (1965). *J. Chem. Phys.* **43**, 2299.
Kaufman, M., Wharton, L., and Klemperer, W. (1965). *J. Chem. Phys.* **43**, 943.
Kelley, K. K. (1960). *U. S. Bur. Mines Bull.* **584**.
Kelley, K. K., and Christensen, A. U. (1961). *U. S. Bur. Mines Rept. Invest.* **5710**.
Kelley, K. K. and King. E. G. (1961). *U. S. Bur. Mines Bull* **592**.
King, E. G. (1957). *J. Am. Chem. Soc.* **79**, 2399.
King, E. G., and Christensen, Jr., A. U. (1958). *J. Am. Chem. Soc.* **80**, 1800.
King, E. G., Weller, W. W., and Pankratz, L. B. (1961). *U.S. Bur. Mines Rept. Invest.* **5857**.
Kirillin, V. A., Sheindlin, A. E., Chekhovski, V. Ya., and Petrov, V. M. (1963). *Zh. Fiz. Khim.* **37**, 2249.
Knacke, O., and Prescher, K. E. (1964). *Z. Erzbergbau Metallhuettenw.* **17**, 28.
Kolesov, V.P., Skuratov, S. M., and Zaikin, I. D. (1959). *Zh. Neorgan. Khim.* **4**, 1237.
Kornilov, A. N., Leonidov, V. Ya., and Skuratov, S. M. (1962). *Dokl. Akad. Nauk. SSSR* **144**, 355.
Kornilov, A. N., Ushakova, I. M., and Skuratov, S. M. (1967). *Zh. Fiz. Khim.* **41**, 200.
Kostryukov, V. N., and Morozova, G. K. (1960). *Zh. Fiz. Khim.* **34**, 1833.
Krikorian, O. H., and Carpenter, J. H. (1965). *J. Phys. Chem.* **69**, 4399.
Kusenko, F. G., and Gel'd, P. V. (1960a). *Izv. Akad. Sib. Otdel. Nauk. SSSR* **2**, 46.
Kusenko, F. G., and Gel'd, P. V. (1960b). *Zh. Obshch. Khim.* **30**, 3847.
Lagerqvist, A., and Huldt, L. (1953). *Z. Naturforsch.* **8a**, 493.
Lagerqvist, A., and Huldt, L. (1954). *Z. Naturforsch.* **9a**, 991.
Lagerqvist, A., and Selin, L. E. (1956). *Arkiv Fysik* **11**, 323.
Lagerqvist, A., and Selin, L. E. (1957). *Arkiv Fysik* **12**, 553.

Lagerqvist, A., and Uhler, U. (1949). *Arkiv Fysik* **1**, 459.
Lagerqvist, A., and Uhler, U. (1953). *Arkiv Fysik* **6**, 95.
Lagerqvist, A., and Uhler, U. (1967). *Z. Naturforsch.* **22b**, 551.
Lagerqvist, A., Lind, E., and Barrow, R. F. (1950). *Proc. Phys. Soc. (London)* **A63**, 1132.
Lagerqvist, A., Nilsson, N. E. L., and Barrow, R. F. (1957). *Arkiv Fysik* **12**, 543.
Lagerqvist, A., Nilsson, N. E. L., and Wigartz, K. (1958). *Arkiv Fysik* **13**, 379.
Lagerqvist, A., Nilsson, N. E. L., and Wigartz, K. (1959). *Arkiv Fysik* **15**, 521.
Laksham, S. V. J. (1960). *Z. Physik* **158**, 367, 386.
Lewis, G. N., Randall, M., Pitzer, K. S., and Brewer, L. (1961). " Thermodynamics," 2nd ed., Appendix 7. McGraw-Hill, New York.
Mack, E., Osterhof, G. G., and Kraner, H. M. (1923). *J. Am. Chem. Soc.* **45**, 617.
McEwan, M. J., and Phillips, L. F. (1966). *Trans. Faraday Soc.* **62**, 1717.
McGrath, W. D., and McGarvey, J. J. (1962). *J. Chem. Phys.* **37**, 1544.
McGrath, W. D., and McGarvey, J. J. (1964). *Proc. Roy. Soc.* **A278**, 490.
Macur, G. J., Edwards, R. K., and Wahlbeck, P. G. (1966). *J. Phys. Chem.* **70**, 2956.
Mah. A. D. (1954). *J. Am. Chem. Soc.* **76**, 3364.
Mah. A. D. (1957). *J. Phys. Chem.* **61**, 1572.
Mah, A. D. (1958). *J. Am. Chem. Soc.* **80**, 3872.
Mah. A. D. (1959). *J. Am. Chem. Soc.* **81**, 1582.
Mah, A. D. (1960). *U. S. Bur. Mines Rept. Invest.* **5600**.
Mah. A. D. (1963). *U. S. Bur. Mines Rept. Invest.* **6171**.
Mah, A. D., and Kelley, K. K. (1961). *U. S. Bur. Mines Rept. Invest.* **5858**.
Mah, A. D., Pankratz, L. B., Weller, W. W., and King, E. G. (1967). *U. S. Bur. Mines Rept. Invest.* **7026**.
Mal'tsev, A. A., Kataev, D. I., and Tatevskii, V. M. (1960). *Opt. i Spektroskopiya* **9**, 713.
Mann, J. B. (1967). *J. Chem. Phys.* **46**, 1646.
Martin, E. V. (1932). *Phys. Rev.* **41**, 167.
Masse, J. L., and Masse-Baerlocher, M. (1967). *J. Chim. Phys.* **64**, 417.
Meinel, H., and Krauss, L. (1966). *Z. Naturforsch.* **21a**, 1878.
Messier, D. R. (1967). *J. Am. Ceram. Soc.* **50**, 665.
Meyer, B. (1965). *J. Mol. Spectry.* **18**, 443.
Meyer, C., Haeusler, C., and Barchewitz, P. (1965). *J. Phys. (Paris)* **26**, 799.
Moore, C. E. (1949). Atomic energy levels, Vol. I. *Natl. Bur. Std. (U.S.) Circ.* **467**.
Moore, C. E. (1952). Atomic energy levels, Vol. II. *Natl. Bur. Std. (U.S.) Circ.* **467**.
Moore, C. E. (1958). Atomic energy levels, Vol. III. *Natl. Bur. Std. (U.S.) Circ.* **467**.
Morozova, M. P., and Stolyarova, T. A. (1960). *Zh. Obshch. Khim.* **30**, 3848.
Mulford, R. N. R., and Lamar, L. E. (1961). *Proc. Intern. Conf. Plutonium Metallurgy 2nd, Grenoble, France*, 1960, p. 411. Cleaver-Hume Press, London.
Munir, Z. A., and Searcy, A. W. (1964). *J. Electrochem. Soc.* **111**, 1170.
NAS-NRC (1963). Table of general physical constants. *Natl. Bur. Std. Tech. News Bull.* **47** (10).
Nesmeyanov, A. N., Firsova, L. P., and Isakova, E. P. (1960a). *Zh. Fiz. Khim.* **34**, 1200.
Nesmeyanov, A. N., Firsova, L. P., and Isakova, E. P. (1960b). *Zh. Fiz. Khim.* **34**, 1699.
Newbury, R. S., Barton, Jr., G. W., and Searcy, A. W. (1968). *J. Chem. Phys.* **48**, 793.
Nikonov, B. P., and Otmakhova, N. G. (1961). *Zh. Fiz. Khim.* **35**, 1494.
Ninomiya, M. (1955). *J. Phys. Soc. Japan* **10**, 829.
Norman, J. H., Staley, H. G., and Bell, W. E. (1964). *J. Phys. Chem.* **68**, 662.
Norman, J. H., Staley, H. G., and Bell, W. E. (1965a). *J. Chem. Phys.* **42**, 1123.

Norman, J. H., Staley, H. G., and Bell, W. E. (1965b). *J. Phys. Chem.* **69**, 1373.
Norman, J. H., Staley, H. G., and Bell, W. E. (1967). *J. Phys. Chem.* **71**, 3686.
Norman, J. H., Staley, H. G., and Bell, W. E. (1968). *Advan. Chem. Ser.* **72**, 101.
Norrish, R. G. W., and Oldershaw, G. A. (1959). *Proc. Roy. Soc.* **A249**, 498.
Novokhatskii, L. A., and Lenev, L. M. (1966). *Zh. Neorgan. Khim.* **11**, 2014.
Oetting, F. L. (1967). *Chem. Rev.* **67**, 261.
Ohse, R. W. (1966). *J. Chem. Phys.* **44**, 1375.
Otto, E. M. (1964). *J. Electrochem. Soc.* **111**, 88.
Otto, E. M. (1966). *J. Electrochem. Soc.* **113**, 643.
Padley, P. J., and Sugden, T. M. (1959). *Trans. Faraday Soc.* **55**, 2054.
Panish, M. B. (1961). *J. Chem. Phys.* **34**, 2197.
Panish, M. B., and Reif, L. (1961). *J. Chem. Phys.* **34**, 1915.
Panish, M. B., and Reif, L. (1962). *J. Chem. Phys.* **37**, 128.
Panish, M. B., and Reif, L. (1964). *J. Chem. Phys.* **38**, 253.
Pankratz, L. B., and Kelley, K. K. (1963a). *U. S. Bur. Mines Rept. Invest.* **6198**.
Pankratz, L. B., and Kelley, K. K. (1963b). *U. S. Bur. Mines Rept. Invest.* **6295**.
Pankratz, L. B., King, E. G., and Kelley, K. K. (1962). *U. S. Bur. Mines Rept. Invest.* **6033**.
Papousek, D. (1961). *Collection Czech. Chem. Commun.* **26**, 1909.
Pardue, W. M., and Keller, D. L. (1964). *J. Am. Ceram. Soc.* **47**, 610.
Paule, R. C., and Margrave, J. L. (1963). *J. Phys. Chem.* **67**, 1896.
Phillips, J. G. (1951). *Astrophys. J.* **114**, 152.
Phillips, J. G. (1952). *Astrophys. J.* **115**, 567.
Phillips, J. G. (1969). *Astrophys. J.* **157**, 449.
Phillips, L. F., and Sugden, T. M. (1961). *Trans Faraday Soc.* **57**, 914.
Phillips, Jr., W. L., (1963). *J. Less-Common Metals* **5**, 97.
Phipps, T. E., Sears, G. W., and Simpson, O. C. (1950). *J. Chem. Phys.* **18**, 724.
Plante, E. R., and Szwarc, R. (1966). *J. Res. Natl. Bur. Std.* **A70**, 175.
Platteeuw, J. C. (1953). Dissertation. Tech. Wetenschap aan de Tech. Hogeschool te Delft, Delft.
Platteeuw, J. C., and Meyer, G. (1956). *Trans. Faraday Soc.* **52**, 1066.
Porter, G. (1950). *Discussions Faraday Soc.* **9**, 60.
Porter, R. F., Chupka, W. A., and Inghram, M. G. (1955). *J. Chem. Phys.* **23**, 1347.
Powell, F. X., and Lide, Jr., D. R. (1964). *J. Chem. Phys.* **41**, 1413.
Rand, M. H., and Kubaschewski, O. (1963). " The Thermochemical Properties of Uranium Compounds." Wiley (Interscience), New York.
Rank, D. H., St. Pierre, A. G., and Wiggins, T. A. (1965). *J. Mol. Spectry.* **18**, 418.
Rao, K. S. (1958). *Can. J. Phys.* **36**, 1526.
Raziunas, V., Macur, G. J., and Katz, S. (1963). *J. Chem. Phys.* **39**, 1161.
Raziunas, V., Macur, G. J., and Katz, S. (1965a). *J. Chem. Phys.* **42**, 2634.
Raziunas, V., Macur, G. J., and Katz, S. (1965b). *J. Chem. Phys.* **43**, 1010.
Richards, A. W. (1956). *Bull. Inst. Mining Met.* **65**, 151.
Richards, W. G., Verhaegen, G., and Moser, C. M. (1966). *J. Chem. Phys.* **45**, 3226.
Richardson, F. D., and Webb, L. E. (1955). *Bull. Inst. Mining Met.* **64**, 529.
Rosen, B. (1951). " Donnees Spectroscopiques Concernant les Molecules Diatomiques." Comite Intern. des Tables Annualles de Constantes et Données Numeriques (Intern. Union of Chem.). Hermann, Paris.
Rossini, F. D., Wagman, D. D., Evans, W. E., Levine, S., and Jaffe, I. (1952). Selected values of chemical thermodynamics properties, *Natl. Bur. Std.* (*U.S.*) *Circ.* **500**.

Rowlinson, H. C., and Barrow, R. F. (1953). *J. Chem. Phys.* **21**, 378.
Ryabova, V. G., and Gurvich, G. V. (1965). *Teplofiz. Vysokikh Temperatur Akad. Nauk SSSR* **3**, 318.
Santharam, C. V., and Rao, P. T. (1962). *Z. Physik* **168**, 553.
Santharam, C. V., and Rao, P. T. (1963). *Indian J. Phys.* **37**, 14.
Savage, W. R., Hudson, D. E., and Spedding, F. H. (1959). *J. Chem. Phys.* **30**, 221.
Scari, O. (1956). *Acta Phys. Acad. Sci. Hung.* **6**, 73.
Schäfer, H., and Liedmeier, F. (1964). *Z. Anorg. Allgem. Chem.* **329**, 225.
Schäfer, H., Schneidereit, G., and Gerhardt, W. (1963). *Z. Anorg. Allgem. Chem.* **319**, 327.
Scheer, M. D., and Fine, J. (1965). *J. Chem. Phys.* **42**, 3645.
Schofield, K. (1967). *Chem. Rev.* **67**, 707.
Semenov, G. A. (1965). *Zh. Neorgan. Khim.* **10**, 2390.
Semenov, G. A., and Ovchinnikov, K. V. (1965). *Zh. Obshch. Khim.* **35**, 1517.
Sen Gupta, A. K. (1934). *Z. Physik* **91**, 471.
Sen Gupta, A. K. (1939). *Indian J. Phys.* **13**, 145.
Sen Gupta, A. K. (1943). *Indian J. Phys.* **17**, 216.
Shchukarev, S. A., and Ryabov, A. N. (1960). *Zh. Neorgan. Khim.* **5**, 1931.
Shchukarev, S. A., and Semenov, G. A. (1957). *Zh. Neorgan. Khim.* **2**, 1217.
Shchukarev, S. A., and Semenov, G. A. (1961). *Dokl. Akad. Nauk SSSR* **141**, 652.
Shchukarev, S. A., Semenov, G. A., and Rat'kovskii, I. A. (1962). *Zh. Prikl. Khim.* **35**, 1454.
Shchukarev, S. A., Semenov, G. A., and Frantseva, K. E. (1966). *Zh. Neorgan. Khim.* **11**, 233.
Sherwood, E. M., Rosenbaum, D. M., Blocher, Jr., J. M., and Campbell, I. E. (1955). *J. Electrochem. Soc.* **102**, 650.
Shetlar, M. D. (1965). UCRL-11979. Lawrence Radiation Lab., Univ. of California, Berkeley, California.
Shin-Piaw, C. (1938). *Ann. Phys. (Paris)* [11] **10**, 173.
Shomate, C. H., and Huffman, E. H. (1943). *J. Am. Chem. Soc.* **65**, 1625.
Singh, N. L. (1959). *Can. J. Phys.* **37**, 136.
Singh, R. B., and Rai, D. K. (1965). *J. Quant. Spectry. Radiative Transfer* **5**, 723.
Smoes, S., Drowart, J., and Verhaegen, G. (1965). *J. Chem. Phys.* **43**, 732.
Spedding, F. H., and Daane, A. H. (1954). *J. Metals* **6**, 504.
Stafford, F. E., and Berkowitz, J. (1964). *J. Chem. Phys.* **40**, 2963.
Storms, E. K. (1965). LA-DC-6953. Los Alamos Scientific Lab., Los Alamos, New Mexico.
Storms, E. K. (1966). *Thermodyn. Proc. Symp., Vienna, July 1965*, pp. 309–343. Intern. At. Energy Agency, Vienna, Austria.
Strassmair, H., and Stark, D. (1967). *Z. Angew. Phys.* **23**, 40.
Stubblefield, C. T., Eick, H., and Eyring, L. (1956). *J. Am. Chem. Soc.* **78**, 3018.
Studier, M. H. (1962). *J. Phys. Chem.* **66**, 189.
Stull, D. R., and Sinke, G. C. (1956). "Thermodynamic Properties of the Elements," Am. Chem. Soc., Washington, D.C.
Swaminathan, T. M., and Krishnamurty, S. G. (1954). *Current Sci. (India)* **23**, 258.
Szwarc, R., Plante, E. R., and Diamond, J. J. (1965). *J. Res. Natl. Bur. Std.* **A69** 417.
Theard, L. P., and Hildenbrand, D. L. (1964). *J. Chem. Phys.* **41**, 3416.
Toerring, T. (1964). *Z. Naturforsch.* **19a**, 1426.
Toerring, T. (1966). *Z. Naturforsch.* **21a**, 287.
Tyte, D. C. (1967). *Proc. Phys. Soc. (London)* **92**, 1134.

Uhler, U. (1953). *Arkiv Fysik* **7**, 125.
Uhler, U. (1954). *Arkiv Fysik* **8**, 265.
Uhler, U., and Akerlind, L. (1956). *Arkiv Fysik* **10**, 431.
Uhler, U., and Akerlind, L. (1961). *Arkiv Fysik* **19**, 1.
Veits, I. V., and Gurvich, L. V. (1956). *Opt. i Spektroskopiya* **1**, 22.
Verhaegen, G., and Richards, W. G. (1966). *J. Chem. Phys.* **45**, 1828.
Vesselovskii, V. K. (1943). *Zh. Prikl. Khim.* **16**, 397.
Wagman, D. D., Evans, W. H., Parker, V. B., Halow, I., Bailey, S. M., and Schumm, R. H. (1968). Selected values of chemical thermodynamics properties, *Natl. Bur. Std. (U.S.) Tech. Note* **270–3**.
Wahlbeck, P. G., and Gilles, P. W. (1967). *J. Chem. Phys.* **46**, 2465.
Walsh, P. N., Goldstein, H. W., and White, D. (1960). *J. Am. Ceram. Soc.* **43**, 229.
Walsh, P. N., Dever, D. F., and White, D. (1961). *J. Phys. Chem.* **65**, 1410.
Washburn, C. A. (1963). UCRL-10991. Lawrence Radiation Lab., Univ. of California, Berkeley, California.
Watson, W. W., and Shambon, A. (1936). *Phys. Rev.* **50**, 607.
Weller, W. W., and King, E. G. (1963). *U. S. Bur. Mines Rept. Invest.* **6245**.
Weltner, Jr., W., and McLeod, Jr., D., (1965a). *Nature* **206**, 87.
Weltner, Jr., W., and McLeod, Jr., D. (1965b). *J. Phys. Chem.* **69**, 3488.
Weltner, Jr., W., and McLeod, Jr., D. (1965c). *J. Chem. Phys.* **42**, 882.
Weltner, Jr., W., and McLeod, Jr., D. (1965d). *J. Mol. Spectry.* **17**, 276.
Wharton, L., and Klemperer, W. (1963). *J. Chem. Phys.* **38**, 2705.
Wharton, L., Kaufman, M., and Klemperer, W. (1962). *J. Chem. Phys.* **37**, 621.
Wise, S. S., Margrave, J. L., Feder, H. M., and Hubbard, W. N. (1963). *J. Phys. Chem.* **67**, 815.
White, D., Walsh, P. N., Goldstein, H. W., and Dever, D. F. (1961). *J. Phys. Chem.* **65**, 1404.
White, D., Walsh, P. N., Ames, L. L., and Goldstein, A. W. (1962). *Thermodyn. Proc. Symp. Vienna, May 1962*, pp. 417–443. Intern. At. Energy Agency, Vienna, Austria.
White, D., Seshadri, K. S., Dever, D. F., Mann, D. E., and Linevsky, M. J. (1963). *J. Chem. Phys.* **39**, 2463.
Wolff, E. G., and Alcock, C. B. (1962). *Trans. Brit. Ceram. Soc.* **61**, 667.

The Matrix-Isolation Technique Applied to High Temperature Molecules

W. Weltner, Jr.

DEPARTMENT OF CHEMISTRY, UNIVERSITY OF FLORIDA, GAINESVILLE, FLORIDA

I. Introduction 85
II. Experimental Procedures 86
III. Optical Spectroscopy 87
A. The VO Molecule 87
B. Metal Oxide Molecules 89
C. Metal Halide Molecules 94
IV. Electron-Paramagnetic-Resonance Spectroscopy 98
V. Orientation of Molecules in Matrices 101
References 103

I. Introduction

To quote from recent contributors to this series (Cheetham and Barrow, 1967), " the arc spectrum of TaO is extremely complicated, and the preliminary study of the transitions observed in the matrix has proved most valuable in directing attention to particular regions of the arc spectrum." Those authors are referring to the spectrum of molecular TaO trapped in a solid inert-gas matrix at 4°K. The simplification in spectra when measured at low temperatures, rather than near 2000°K, is one of the benefits of the application of matrix isolation to the study of high temperature molecules.

Matrix isolation is a technique for trapping molecules as isolated entities in an inert solid, or matrix, in order to investigate their properties, usually by optical means. It has been applied to the study of a large number of free radicals, and its usefulness has expanded from optical spectroscopy into the area of electron paramagnetic resonance (EPR). This discussion will be limited to those molecules that are more or less associated only with high temperature systems and are present in the saturated vapor over refractory solids. There are many species, such as OH, CN, or CH_3, which occur in a variety of environments, but do not fall into this category. References to the numerous radicals investigated in matrices by Pimentel, Robinson, Broida, Milligan and Jacox, Jen, and others may be found in earlier reviews [see Bass and Broida (1960), Jacox and Milligan (1964), Weltner (1967)].

II. Experimental Procedures

In most instances high temperature molecules are produced by the vaporization of a refractory solid and effuse from a Knudsen cell into a vacuum. This molecular beam is allowed to collide with a cloud of inert-gas atoms just prior to quenching on a window cooled to 4°K or 20°K. The inert gas solidifies on the window, and being in excess, it serves to trap and isolate the molecules of interest in a rigid matrix. Usually the ratio of matrix to trapped molecules is greater than 1000. Figure 1 shows one arrangement of furnace and dewar that can be used. High temperatures can be attained by induction heating, resistance heating, or by electron bombardment of the sample cell. The cooled window can be turned to allow absorption or emission studies to be made through other windows on the outside of the dewar.

The matrix material can be any gas (or vapor) that will not react with the trapped molecule and that can be conveniently and rigidly solidified. The rare gases are usually used because they are relatively inert and are transparent to light over a wide wavelength region. Neon can be used only with

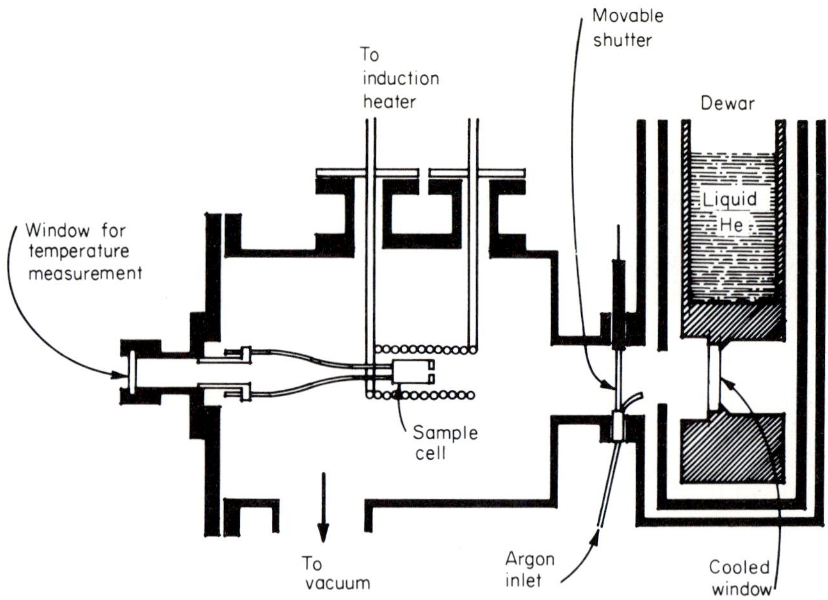

Fig. 1. An apparatus for the matrix isolation of high temperature molecules at 4°K. The sample is inductively heated in vacuum, and the vaporizing molecules are trapped in solid argon on the cooled window.

liquid helium at 4°K as the coolant for the window, since it melts at 24°K and is not rigid enough to prevent solid-state diffusion if the temperature rises above 12°K. Argon, krypton, and xenon can all be used with liquid hydrogen (boiling point, 20°K) as the refrigerant. It is found that, in general, the heavier rare gases perturb the trapped molecule more than argon or neon, and therefore are less desirable as matrix media.

A similar arrangement to that in Fig. 1 can be used to prepare matrices for EPR investigations. In the apparatus of Jen *et al.* (1958) the matrix is quenched at 4°K on a sapphire rod, which is then lowered into the microwave cavity. The cavity is part of the dewar and is cooled to 77°K so as to provide some heat shielding. The lower part of the dewar containing the cavity is suspended between the poles of the magnet. Other experimental arrangements for EPR studies have been described by Kupferman and Pipkin (1968), Farmer *et al.* (1964), and Bennett and Thomas (1964).

III. Optical Spectroscopy

A. The VO Molecule

The VO molecule is representative of the diatomic transition-metal oxides of importance at high temperatures, particularly in stellar atmospheres. It is easily produced in the gas by the vaporization of a mixture of solid V_2O_5 and vanadium metal at about 1800°C. An unpublished investigation of its matrix absorption spectrum at 4°K, carried out in collaboration with McLeod (1966) will be used here to illustrate the relationship between gas-phase and solid-state spectroscopy. Later on, the EPR work of Kasai (1968) on VO will be discussed. That research and the optical absorption study of Richards and Barrow (1968) have established the ground state as $^4\Sigma^-$.

In the gas phase three systems of bands have been observed: one in the infrared near 7900 Å, one in the visible at 5700 Å, and a third less well-defined system near 10,500 Å. The first two systems have been seen in absorption (Richards and Barrow, 1968) and emission (Keenan and Schroeder, 1952; Lagerqvist and Selin, 1957) and the third only in emission (Kleman and Liljeqvist, 1955; Lagerqvist, 1957). The strong visible system exhibits irregularities due to perturbations from a nearby electronic state.

All of these systems also appear in the absorption spectrum of VO in a neon matrix, including evidence of vibrational perturbations of the strong visible bands. All molecules are expected to be in the zero-point vibrational level of the ground electronic state at 4°K, so that the observation of these transitions indicates that the corresponding gas bands are not " hot " bands, but are terminating in the ground state. In addition, a progression of relatively weak, sharp bands beginning at 4185 Å is observed in the matrix spectrum and has apparently not been detected in the gas phase. Three of these matrix progressions are shown in Fig. 2. The fourth lies at about 8900 Å and has only been seen for isotopically substituted V ^{18}O because the spectrum was not measured that far into the infrared for V ^{16}O. V ^{18}O molecules are prepared by passing $^{18}O_2$ over hot vanadium metal in a tungsten effusion cell.

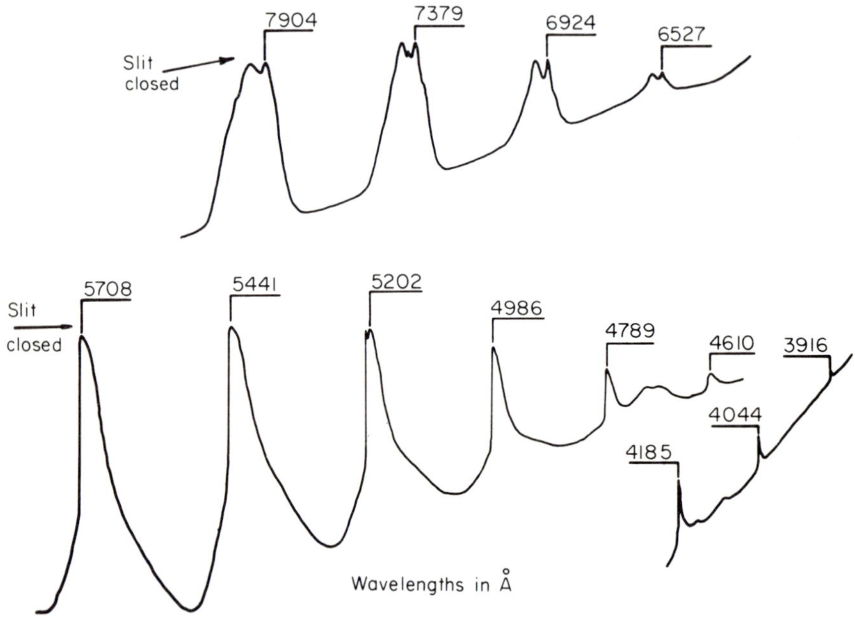

Fig. 2. The electronic spectrum of the V ^{16}O molecule trapped in a neon matrix at 4°K.

All of the matrix bands and derived vibrational frequencies for V ^{16}O and V ^{18}O are listed in Tables I and II, and a comparison of the isotopic molecules is given in Table III. The agreement of ν(V ^{18}O)/ν(V ^{16}O) in column 6 of Table III with the theoretical ratio of 0.961 is considered satisfactory, since the band widths in the matrix are considerably larger than in the gas and the peaks are less well-defined. Usually the band widths in matrices are due to overlapping bands arising from molecules in different sites in the matrix; most trapped molecules are not able to rotate in inert-gas solids. The C–X system between 7960 and 6200 Å illustrates the broad bands that can be encountered and shows the reason for giving the positions of the two subpeaks in Tables I and II.

The two bands that are observed near 8900 Å for V ^{18}O (see Table II) are presumably transitions to high v' levels among a progression of bands whose (0, 0) is at longer wavelengths. This system might then be that near 10,500 Å observed by Kleman and Liljeqvist (1955) in flames, and verified by Lagerqvist (1957); or it may be a new system.

Measurements of the infrared spectra (3 to 40 μ) of trapped VO yield the vibrational frequencies for the two isotopic molecules in the ground electronic state as 963 and 1005 cm^{-1}, as given in Table III. As expected, only one strong band appeared in each case. The frequency for V ^{16}O is in good agreement with the gas phase value of $\Delta G''_{1/2} = 1001.6$ cm^{-1} (Lagerqvist and Selin, 1957). Agreement within a few cm^{-1} has been found for all of the transition-metal oxide molecules studied. In argon, the vibrational frequencies may be lowered by about 20 cm^{-1} from those in the gas phase. The excellent agreement with the gas phase values of the electronic transitions and vibrational frequencies in the excited states of VO is shown in Table IV.

B. Metal Oxide Molecules

1. *Transition-Metal Oxides*

In a manner similar to that for VO, the optical spectroscopy of the matrix-isolated diatomic oxides of almost all the transition metals containing up to four d electrons have now been investigated. They are probably the most important high temperature species among these diatomic oxides. Only CrO, NbO, and MoO have not been observed in matrices. Although the ground states of CrO and MoO are in doubt (MoO has not been seen spectroscopically in the gas), NbO is almost surely $^2\Delta$ from the work of Uhler (1954) and from analogy with TaO. The matrix work on all of these oxides has been reviewed recently elsewhere (Weltner, 1967), and they have been considered rather thoroughly in Volume 1 of this series by Cheetham and Barrow (1967) and Carlson and Claydon (1967).

MO_2 molecules have been observed mass spectrometrically from many high temperature sources, and recent work in that area has been directed toward establishing whether they are bent or linear by use of an electrostatic quadrupole deflecting field (Kaufman *et al.*, 1967). This determines whether the molecule has a permanent dipole moment. It was found that the oxide molecules TiO_2, ZrO_2, CeO_2, ThO_2, TaO_2, and UO_2 are polar and therefore bent; SiO_2 is linear. Earlier matrix work had also indicated that ZrO_2, ThO_2, TaO_2, and WO_2 are bent in the ground electronic state. This can be inferred in several ways from optical matrix spectra, sometimes with the help of thermodynamic data.

A linear symmetric molecule will exhibit only two infrared-active vibrational frequencies so that the simple count of observed matrix bands can give the answer. However, the intensity of the symmetric stretching frequency (ν_1) may be weak if the molecule is only slightly bent, so that this is not a sure test. Also, in matrices care must be taken that isolation is occurring and

TABLE I. ABSORPTION BANDS OF $V^{16}O$ IN A NEON MATRIX AT 4°K

v'	λ (Å)	ν (cm^{-1})	$\Delta G_{v'+1/2}$ (cm^{-1})
C–X system			
0	7956	12,566	
	7904	12,648	
			900
1	7428	13,459	
	7329	13,548	
			891
2	6965	14,354	
	6924	14,439	
			879
3	6567	15,224	
	6527	15,318	
			864
4	6217	16,081	
	6178	16,182	
A–X system			
0	5706.8	17,518	
			856
1	5441	18,374	
			845
2	5201.7[a]	19,219	
			827
3	4987.2	20,046	
			821
4	4790.9	20,867	
D–X system (new)			
0	4185	23,888	
			833
1	4044	24,721	
			811
2	3916	25,532	
			790
3	3798	26,322	

[a] Perturbed.

TABLE II. ABSORPTION BANDS OF V ^{18}O IN A NEON MATRIX AT 4°K

v'	λ (Å)	ν (cm^{-1})	$\Delta G_{v'+1/2}$ (cm^{-1})
C–X system			
0	7959	12,560	
	7904	12,648	
			865
1	7446	134,626	
	7398	13,513	
			850
2	7004	14,273	
	6960	14,363	
			844
3	6611	15,121	
	6574	15,207	
			834
4	6265	15,958	
	6232	16,041	
A–X system			
0	5705.8	17,521	
			817
1	5451.6[a]	18,338	
			809
2	5221.3	19,147	
			794
3	5013.3	19,941	
			790
4	4822.3	20,731	
D–X system (new)			
0	4184	23,893	
			797
1	4049	24,690	
			777
2	3926	25,467	
			751
3	3813	26,218	
Infrared system			
?	8951	11,169	
	8924	11,203	
			848
?	8319	12,017	
	8296	12,051	

[a] Perturbed.

that bands due to dimers, etc., are eliminated. If isotopic substitution of the central metal atom is convenient, one can use the infrared data for these additional molecules to calculate the apex angle. The calculated angle is insensitive to the relatively small mass changes, so that usually the frequencies are not known accurately enough to establish the angle with certainty. [Mann *et al.* (1967), have, however, measured the matrix frequencies of MgF_2 to 0.01 cm^{-1} so as to avoid this objection (see below)]. One can also compare calculated thermodynamic functions, using measured vibrational frequencies and a particular molecular structure, with experimental thermodynamic data.

TABLE III. Comparison of Matrix Spectra of $V\,^{16}O$ and $V\,^{18}O$

Band[a] System	(0, 0)[a] Band		$\Delta G'_{1/2}$ (cm^{-1})[b]		
	λ (Å)	ν (cm^{-1})	$V\,^{18}O$	$V\,^{16}O$	$\left(\frac{\nu^{18}}{\nu^{16}}\right)^d$
$C^4\Pi - X^4\Sigma^-$	7956	12,566	865	900	0.950
$A^4\Sigma^- - X^4\Sigma^-$	5707	17,518	817	856	0.954
D $- X^4\Sigma^-$	4185	23,888	797	833	0.957
X	0	0	963[c]	1005[c]	0.958

[a] See Tables I and II.
[b] Vibrational frequencies in the excited electronic states.
[c] Vibrational frequencies in the ground electronic states from infrared spectra.
[d] Theoretical value=0.961.

TABLE IV. Comparison of the Gas Phase and Neon Matrix Spectra of $V\,^{16}O$

Band System	Gas		Neon matrix	
	λ (Å)[a] ν (cm^{-1})	$\Delta G'_{1/2}$ (cm^{-1})[b]	λ (Å)[a] ν (cm^{-1})	$\Delta G'_{1/2}$ (cm^{-1})[b]
$A^4\Sigma^- - X^4\Sigma^-$	5739[c] 17,419	852.7	5707 17,517	856
$C^4\Pi - X^4\Sigma^-$	7973[d] 12,539	897.6	7956 12,566	900

[a] (0, 0) bands.
[b] Vibrational interval in excited electronic state.
[c] From Lagerqvist and Selin (1957) and Richards and Barrow (1968).
[d] Chosen from Keenan and Schroeder's (1952, Table 1) as closest (0, 0) band to matrix value.

The ground-state vibrational spectra of ZrO_2 and ThO_2 have been measured in argon by Linevsky (1965). The observation of two stretching frequencies (see Table V) and the substitution of isotopic Zr or Th indicates bent structures with angles of about 110°. Two stretching frequencies were also found for TaO_2 and WO_2 (Weltner and McLeod, 1965a,b). In this case, electronic absorption spectra were also observed at 4°K and are suggestive of the ease of bending in these molecules. The WO_2 spectrum is shown in Fig. 3; the TaO_2 spectrum is similar, except that it lies between 8500 and 7000 Å. The extensive progressions of bands separated by 300 cm^{-1} indicates a change of the O—W—O angle during this electronic excitation and the frequency

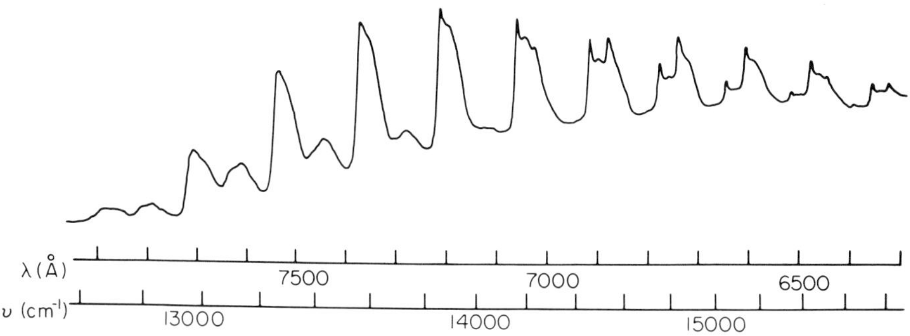

Fig. 3. The electronic spectrum of the WO_2 molecule trapped in a neon matrix at 4°K. (Reproduced with the permission of the *J. Mol. Spectry.*)

TABLE V. PROPERTIES OF METAL DIOXIDES OBTAINED FROM MATRIX SPECTRA

	ThO_2	ZrO_2	TaO_2	WO_2 [a]
Electronic state designation		$^1A_1$	$^2B_1$	$^1A_1$
(Ground state)	$\nu_1''=787$ cm^{-1}	$\nu_1''=884$ cm^{-1}	$\nu_1''=998$ cm^{-1}	$\nu_1''=992$ cm^{-1}
	$\nu_2''=?$	$\nu_2''=?$	$\nu_2''=?$	$\nu_2''=?$
	$\nu_3''=735$	$\nu_3''=819$	$\nu_3''=927$	$\nu^{3\prime\prime}=928$
Electronic state designation	—	—	$^2A_1$ or $^2A_2$	^1A or ^1B
(Excited state)	—	—	$\nu_1'=937$ cm^{-1}	$\nu_1'=972$ cm^{-1}
	—	—	$\nu_2'=285$	$\nu_2'=300$
	—	—	$\nu_3'=?$	$\nu_3'=?$

[a] Electronic states of WO_2 were incorrectly assigned as triplet (Weltner and McLeod, 1965b), as pointed out by Prophet.

interval corresponds to the bending motion ν_2' in the excited electronic state. There are several series of these progressions, not easily discernible from the figure, separated by the symmetric stretching frequency ν_1' in the excited state.

A summary of the data for these four molecules is given in Table V. The designations of the electronic states of ZrO_2, WO_2, and TaO_2 are proposed from considerations similar to those of Walsh (1953), but now including d orbitals (Weltner and McLeod, 1965a, b). A gas-phase study of these molecules is needed to obtain accurate molecular structure data.

2. *Other Oxides*

The infrared spectra of LiO, Li_2O, and Li_2O_2 have been investigated in matrices by White *et al.* (1963), and Li_2O in more detail by Seshadri *et al.* (1966). Li_2O was found to be linear or very nearly so. This again supports electric deflection measurements (Büchler *et al.*, 1963). LiO_2 bands were discovered, and it also is nearly linear.

Matrix-isolated B_2O_3 was studied by Weltner and Warn (1962) and by Sommer *et al.* (1963). Both groups of workers concluded from infrared spectra that the molecule is a V-shaped O=B—O—B=O. It is gratifying to see that the electric deflection method recently found it to be polar and therefore in accord with this model (Kaiser *et al.*, 1968).

C. Metal Halide Molecules

1. *Lithium Fluoride and Aluminum Fluoride*

The molecules vaporizing from solid lithium fluoride have been investigated more extensively via matrix-isolation than any other species. The original infrared study of matrix-isolated LiF (Linevsky, 1961) has led to the further determination of its spectrum in all of the rare gases and in N_2, CO, and CH_4 (Snelson and Pitzer, 1963; Schlick and Schnepp, 1964; Linevsky, 1963, 1965; Redington, 1966; Snelson, 1967a).

The vibrational band of diatomic LiF occurs at 868 cm^{-1} in neon at 4°K, a shift of -22 cm^{-1} from the gas value. This perturbing effect in the solid is even larger in the heavier rare gases (see Table VI) and arises from the matrix interaction with the large dipole moment (6.33 *D*) of LiF (Linevsky, 1961; Redington, 1966). This extreme behavior and the many other bands observed in the matrices have been the reasons for the thorough infrared investigations of this system. The above authors have been largely concerned with the identification of polymer bands and their separation into those of the dimer, trimer, etc. The "double-oven" technique (Linevsky, 1964; Snelson,

TABLE VI. VIBRATIONAL FREQUENCIES OF METAL HALIDE MOLECULES FROM MATRIX INFRARED SPECTRA

Molecule	Vibrational frequencies (cm^{-1})		
	Neon	Argon	Krypton
LiF[a]	868	835	820
AlF[b]	785	774	768
AlF_3 [b]	$\nu_2=280$ $\nu_3=950$ $\nu_4=254$	$\nu_2=260$ $\nu_3=935$ $\nu_4=247$	$\nu_2=244$ $\nu_3=930$ $\nu_4=244$
BeF_2 [c] (linear)	$\nu_2=330$ $\nu_3=1540$	$\nu_2=309$ $\nu_3=1528$	$\nu_2=302$ $\nu_3=1524$
$BeCl_2$ [c] (linear)	$\nu_2=238$ $\nu_3=1122$	$\nu_3=1108$	
$BeBr_2$ [f] (linear)	$\nu_2=207$ $\nu_3=993$	$\nu_3=985$	
BeI_2 [f] (linear)	$\nu_3=872$	$\nu_3=877$	
MgF_2 (bent?)	$\nu_2=254$[c] $\nu_3=862$	$\nu_3=842$[d]	$\nu_1=478$[a] $\nu_2=242$ $\nu_3=837$
CaF_2 (bent)	$\nu_1=504$[c] $\nu_3=581$	$\nu_1=489$ $\nu_3=561$	$\nu_1=487$[c] $\nu_2=164$[e] $\nu_3=555$[c]
SrF_2 (bent)	$\nu_1=468$[c] $\nu_3=471$	$\nu_1=447$[c] $\nu_3=450$	$\nu_1=439$[c] $\nu_2=85$? [e] $\nu_3=443$[c]
BaF_2 (bent)	$\nu_1=437$[c] $\nu_3=413$	$\nu_1=421$[c] $\nu_3=398$	$\nu_1=416$[c] $\nu_2=65$?[e] $\nu_3=392$[c]

[a] Taken from Snelson (1966). Gas phase value = 890 cm^{-1}.
[b] Taken from Snelson (1967b). Gas phase value for AlF = 798 cm^{-1}.
[c] Snelson (1966).
[d] Mann *et al.* (1967).
[e] Mann (1965).
[f] Snelson (1968).

1967a), which allows the superheating of the saturated vapor so as to break down the polymers, has been helpful in distinguishing the various molecules.

The dimer is present in the saturated vapor in higher concentration (Porter and Schoonmaker, 1958) than LiF. It is expected to be of rhombus

(V_h) symmetry, that is, a planar diamond shape, as has been found for Li_2Cl_2 in the gas (Bauer *et al.*, 1960). This species has only three infrared-active frequencies out of its total of six; hence, three of those given in Table VI are approximate calculated values (Snelson, 1967a). The more uncertain trimer assignments have not been included in the table.

AlF_3 has been observed in the infrared by Linevsky (1964) and AlF, AlF_3, and Al_2F_6 by Snelson (1967b). The vibrational frequencies of AlF and AlF_3 in various matrices are also given in Table VI.

2. *Dihalides*

Recently, some alkaline-earth dihalides have provoked controversy because of the surprising result that they are nonlinear. Apparantly this should have been expected, since Skinner (1955) had earlier pointed out the importance of d orbitals and possible nonlinearity in the heavier dihalides. Experimentally, this was first revealed by the electric deflection technique developed by Klemperer and co-workers [see Wharton *et al.* (1963), Büchler *et al.* (1964a, b)]. It was found that CaF_2, SrF_2, $SrCl_2$, and all of the Ba dihalides are polar and bent, whereas all other Group IIA dihalides are linear. Matrix spectral data, except for perhaps MgF_2, are in agreement with those conclusions.

In Table VI, taken from the extensive work of Snelson (1966, 1968), White, Mann, Linevsky and co-workers [see Mann (1965), Mann *et al.* (1967)], are given the vibrational frequencies observed in the infrared for the various molecules in neon, argon, and krypton matrices. For many of them ν_2, the bending frequency, has not yet been observed. There is a disagreement for MgF_2, since Mann *et al.* (1967) believe that it is nonlinear. Their analysis of the infrared spectrum, including Mg isotope bands, yields two stretching frequencies and a calculated angle of about 150°. This also does not agree with Snelson's (1966) conclusions.

Kaufman *et al.* (1967) have pointed out an interesting isoelectronic analogy between the bent alkaline-earth difluorides and the bent metal dioxides. This is in support of Walsh's ideas (1953), which have been extended to include d orbitals by Hayes (1966) in a discussion of the dihalides.

The transition-metal dihalides involving 3d electrons from Mn through Cu have been found to be linear. Electric deflection and matrix infrared spectra again agree on that. Their vibrational frequencies in an argon matrix at 4°K are given in Table VII. The bending frequencies are all taken from very recent work by Thompson and Carlson (1968). Because of their linearity, the symmetric stretching frequencies are generally unknown, except for $NiCl_2$, where it was determined from the fluorescence spectrum of an argon

TABLE VII. VIBRATIONAL FREQUENCIES OF LINEAR TRANSITION-METAL DIHALIDES FROM ARGON MATRIX SPECTRA

Molecule	Vibrational frequencies (cm^{-1})	
	Argon matrix[a]	Gas[b]
NiF_2	$\nu_3 = 780$[c]	
CuF_2	$\nu_3 = 769$[d]	
$NiCl_2$	$\nu_1 = 360 \pm 5$[e]	
	$\nu_2 = 85$	
	$\nu_3 = 521$[c]	515 ± 15
$NiBr_2$	$\nu_2 = 69$	
	$\nu_3 = 414$	
$CoCl_2$	$\nu_2 = 95$	
	$\nu_3 = 492$	493 ± 10
$FeCl_2$	$\nu_2 = 88$	
	$\nu_3 = 494$	492 ± 10
$MnCl_2$	$\nu_2 = 83$	
	$\nu_3 = 477$	467 ± 5

[a] All frequencies, except those indicated, are from Thompson and Carlson (1968).
[b] Leroi *et al.* (1962).
[c] Milligan *et al.* (1965).
[d] Kasai *et al.* (1966).
[e] Gruen *et al.* (1968). This frequency is from fluorescence spectra of an argon matrix.

matrix (Gruen *et al.*, 1968). The geometry and bonding of these dihalides has been discussed by Charkin and Dyatkina (1965) in terms of the valence states of the metals (Skinner, 1955). Knowledge of the heavier metal dihalides is lacking. Reference should also be made to the chapter by Hildenbrand (1967) in Volume 1 of this series for a discussion of the fluorides and their bonding.

DeKock and Gruen (1966, 1967) have measured the electronic spectra of some of the gaseous 3d transition-metal halides and analyzed them in terms of ligand-field theory. The spectra are interesting in that they clearly involve d–d transitions and can be interpreted in terms of an intermediate-field model.

IV. Electron-Paramagnetic-Resonance Spectroscopy

Matrix-isolation is especially applicable to EPR research on high temperature molecules, since such spectra can be observed in no other way. In other words, gas-phase or solution EPR investigations are not likely to become feasible for molecules produced at 1000 to 2000°K. Although there has been increasing interest in spin-resonance studies of matrix-isolated radicals in recent years [see, for example, Atkins and Symons (1967)], very few of them fall into our classification of high temperature molecules. Only CuF_2, ScO, YO, LaO, and VO can be considered as being in this category; their magnetic parameters are collected in Table VIII. Work on the first four of these

TABLE VIII. Magnetic Properties of High Temperature Molecules

Molecule	Ground electronic state	g tensor elements	Hyperfine-coupling tensor elements (MHz)
CuF_2	$^2\Sigma$	$g_{\parallel}=1.913\pm0.001$ $g_{\perp}=2.601\pm0.001$	$A_{\parallel}(^{63}Cu)=1933\pm5$ $A_{\perp}(^{63}Cu)=2090\pm5$ $A_{\parallel}(F)=308\pm1$ $A_{\perp}(F)=206\pm1$
ScO	$^2\Sigma$	$g=2.00\pm0.01$	$A=2010\pm10$
YO	$^2\Sigma$	$g=2.003\pm0.002$	$A=803\pm3$
LaO	$^2\Sigma$	$g=2.01\pm0.01$	$A=3890\pm10$
VO	$^4\Sigma$	$g_{\parallel}=2.0023\pm0.001$ $g_{\perp}=1.9804\pm0.001$	$A_{\parallel}=714.7\pm0.5$ $A_{\perp}=837.1\pm0.5$

molecules has been reviewed earlier by the author (1967), but VO has just recently been completed by Kasai (1968).

EPR spectroscopy, concerned as it is with transitions between spin states or magnetic energy levels of molecules, reveals the properties of the wave functions of the unpaired electrons through the measured g and hyperfine-coupling tensors. The magnetic properties of a molecule, as given by its g tensor, are fixed with respect to the frame of the molecule and have principal values along its symmetry axes. For example, in a linear molecule, there are two principal g values: $g_{\parallel}$, along the axis, and $g_{\perp}$, perpendicular to the axis. However, when the molecules are not parallel or perpendicular to the magnetic field, but randomly oriented as in matrices, one is measuring an effective g value that is a function of $g_{\parallel}$, $g_{\perp}$, and the angle of orientation. This causes the spectrum to be more smeared out, and absorption of energy does not occur just when the magnetic field $H = h\nu/g_{\parallel}\beta$ or $h\nu/g_{\perp}\beta$, but also at essentially all fields in between. (Here ν would be fixed at, say, 9.5 GHz, β = Bohr magneton and h = Planck's constant). This yields a characteristic powder pattern that can usually be analyzed for $g_{\parallel}$ and $g_{\perp}$, perhaps by means of a computer if the spectrum is complicated by hyperfine structure. Figure 4a shows an EPR absorption spectrum of a matrix containing a randomly

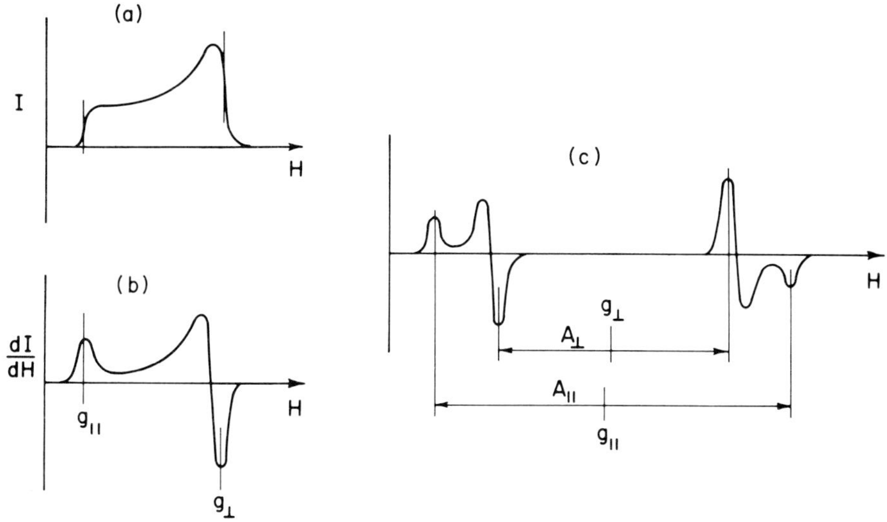

Fig. 4. The EPR spectra of a linear molecule ($^2\Sigma$ ground state) randomly oriented in a solid matrix. (a) Absorption spectrum, where the magnetic field H is the abscissa. (b) Spectrum actually measured, the first derivative of (a). (c) Derivative spectrum obtained if the molecule also contains a magnetic nucleus with spin $\frac{1}{2}$, illustrating hyperfine splittings.

oriented linear molecule with one unpaired electron. The first derivative of this spectrum, shown in Fig. 4b, is the one actually measured. The absorption spectrum has this shape because absorption by molecules oriented perpendicular to the magnetic field is more favored than other orientations when an average is taken over all angles. One can see that $g_{\parallel}$ and $g_{\perp}$ are obtained directly from the magnetic field measurements, with knowledge, of course, of the fixed microwave frequency.

In an orbitally nondegenerate ground state, i.e., a Σ state, $g_{\parallel}$ and $g_{\perp}$ differ from the free electron value ($g_e = 2.0023$) because of spin–orbit interaction between the ground state and excited states of the molecule. Usually this interaction is not large so that $g_{\parallel}$ and $g_{\perp}$ are not very different from g_e and a powder spectrum similar to that in Fig. 4b can be observed. If $g_{\parallel}$ and $g_{\perp}$ differ very much, as may occur in a molecule with an orbitally degenerate ground state, the spectrum will be so spread out as to be unobservable. There is, however, the possibility of trapping molecules in an *oriented* position with respect to the condensing surface, which would in principle make it possible to investigate linear molecules with Π, Δ, etc., ground states. This will be discussed in Section V.

The interaction of the unpaired electron spins with a nucleus having a magnetic moment leads to hyperfine splitting, characterized by a tensor A, and this splitting yields information about the electronic distribution in the molecule and the bonding. For example, the spectrum in Fig. 4b might become that shown in Fig. 4c if the molecule contained a nucleus of spin $\frac{1}{2}$ and the odd electron interacted rather strongly with the nucleus. If the odd electron is in an s state, this interaction is isotropic, that is, independent of the orientation of the molecules in the magnetic field and $A_{\parallel}=A_{\perp}$, However, addition of p or d character makes the hyperfine splitting dependent upon orientation. Then from knowledge of $A_{\parallel}$ and $A_{\perp}$ for a linear molecule, as obtained from Fig. 4c, one can distinguish the s from the p or d character of the odd electron.

In the case of ScO, YO, and LaO, where the elemental metal atoms are of electronic configuration s^2d^1, EPR revealed that the situation is quite different in the oxide molecule. The d_{z^2} orbital of the metal atom has become a bonding d_{σ} orbital, and the unpaired spin in the molecule is almost completely contributed by the nonbonding s electron on the metal atom. This electron then interacts strongly and isotropically with the nuclear moment of Sc, Y, or La. The hyperfine splittings observed in the EPR spectra could then be shown to account for splittings observed by Åkerlind (1962) in the high dispersion optical spectra in the gas phase. The EPR analysis showed clearly that the ground state of these molecules is $^2\Sigma$ (Weltner *et al.*, 1967).

Kasai's (1968) analysis of the EPR spectrum of matrix-isolated VO shows unequivocally that its ground state is $^4\Sigma$ and not $^2\Delta$, the most likely alternative. This result is in accord with the calculations of Carlson and Moser (1966) and the gas-phase spectral analysis of Richards and Barrow (1968) mentioned earlier. The molecule contains three unpaired electrons in a $\sigma\delta^2$ configuration where the σ and δ orbitals are essentially the nonbonding 4s and $3d_{xy}$ atomic orbitals of vanadium. The interaction between these electrons leads to a splitting of the ground-state electronic energy level even in the absence of a magnetic field (zero-field splitting, D). In this diatomic molecule the interaction is large (i.e., $D \gg g\beta H$) because the spins are confined to a small molecular volume and at 4°K only the lowest of the levels is populated. Only one allowed transition is then observed between $M=\pm\frac{1}{2}$ levels, yielding a spectrum basically similar to Fig. 4b but with $g_{\parallel}\cong g_e$ and an " effective " value of $g_{\perp}\cong 2g_e$, As in Fig. 4c, hyperfine coupling causes further structure to appear in the spectrum due to the spin of 7/2 of the vanadium nucleus. The final g and A parameters found by Kasai are given in Table VIII.

The confirmation of a $^4\Sigma(\sigma\delta^2)$ ground state for VO demonstrates an interesting analogy in the electronic properties of the two series of molecules TiO, ZrO, HfO and VO, NbO, TaO. Whereas the ground state of TiO is $^3\Delta(\sigma\delta)$, that of ZrO and HfO is $^1\Sigma(\sigma^2)$; the same sort of shift in the filling of levels occurs in the VO series, since the ground state of NbO and TaO is $^2\Delta(\sigma^2\delta)$.

V. Orientation of Molecules in Matrices

It has been found from EPR investigations of $Cu(NO_3)_2$, CuF_2, and NO_2 (Kasai *et al.*, 1966) that molecules can be oriented in matrices with respect to the substrate on which condensation is occurring. In those cases the substrate was a flat sapphire surface at 4°K, and it occurred in neon or argon matrices. Apparently, the phenomenon of orientation has also been observed by Snelson (1967b) during measurements of the infrared spectrum of trapped Li_2F_2 and BF_3, where the substrate was a CsI window at 4°K.

Orientation is detected by turning the matrix in the magnetic field, or in the light beam, and finding a change in the observed spectrum. If, for example, a linear molecule is oriented in the matrix with its axis parallel to the condensing surface, one would observe the effects of both $g_{\parallel}$ and $g_{\perp}$ when the surface is parallel to the magnetic field, but only the effects of $g_{\parallel}$ when the surface is perpendicular to the field. Figure 5 is an attempt to illustrate this. The EPR spectrum undergoes a dramatic change when the surface is turned by 90°. One observes the $g_{\parallel}$ band in Fig. 4b to decrease in intensity and the $g_{\perp}$ band to increase in intensity when the condensing surface is turned to face the magnetic field. If all molecules were oriented parallel to the flat sapphire surface, the $g_{\parallel}$ band would completely disappear and only one symmetric line at $H=h\nu/g_{\perp}\beta$ would remain. Whipple [see Kasai *et al.*, (1966)] has, in fact, derived from the change in the spectrum of $Cu(NO_3)_2$ in neon the distribution or degree of orientation of these molecules in the matrix. It is found that this large, flat molecule is strongly oriented with its plane of symmetry parallel to the sapphire surface, and that there is only a small proportion of molecules with their planes perpendicular to the surface. NO_2, even though is is not very planar, acts like $Cu(NO_3)_2$ (Kasai *et al.*, 1965), whereas CuF_2 orients as in the above example of a linear molecule.

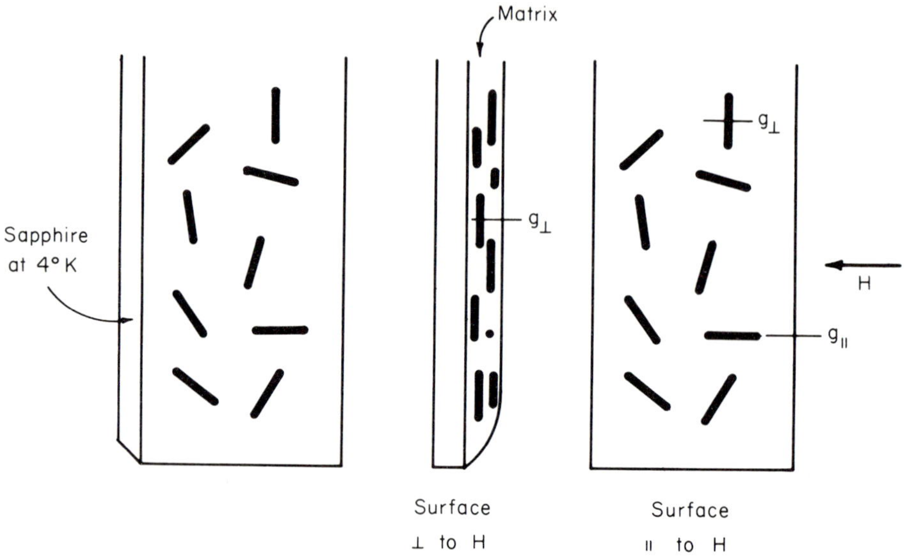

Fig. 5. Linear molecules oriented in a matrix with their axes parallel to the flat surface of the sapphire rod. The two views on the right illustrate the positions of the sapphire surface perpendicular and parallel to the magnetic field *H*.

The mechanism for orientation is not entirely clear, and it is likely that it varies for different molecules, depending upon their size and shape. A large molecule such as $Cu(NO_3)_2$, vaporizing at 150°C, must "lie down" preferentially on an atomically smooth neon plane parallel to the sapphire surface, and it is possible that this could be a crystallographic surface in the polycrystalline noble-gas solid. Evidence from x-ray diffraction work shows that preferential orientation of such crystallites occurs when the pure rare gases are condensed at low temperatures (Peiser, 1960).

X-ray work also indicates that the trapping of free radicals augments the crystallite orientation, presumably because of some annealing produced by partial recombination of the reactive species. It is significant that most cases where marked orientation occurs have been in the less rigid neon matrices at 4°K. There is evidence from the trapping of NO_2 that the temperature of the condensing gas molecules affects the degree of molecular orientation during quenching, again suggesting that a certain amount of local diffusion or annealing is necessary to cause preferential orientation. During the isolation of CuF_2, which vaporizes at 900°C, orientation occurred in argon, but not in neon, which can be interpreted as support for this effect.

The EPR investigation of VO (Kasai, 1968) revealed that it was oriented preferentially with its axis perpendicular to the sapphire surface, although there was a less populated site where the orientation was parallel. Such a small molecule might be expected to seek a substitutional site in a solid argon lattice. This is probably also the case for ScO, YO, and LaO where some preferential orientation was detected. However, these latter molecules are almost isotropic in their magnetic properties, i.e., they are similar to those of a $^2S_{1/2}$ atom, so that their spectra are less sensitive than VO to orientation effects. In the ground state, VO has an interatomic distance of 1.59 Å (Lagerqvist and Selin, 1957), so that its van der Waals' length is less than the diameter of an argon atom, which is 3.76 Å in the pure solid at 4°K. An interstitial site in argon would have to be considerably distorted to contain VO, since its diameter is only 1.56 Å. However, neither of these possibilities explains why the molecule orients in a particular direction in a cubic solid where there are equivalent environments in several directions in the lattice. It seems likely that the large temperature gradient and the directional gas flow occurring during the condensation process are the driving forces in causing a preferential orientation of crystallites and of trapped molecules, as suggested by Peiser (1960).

ACKNOWLEDGMENT

The support of the National Science Foundation during the period when this review was written is gratefully acknowledged. The author is also grateful to P. H. Kasai, D. McLeod, Jr., and E. B. Whipple for their part in making this review possible.

REFERENCES

Åkerlind, L. (1962). *Arkiv Fysik* **22**, 65.

Atkins, P. W., and Symons, M. C. R. (1967). "The Structure of Inorganic Radicals." Am. Elsevier, New York.

Bass, A. M., and Broida, H. P., eds. (1960). "Formation and Trapping Free Radicals." Academic Press, New York.

Bauer, S. H., Ino, T., and Porter, R. F. (1960). *J. Chem. Phys.* **33**, 685.

Bennett, J. E., and Thomas, A. (1964). *Proc. Roy. Soc.* **A280**, 123.

Büchler, A., Stauffer, J. L., Klemperer, W., and Wharton, L. (1963). *J. Chem. Phys.* **39**, 2299.

Büchler, A., Stauffer, J. L., and Klemperer, W. (1964a). *J. Chem. Phys.* **40**, 3471.

Büchler, A., Stauffer, J. L., and Klemperer, W. (1964b). *J. Am. Chem. Soc.* **86**, 4544.

Carlson, K. D., and Claydon, C. R. (1967). *Advan. High Temp. Chem.* **1**, 43.
Carlson, K. D., and Moser, C. M. (1966). *J. Chem. Phys.*, **44**, 3259.
Charkin, O. P., and Dyatkina, M. E. (1965). *Zh. Strukt. Khim.* **6**, 579.
Cheetham, C. J., and Barrow, R. F. (1967). *Advan. High Temp. Chem.* **1**, 7.
DeKock, C. W., and Gruen, D. M. (1966). *J. Chem. Phys.* **44**, 4387.
DeKock, C. W., and Gruen, D. M. (1967). *J. Chem. Phys.* **46**, 1096.
Farmer, J. B., Gerry, M. C. L., and McDowell, C. A. (1964). *Mol. Phys.* **8**, 253.
Gruen, D. M., Clifton, J. R., and DeKock, C. W. (1968). *J. Chem. Phys.* **48**, 1394.
Hayes, E. F. (1966). *J. Phys. Chem.* **70**, 3740.
Hildenbrand, D. L. (1967). *Advan. High Temp. Chem.* **1**, 193.
Jacox, M. E., and Milligan, D. E. (1964). *Appl. Opt.* **3**, 873.
Jen, C. K., Foner, S. N., Cochran, E. L., and Bowers, V. A. (1958). *Phys. Rev.* **112**, 1169.
Kaiser, E. W., Muenter, J. S., and Klemperer, W. (1968). *J. Chem. Phys.* **48**, 3339.
Kasai, P. H. (1968). *J. Chem. Phys.*, **49**, 4979.
Kasai, P. H., Weltner, Jr., W., and Whipple, E. B. (1965). *J. Chem. Phys.* **42**, 1120.
Kasai, P. H., Whipple, E. B., and Weltner, Jr., W. (1966). *J. Chem. Phys.* **44**, 2581.
Kaufman, M., Muenter, J., and Klemperer, W. (1967). *J. Chem. Phys.* **47**, 3365.
Keenan, P. C., and Schroeder, L. W. (1952). *Astrophys. J.* **115**, 82.
Kleman, B., and Liljeqvist, B. (1955). *Arkiv Fysik* **9**, 377.
Kupferman, S. L., and Pipkin, F. M. (1968). *Phys. Rev.* **166**, 207.
Lagerqvist, A. (1957). *Mem. Soc. Roy. Sci. Liege* **18**, 550.
Lagerqvist, A., and Selin, L. E. (1957). *Arkiv Fysik* **12**, 553.
Leroi, G. E., James, T. C., Hougen, J. T., and Klemperer, W. (1962). *J. Chem. Phys.* **36**, 2879.
Linevsky, M. J. (1961). *J. Chem. Phys.* **34**, 587.
Linevsky, M. J. (1963). *J. Chem. Phys.* **38**, 658.
Linevsky, M. J. (1964). *Proc. Meeting Interagency Chem. Rocket Propulsion Group, Thermochem. 2nd, Alexandria, Virginia*, 1964, Johns Hopkins Appl. Phys. Lab., Chem. Propulsion Inform. Agency Publ. No. 54U, p. 87.
Linevsky, M. J. (1965). *Proc. Meeting Interagency Chem. Rocket Propulsion Group, Thermochem. 3rd, El Segundo, California*, 1965, Johns Hopkins Appl. Phys. Lab., Chem. Propulsion Inform. Agency Publ. No. 82(U), p. 71.
Mann, D. E. (1965). *Proc. Meeting Interagency Chem. Rocket Propulsion Group, Thermochem. 3rd, El Segundo, California*, 1965, Johns Hopkins Appl. Phys. Lab., Chem. Propulsion Inform. Agency Publ. No. 82(U), p. 79.
Mann, D. E., Calder, G. V., Seshadri, K. S., White, D., and Linevsky, M. J. (1967). *J. Chem. Phys.* **46**, 1138.
Milligan, D. E., Jacox, M. E., and McKinley, J. D. (1965). *J. Chem. Phys.* **42**, 902.
Peiser, H. S. (1960). In " Formation and Trapping of Free Radicals " (A. M. Bass and H. P. Broida, eds.), Chapter 9. Academic Press, New York.
Porter, R. F., and Schoonmaker, R. C. (1958). *J. Chem. Phys.* **29**, 1070.
Redington, R. L. (1966). *J. Chem. Phys.* **44**, 1238.
Richards, D., and Barrow, R. F. (1968). *Nature* **217**, 842.
Schlick, S., and Schnepp, O. (1964). *J. Chem. Phys.* **41**, 463.
Seshadri, K. S., White, D., and Mann, D. E. (1966). *J. Chem. Phys.* **45**, 4697.
Skinner, H. A. (1955). *Trans. Faraday Soc.* **51**, 1036.
Snelson, A. (1966). *J. Phys. Chem.* **70**, 3208.
Snelson, A. (1967a). *J. Chem. Phys.* **46**, 3652.

Snelson, A. (1967b). *J. Phys. Chem.* **71**, 3202.
Snelson, A. (1968). *J. Phys. Chem.* **72**, 250.
Snelson, A., and Pitzer, K. S. (1963). *J. Phys. Chem.* **67**, 882.
Sommer, A., White, D., Linevsky, M. J., and Mann, D. E. (1963). *J. Chem. Phys.* **38** 87.
Thompson, K. R., and Carlson, K. D. (1968). *J. Chem. Phys.* **49**, 4379.
Uhler, U. (1954). *Arkiv Fysik* **8**, 265.
Walsh, A. D. (1953). *J. Chem. Soc.* p. 2266.
Weltner, Jr., W. (1967). *Science* **155**, 155.
Weltner, Jr., W., and MacLeod, Jr., D. (1965a). *J. Chem. Phys.* **42**, 882.
Weltner, Jr., W., and McLeod, Jr., D. (1965b). *J. Mol. Spectry.* **17**, 276.
Weltner, Jr., W., and McLeod, Jr., D. (1966). Unpublished work.
Weltner, Jr., W., and Warn, J. R. W. (1962). *J. Chem. Phys.* **37**, 292.
Weltner, Jr., W., McLeod, Jr., D., and Kasai, P. H. (1967). *J. Chem. Phys.* **46**, 3172.
Wharton, L., Berg, R. A., and Klemperer, W. (1963). *J. Chem. Phys.* **39**, 2023.
White, D., Seshadri, K. S., Dever, D. F., Mann, D. E., and Linevsky, M. J. (1963). *J. Chem. Phys.* **39**, 2463.

Molecular Beams in High Temperature Chemistry

Thomas A. Milne and Frank T. Greene

MIDWEST RESEARCH INSTITUTE, KANSAS CITY, MISSOURI

I. Introduction 107
II. Beam Formation from High Pressures 108
A. Ideal Free-Jet Expansion 109
B. Phenomena Involved in Free-Jet Expansions 111
III. Molecular Beam Techniques 113
A. Sources 113
B. Detectors 115
C. Beam Modulation 116
D. Velocity Selection and Analysis 117
IV. Molecular Beam Applications 120
A. Direct Sampling at High Pressures 120
B. Studies Using High Pressure Beam Properties 130
C. Electric and Magnetic Deflection Experiments 135
D. Miscellaneous Applications of Molecular Beams 139
References 140

I. Introduction

The use of molecular beams in the study of atomic and molecular properties has a long history. The early work, which dealt mainly with the behavior of individual species in interaction with electric and magnetic fields, is amply reviewed (Fraser, 1931, 1937; Estermann, 1946, 1959; Smith, 1955; Ramsey, 1956) and serves as the foundation for many techniques in use today. More recently, development of techniques and supporting theory has opened up the possibility of using molecular beams in interaction with other beams, with gases, or with surfaces to examine on the most fundamental level chemical kinetics, intermolecular forces, and surface properties. These aspects of molecular beam studies, although they have considerable relevance to high temperature chemistry, are the subject of several good reviews and are not included here (Datz and Taylor, 1959; Fite and Datz, 1963; Zorn, 1964; Pauly and Toennies, 1965; Herschbach, 1966; Ross, 1966; Bernstein and Muckermann, 1967; Stickney, 1968).

One of the most recent developments in molecular beam technology has great relevance to high temperature chemistry and forms the major subject of this chapter. This development involves the formation of molecular beams from high pressure systems and their introduction into detection devices, particularly mass spectrometers. Such an application is a logical extension of the very fruitful coupling of Knudsen cells and their effusive beams with mass spectrometers, which has constituted a main thrust of high temperature chemistry in the last 15 years (Inghram and Drowart, 1960; Drowart and Goldfinger, 1967; Berkowitz and Walter, 1968; Zmbov *et al.* 1968).

In this review we try to portray the main features of molecular beam formation from high pressures, the important phenomena accompanying such expansions, and the techniques for production, detection, and diagnosis of such beams. We then discuss applications involving the sampling of high temperature systems (very broadly interpreted) and the use of these beams as a means of generating and studying cold molecules from hot systems, observing nucleation processes, and obtaining the properties of unusual clusters of molecules and atoms. Also included is a discussion of several types of high temperature studies, using high pressure beams or effusive beams, which are of special interest to us.

II. Beam Formation from High Pressures

A sufficient reason for the high temperature chemist to pursue direct observations of species in high pressure environments is contained in the classic discussion by Brewer (1950), who argued that the complexity of saturated vapors will tend to increase with increasing temperature. There is thus always the possibility that the species important in high pressure, high temperature systems of scientific or practical concern will not be revealed by laboratory studies on these systems at lower pressures. A current example is provided by carbon, where heretofore unstudied carbon clusters may dominate the vapor at the triple point. There is also increasing interest in the ability to directly sample high pressure, high temperature systems of practical concern, such as rocket chambers, internal combustion engines, or the atmosphere of Venus. Since the most practical universal detector for high temperature gaseous species currently available, the mass spectrometer, requires a molecular beam in at least a formal sense to transport condensable species into the ionizing region, the question of high pressure beam formation is of importance. In many cases beam formation phenomena will very likely prove of crucial importance in interpreting results. Finally, as discussed below, the process of high pressure expansion to molecular flow itself provides a unique environment for studying nucleation, relaxation processes, and the properties of clustered species, and for obtaining high temperature molecules with reduced internal energy. A discussion of some of these uses of high pressure beams has been given by Milne and Greene (1968).

This section presents a simple description of the process of beam formation from pressures such that continuum flow exists during the initial expansion. A brief mention of the ill-defined transition region from molecular flow is also included. We omit here a discussion of effusive flow and the many problems associated with nonideal orifices (Freeman and Edwards, 1967), evaporation and condensation coefficients (Rosenblatt and Lee, 1968), cell reactions (Ward *et al.*, 1967), surface migration (Winterbottom, 1968), and angular effusion anomalies (Wang and Wahlbeck, 1968).

A. Ideal Free-Jet Expansion

When the mean free path of a gas passing through an orifice is much less than the narrowest orifice dimension (that is, the Knudsen number, λ/d, is much less than 1), then the behavior is dominated by collisions in the initial continuum expansion, which occurs downstream (and slightly upstream) of the orifice. According to the simplest model of beam formation, if the pressure downstream from the orifice is kept quite low, typically 10^{-5}–10^{-6} of the source pressure, then the gas history may be approximated by assuming an isentropic expansion, which is terminated abruptly by a transition to collisionless flow and followed by collimation of this gas into a molecular beam. If one takes care to avoid attenuation, preferential scattering of the beam, or the possible entrainment of background gas in the jet, then the main processes of interest occur in the initial expansion.

Such expansions, when occurring though thin-edged orifices or converging nozzles without a flow-controlling diverging section, are called free jets. Free jets are currently the subject of intensive study by fluid dynamicists because they form the basis of the Kantrowitz–Grey (1951) type of supersonic molecular beam for the production of high energy, high intensity, narrow velocity distribution beams and because they can provide a supersonic flow field of known properties for various aerodynamic studies. The first 10 years or so of study of free jets and beam formation have been covered in several recent reviews (Leonas, 1964; Knuth, 1964; French, 1965, 1966; Anderson *et al.*, 1965a, 1966) and in a series of international symposia on rarefied gas dynamics (Talbot, 1961; Laurmann, 1963; de Leeuw, 1965; Brundin, 1967). The most recent discussions of current work with free jets and related problems are contained in the latest of these symposia (Trilling and Wachman, 1969). Emphasis in this section will be on the simplest picture of the phenomena involved, with specific references to complications of concern to potential high temperature applications.

Two common situations in beam formation may be distinguished. In both, pressures downstream of the orifice are kept sufficiently low that either no distinguishable Mach disk forms or its location is well downstream of the position of the first collimating slit or skimmer. In the first case this skimmer is placed downstream of the region of transition from continuum to molecular flow so that it intercepts a nearly collisionless flow field. In the second case, the skimmer is placed in the continuum flow, at some region of desired Mach number, and collisionless flow is supposed to occur inside and beyond the skimmer. The history of the gas during the expansion is determined in either case, according to the simplest model, by the Mach number it achieves and the time involved before collisions cease.

Direct experimental data for small orifices and simple gases give the Mach number of the gas during the continuum expansion (Anderson *et al.*, 1965b) and can be fit to an effective terminal Mach number at which collisions essentially cease (Scott and Phipps, 1967; Abuaf *et al.*, 1967b). Over the continuum portion of the expansion these results agree well with the calculations of Sherman (1963). Ashkenas and Sherman (1965) give the calculated Mach number as a function of distance downstream as

$$M = A\left(\frac{x-x_0}{d}\right)^{\gamma-1} - \frac{1}{2}\left(\frac{\gamma+1}{\gamma-1}\right)\Big/ A\left(\frac{x-x_0}{d}\right)^{\gamma-1} \tag{1}$$

where γ is the ratio of specific heats, d is the orifice diameter, and x is the distance downstream from the orifice. A and x_0/d have the values for different γ's which are shown in the tabulation. At large distances $M \cong A(x/d)^{\gamma-1}$

γ	A	x_0/d
$\frac{5}{3}$	3.26	0.075
$\frac{7}{5}$	3.65	0.40
$\frac{9}{7}$	3.96	0.85

$$\rho/\rho_0 = (1+\tfrac{1}{2}(\gamma-1)M^2)^{-1/\gamma-1} \cong (\tfrac{1}{2}(\gamma-1))^{-1/\gamma-1}A^{-2/\gamma-1}(x/d)^{-2}$$

which for $\gamma = \frac{5}{3}$ reduces to

$$\rho/\rho_0 = 0.16/(x/d)^2 \tag{2}$$

where ρ_0 is the source density and ρ is the density at a distance x from the orifice of diameter d. (Note that the expansion is scaled in orifice diameters.) It is interesting that the flow is sourcelike and differs relatively little in the rate of density decrease from the rate given by the Knudsen effusion expression, were this relationship to apply over the same source conditions (Taylor, 1963):

$$(\rho/\rho_0)_{\text{Knudsen}} = 0.0625/(x/d)^2$$

Standard expressions for isentropic expansion relate the Mach number or density to the local pressure, temperature, and velocity. Typical expansions take place in a few microseconds and involve several thousand binary collisions. Resultant velocity distributions can be extremely narrow. Expansion to rather large Mach numbers has been achieved in simple gases. Anderson and Fenn (1965) give the following expression based on experimental velocity distribution measurements for argon:

$$M_{\text{Terminal}} = 1.17(\lambda/d)^{(\gamma-1)/\gamma}$$

where γ is the ratio of specific heats, λ is the mean free path of the source gas, and d is the orifice diameter. The achievement of Mach numbers in the range of 10 to 20 results in enormous cooling and often supersaturation of the gas. The consequences of this cooling figure prominently in our later discussion of sampling and other applications.

Prediction of the theoretical beam intensity at a large distance from the orifice, with intervening slits, requires consideration of the terminal Mach number and the geometry of the skimmer. With a sufficiently large skimmer, Anderson and Fenn (1965) have argued that Ashkenas and Sherman's source-flow expression, Eq. (2), can be applied at any distance, even into the essentially collisionless regime. Thus an upper limit to attainable beam intensities is provided if departures from ideality and scattering are negligible. Anderson *et al.* (1966) have given an expression for intensity in the case of a very small skimmer and for a given terminal Mach number M_T. Likewise, Anderson and Fenn (1965) give an expression for large skimmers. A detailed discussion of beam intensities is also presented by Hagena and Morton (1967), who consider both of the above mentioned beam formation cases. In very recent work (Bossel *et al.*, 1968) it has been shown that it is possible in practice to achieve beams of Ar and N_2 with very nearly the theoretical intensity and translational energy when the skimmer is immersed in the free-jet flow field. The question of velocity distributions in the beam is dealt with by Amend and Hurlbut (1968) and Hagena and Morton (1967), among others.

B. Phenomena Involved in Free-Jet Expansions

Usually many complications arise to cause deviations from predictions using the simple picture just noted. It is not our purpose to discuss in detail these deviations from the ideal expansion, but to point out the consequences of the first order effects of even an ideal expansion. The above mentioned reviews on free jets provide detailed discussions. Ashkenas and Sherman (1965), Anderson *et al.* (1965b, 1968) and Busol *et al.* (1967) discuss the question of effective orifice size and the scaling of the expansion. The effects of shock waves at the skimmer, flow inside the skimmer, background gas penetration in the free jet, collisions within the beam, and similar perturbations are treated by Fenn and Anderson (1965), McMichael and French (1966), Abuaf *et al.* (1967b), Rebrov and Sharafutdinov (1968), Govers *et al.* (1968), LeRoy *et al.* (1968), Campargue (1968), and others.

For pure monatomic gases, the expansion is relatively simple in that only translational degrees of freedom exist, except in cases involving nucleation as will be discussed later. Even for these gases, however, the expansion cannot be truly isentropic and, in fact, the translational relaxation is anisotropic with parallel and perpendicular temperatures being distinguished in the resulting molecular beam (Hamel and Willis, 1966; Muntz, 1967; Hamel, 1968; Gazzola *et al.*, 1968).

With polyatomic gases one must consider the relaxation of the various internal degrees of freedom during the expansion. Direct experimental data on the translational and rotational states during expansion have been obtained by velocity distribution measurements (Anderson and Fenn, 1965) for the former and by optical spectroscopy for the latter (Robben and Talbot, 1966). The vibrational behavior has been inferred from the apparent γ, which describes the measured free-jet properties (Hagena and Henkes, 1960). As a first approximation one can expect translational relaxation to the terminal Mach number condition (a matter of definition). Rotational relaxation will be considerable, but final rotational temperatures are substantially higher than translational temperatures. Vibrational relaxation will be much less complete than rotational. For excited electronic states, radiative processes will probably dominate over collisional deexcitation. When collisions have essentially stopped, the state of the molecules will remain frozen, with the possible exception of unimolecular reactions such as those postulated by Leckenby and Robbins (1966), in which weakly bound polyatomic clusters essentially rotate themselves apart in the absence of collisions. Additional references to these relaxation phenomena are given in the section on free-jet isolation spectroscopy.

Additional problems enter with the expansion of gaseous mixtures and with the passage of reactive or condensable species through cold orifices. These problems are discussed below in the context of their effect on the interpretation of sampling and related experiments.

The properties of beams formed under transition flow at the orifice are poorly understood compared to the limiting cases of Knudsen effusion and high pressure free-jet formation. Recent papers by Wahlbeck and co-workers [see Wang and Wahlbeck (1968), and references therein] discuss the transition from molecular flow to continuum flow through long and short cylindrical orifices in terms of angular number distributions. Discharge characteristics through orifices are treated by Smetana *et al.* (1967), and the effect of the onset of collisions at the orifice on velocity distributions has been documented by Scott and Phipps (1967), Polak and Trilling (1968), and Scott *et al.* (1967). Nutt *et al.* (1959) discuss beam formation in the transition region. However, even in an ideal case one can make no specific predictions of the dependence of beam intensity on pressure because of the primary lack of a quantitative model for flow properties in this region. In addition, any prediction of the behavior of the beam intensity with changing source pressure must include consideration of slit geometry, scattering, effective heat capacity of the gas, nucleation and mass separation. In a system where shocks and scattering are avoided, however, several qualitative changes would be expected and have been observed. For thin-edged orifices, as Knudsen numbers decrease toward unity, slow molecules will be lost and velocity distributions will narrow and shift toward higher velocities. Mass separation has not been systematically studied as a function of pressure but, in a typical sampling situation, has been shown to increase steadily with pressure to the expected limiting value (Greene *et al.*, 1964). Heavy-component intensities will steadily increase, but intensities of lighter components may fall. Nucleation is very sensitive to pressure, so that with highly condensable species, clustering may ensue early in the transition to continuum flow (Milne and Greene, 1967e). Consequently, unless one is willing to invest considerable study of the particular system in use, it appears that beam experiments should be performed under conditions of beam formation approximating one of the two limiting cases.

III. Molecular Beam Techniques

A. Sources

Sources for the production of molecular beams have traditionally consisted of a container for the gas to be studied and a slit or orifice that is in thermal equilibrium with the gas and through which the gas effuses or expands in the first step in beam formation. The problems encountered in this sort of arrangement for a high temperature beam source are largely those common to most high temperature chemistry: materials compatibility and achieving uniform high temperatures. However, the gas need not be in thermal equilibrium with the source aperture. Although this situation can give rise to special problems as discussed in Section IV, it can in some cases avoid insoluble materials problems. Furthermore, there need be no slit at all if the gas source is limited in some other manner. For example, the beam source might be a small laser-heated spot or narrow filament. A general discussion of a variety of molecular beam sources has been given by Anderson *et al.* (1965a), Fenn (1967), and Leonas (1964).

From the standpoint of gas dynamics, molecular beam sources can be divided into those for effusive beams, in which the gas expands through the source aperture in molecular flow, and those for nozzle beams, in which the expansion is of a continuum character. Effusive sources may be of the single aperture or multichannel " beam-shaping " type. Discussions of the single aperture type are included in the work of Ramsey (1956), Fraser (1931), Kusch and Hughes (1959), and Trischka (1962). An almost standard design of a cell used at relatively low temperatures has been generally attributed to Kusch [see Miller and Kusch (1955)]. Multichannel sources use a series of very small, long channels or tubes in place of a single aperture. This arrangement gives a greater beam density than does a single aperture for either a given source chamber pumping speed or for a given pressure while maintaining molecular flow conditions. However, the velocity distribution will be perturbed (Hostettler and Bernstein, 1960). A number of these multichannel sources have been discussed in the literature (Minten and Osberghaus, 1958; Hanes, 1960; Giordmaine and Wang, 1960; Datz *et al.*, 1961; Becker, 1961; Gersing *et al.*, 1963; Angenmaier, 1966).

A nozzle source consists simply of an orifice or nozzle, through which the gas expands in continuum flow, and may be considered to include the skimmer. Some discussion of nozzle beam sources has been included in Section II. A detailed discussion and comparison with other sources has been given by Anderson *et al.* (1966). Nozzle and skimmer geometry and their effect on the free-jet expansion are discussed in Section II. In general, if an approach to ideal free-jet behavior is required, either a converging nozzle or a thin (small length-to-diameter) orifice is required. Additional discussion of these factors is given by Knuth (1964), French (1965, 1966), Anderson *et al.* (1965a), Scoles (1965), and Leonas (1964). The optimization of a nozzle beam system has been discussed by Bier and Hagena (1965). Recent descriptions of major beam systems are given by Scoles and Torello (1968), Stair (1968), Miller (1968b), Brown and Heald (1967), Skofronick (1967), and Audit and Rouault (1967b). Cryogenic sources have been incorporated by Vyse *et al.* (1968) and Zapata *et al.* (1968). Shock- and arc-driven sources are described below. Finally, a revival of interest in minibeam systems, using very small pumps, is indicated by Campargue (1967) and Skofronick and McArdle (1968).

B. Detectors

A variety of methods has been used to detect molecular beams. Thermocouples, photographic film, torsion devices, mass collection, and several types of ionization detectors have been employed with various degrees of success. With the possible exception of mass collection followed by radiochemical analysis, only the ionization devices appear to be of much value in current molecular beam research. For most problems in high temperature chemistry a mass spectrometer is desirable.

Two principal methods of ionization have been used in molecular beam detectors: surface and electron bombardment. Surface ionization detectors are the classical molecular beam detector, and are discussed in detail in the standard references on molecular beams. They have the advantage of relative simplicity and nearly unit efficiency. However, surface ionization is restricted to cases in which the surface has a work function greater than the ionization potential of the molecule to be ionized. This characteristic has limited surface ionization detection largely to the alkali metals and their compounds, although there have been reported extensions of this technique to the alkaline earths (Hall *et al.*, 1968), their oxides (Wharton *et al.*, 1962), and other selected species (Pauly and Toennies, 1965). Although electron bombardment provides an essentially universal technique for ionizing beams, the maximum efficiency of a mass spectrometer equipped with an electron bombardment ionizer was, until recently, about 10^{-4} to 10^{-5} or less. Detectors without mass analysis were not much better. As a consequence, molecular beam research was largely limited to study of reactions involving the alkali metals and their compounds and a few other elements. This created a limit on the types of molecules that could be studied that was called the " alkali barrier." The alkali barrier has now been largely surmounted by advances in the general area of high efficiency ion sources, cleaner vacuum systems, and improvements in signal conditioning. Pauly and Toennies (1965) review recent developments in detection in terms of scattering experiments.

One of the first high efficiency sources was that of Weiss (1961), who claimed fractional ionization as high as $\frac{1}{60}$. Other investigators have built similar sources which, although they have not obtained as high an efficiency, have shown substantial improvement over conventional sources. Ion sources of this type have been used satisfactorily with quadrupole and magnetic mass analyzers as molecular beam detectors (Brink, 1966; Kaufman, 1964; Bennewitz and Wedemeyer, 1963).

An additional technique that shows considerable promise is that of field ionization. Over the active ionization area a field ionization source can have a high efficiency. In order to have a high efficiency molecular beam detector.

the problem is largely one of making the active area a major portion of the geometrical area of the detector. Progress toward this goal has been reported by Woods and Fenn (1966) and Johnston and King (1966a, b).

An alternative to the improvement of the ionizer has been demonstrated by Greene and Milne (1967), who used a quadrupole mass filter with a standard ion source to successfully detect very weak beams. They found that by employing molecular beam modulation, ion counting or low noise amplifiers, and a bakeable beam system containing low levels of interfering residual gases, a good signal-to-noise ratio could be obtained with a standard commercial quadrupole.

C. Beam Modulation

There are generally significant advantages in using modulated (pulsed) beams together with phase-sensitive detection. These advantages can include an improved signal-to-noise ratio, discrimination against residual gases and gas scattered from the beam, and discrimination against products of reactions of the beam gas with the ion source. In some cases, as discussed below, mass discrimination and velocity measurement can also be made without significant loss in beam intensity. The improvement in signal-to-noise ratio due to beam modulation will almost always more than compensate for the factor-of-2 loss in duty cycle when the beam is pulsed on half the time. Furthermore, in many experiments beam modulation is necessary to avoid serious errors.

Beam modulation has been used in a number of mass spectrometric beam systems, including those of Foner and Hudson (1953), Fite and Brackmann (1958), Roberts and McElligott (1963), Bennewitz and Wedemeyer (1963), Greene and Milne (1966), and Boyer *et al.* (1968). Analyses of phase-sensitive detection with pertinence to beam modulation mass spectrometry have been given by Rutgaizer (1967), Boyer *et al.* (1968), Yamamato and Stickney (1967), and Harrison *et al.* (1964). A phase-sensitive detector will discriminate against any random signal and detect only a signal having a fixed phase relationship with a reference signal. Consequently, discrimination will be obtained against not only noise in the detection system but " signal " from nonmodulated gas in the molecular beam system. Molecules reflected from a surface directly back into the ionizer will, however, still be detected, since they have a definite phase relationship with that of the beam. They must be eliminated by proper geometry.

In order to obtain discrimination against random gas introduced into the region of the detector by the beam, and chemical reactions of this gas with the ion source (but not ion–molecule reactions), there is an additional condition, which was pointed out in the pioneering modulated beam study of Foner and Hudson (1953). This condition is that the chopping frequency be fast compared to the time constant for the pump-out of any noncondensable species introduced into the detector region. In other words, the beam must be pulsed sufficiently rapidly that the density of noncondensable gases in the ion source does not change significantly during the chopping cycle. The exact criterion for the ratio of chopping frequency to pumping time constant does, of course, depend on the level of discrimination required, and can probably be best determined experimentally. One method is to follow the ratio of an unstable or reactive species to a stable one as a function of chopping frequency. Milne and Greene (1968) used this method and found that a chopping frequency of about 15 Hz for the ion source of their Bendix mass spectrometer was adequate to obtain discrimination against scattered gas to within a few percent. This problem is of particular concern in mass spectrometric vaporization studies, where serious errors can be introduced in either the absolute measurement of the intensity of a noncondensable gas or in the ratios of condensable to noncondensable species if modulation is not used or if the modulation frequency is too low.

The implementation of beam modulation requires only a method of chopping the beam and inclusion in the detector output circuitry of a phase-sensitive device that is phase locked to the beam chopper. Molecular beam choppers have included tuning forks, vibrating reeds, and motor driven sectors. Some of these devices are commercially available. A vibrating reed for use in bakeable and ultraclean systems has been described by McElligott *et al.* (1963). Other molecular-beam choppers have been reported by Schwarz and Madip (1968), Omrod and Patterson (1967), and Andresen (1966). Several commercial lock-in amplifiers are available that are very useful with analog signals. A signal-averaging computer can also be used for demodulation, with the sweep synchronized with the chopping frequency. If ion counting is used instead of analog signal conditioning, a convenient approach seems to be a scheme such as that described by Greene and Milne (1968b). Alternatively, a multichannel analyzer could be used in a manner analogous to the signal-averaging computer.

D. Velocity Selection and Analysis

There are a number of applications of velocity selection or analysis in molecular beam studies. The advantage in using velocity selection or analysis in velocity sensitive experiments, such as electric and magnetic deflection experiments, are described in a later section. In mass spectrometric studies of effusive beams, velocity analysis can function as a very low resolution mass spectrometer that operates on the neutral molecules and is in tandem with the ion mass spectrometer. This allows species identification independent of fragmentation effects. In principle, similar information can also be obtained in the case of nozzle beams, although much higher resolutions will be required and departures from ideal free-jet expansions may lead to ambiguities.

Velocity analysis techniques are of two principal kinds: those that use velocity selection and those that distinguish between velocities by a difference in arrival times at a detector. The velocity selectors in use employ multiple slotted disks or cylinders with helical slots driven at large angular velocities. The theory of these selectors and current practice have been discussed by Hostettler and Bernstein (1960), Grosser *et al.* (1963), Grosser (1966), Kinsey (1966), and Pauly and Toennies (1965). Some discussion is also given in the general references on molecular beam techniques. All of these velocity selectors are relatively complex and are often not very adaptable to many mass spectrometric systems. They have the advantage over time-of-flight techniques of a much higher duty cycle, and are necessary in experiments where only molecules in a specific velocity range can be present.

Time-of-flight techniques are generally more easily adaptable to high temperature problems. Applications of this basic technique can be somewhat arbitrarily divided into two general types: those in which beam intensity as a function of time is measured directly and those in which the flight time is determined as a phase shift by means of a phase-sensitive detector. The direct measurement technique is the older and probably the more generally useful. Consider an idealized situation in which a molecular beam is pulsed on for an infinitesimal time interval. At times between t and $t+dt$ the number of molecules $n(t)$ arriving at the detector have velocities between L/t and $L/(t+dt)$, where L is the distance from the beam pulser to the detector. If the total number of molecules is N, then $n(t)/N$ is the fraction of molecules with velocity between v and $v+dv$ and the velocity distribution can be determined by measuring $n(t)$ as a function of t. It is practical to make the time interval in which the beam is on sufficiently short compared to the flight time of the molecules to satisfactorily approach the idealized condition. Alternatively, longer open times may be used and a correction made for the contribution to $n(t)$ of molecules with a finite range of velocities having the same arrival time at the detector. Corrections may also be required for several other complications, including finite detector depth and finite resolution of the detection circuitry. A good general discussion of time-of-flight velocity analysis has been given by Hagena and Varma (1968) and

Scott (1965). A scheme for use with a Bendix time-of-flight mass spectrometer, which gives exceptional detector resolution and an improved signal-to-noise ratio, has been described by Greene (1968) and implemented by Greene *et al.* (1968b). It could also be applied to conventional mass spectrometers.

Direct time-of-flight measurements have been applied to a variety of problems. Anderson and Fenn (1965), Vidal *et al.* (1967), Muntz (1967), French and Locke (1967), Phipps *et al.* (1963) and Amend and Hurlbut (1968) have measured velocity distributions in nozzle beams. The direct time-of-flight technique has also been applied to atoms scattered from a hot surface (Moran *et al.* 1967), sputtered (Stuart *et al.*, 1963), evaporated from liquid He (Johnston and King, 1966a, b), and produced by dissociative excitation (Levethal and Robiscoe, 1967), and to the identification of metastable molecular states (Freund and Klemperer, 1967).

Although the phase shift velocity analysis method is, in practice, inferior to the direct method with respect to the precision with which a complete velocity distribution can be measured, it has two potentially very useful applications: the measurement of most-probable velocities and the discrimination against a species in an effusive beam that has a different mass than that being studied but gives an ion with the same charge-to-mass ratio. Fundamentally, the phase shift method is simply an extension of the direct time-of-flight method to the case of long beam-on times and low detector resolution. The consequence of this arrangement is that the distribution of velocities tends to be obscured by the poor time resolution of the detector and the broad beam-on pulse, although a complete velocity distribution can be obtained (Boyer *et al.*, 1968). However, the most probable velocity can be readily detected as a delay in the mean time of arrival of the pulse, which will appear as a shift in phase if the chopping frequency is sufficiently rapid. The conversion of the observed phase shift into a most probable velocity has some complications, which are discussed by Boyer *et al.* (1968), and Harrison *et al.* (1964). A knowledge of the most probable velocity can be extremely valuable, as discussed later. The important thing to be noted in connection with this method is that the most probable velocity can generally be determined with little loss in intensity and using only very simple equipment. The only loss will be due to the increased spreading of the beam pulses at the higher frequencies required to obtain a measurable phase shift, and to a reduction in the beam-on to beam-off ratio that may be required to avoid serious overlap of the arrival times of adjacent pulses at the detector. The only equipment required in addition to the phase-sensitive device is a phase meter or calibrated phase shift.

A potentially important application of the phase shift method arises from the fact that for a phase-sensitive detector

$$I = I_0 \cos \theta$$

where I is the observed intensity, I_0 is the maximum intensity observed, and θ is the phase difference between the reference signal and beam signal. It is clear that for $\theta = 90°$, the detected intensity will be zero. In an effusive beam, species with different masses will have different mean velocities. Consequently, θ for each species will be different. By a proper choice of chopping frequency the two species can, in principle, be made to appear 90° apart in phase, so that one species will not be detected while the other will be detected at maximum intensity. For reasonably light molecules and adequately large mass differences the chopping frequencies required are experimentally accessible. This technique, which is analogous to that of Yealland *et al.* (1967), should be useful in cases in which an ion has two neutral precursors, as occurs in studies of monomer–dimer equilibria or free radicals.

IV. Molecular Beam Applications

In the following sections we discuss a number of applications of potential interest to high temperature chemists. With regard to the high pressure molecular beam applications, it is useful to distinguish those that involve the beam formation process only passively, such as direct sampling, and those that take advantage of changes in the gas due to expansion, such as nucleation studies. To the former, the processes of interest to the latter are mainly a necessary evil.

A. Direct Sampling at High Pressures

One of the chief applications of high pressure molecular beams is the extension of the range of source pressures that can be directly sampled by mass spectroscopy from a few Torr to pressures of 10^4 to 10^6 Torr. We review the effects expected to be encountered, some current limitations, and finally a list of applications known to us.

1. *Principal Effects Involved in Interpretation of Beam Composition*

Based on the simple picture of free-jet expansion presented above, a number of potentially serious problems can be anticipated. It is assumed that effects due to shocks or scattering have been eliminated by judicious choice of geometry and pumps, so that only phenomena inherent in the expansion are of concern. The phenomena are categorized here as (a) mass separation,

(b) nucleation and shifting equilibrium, (c) change of internal state, and (d) mutual interaction of system and probe.

(a) An early observation (Becker *et al.* 1955) in the use of nozzle beams was that of the concentration of heavier species relative to light on the axis of the beam. This phenomenon and various contributions to it have been extensively studied and are reviewed by Fenn and Anderson (1965). More recent studies involve mass separation effects in the free jet itself (Sherman 1965; Rothe, 1966; Anderson, 1967b; Sebacher *et al.*, 1968; Schügerl, 1968; Wang and Bauer, 1968) and in shock waves (Sebacher, 1968). Velocity distributions of individual components of mixtures are now being measured to further elucidate the behavior of mixtures (Miller and Andres, 1968; Fisher and Knuth, 1968; Anderson and Cauley, 1968; Greene *et al.*, 1968b). The problem of mass separation appears manageable if proper calibrations for the flow conditions of interest are carried out. The simple thermal spreading of components of a free jet, all of which achieved the same stream velocity but have different random thermal velocities (Stern *et al.*, 1960), will alone account for a first power of the molecular weight enrichment of heavy species, and this correction is suggested at high pressure if calibrations are unavailable (Greene *et al.*, 1964).

(b) As the temperature and pressure change during free-jet expansion, many systems can change in composition due to chemical reaction. For example, any system can nucleate or associate to form clusters as part of the early stages of condensation. Common chemical reactions involving positive activation energies stand a good chance of being frozen at nearly their initial state due to the extreme cooling and short time scale involved in free jets. Presumably, calculations of shifting chemical equilibrium could be carried out for free jets, either with or without considering the effects of the reaction on the flow parameters, in a fashion analogous to that used in rocket nozzles (Westenberg and Favin, 1963).

Homogeneous nucleation in free jets was first inferred from measurements by Becker *et al.* (1956), and later observed directly through mass spectrometric observations by Henkes (1961, 1962), Bentley (1961), Turnbull (1962), and Greene and Milne (1963). More recent observations, and studies of nucleation and the clusters formed, are presented below. The implication of the presence of nucleation to the interpretation of direct sampling observations are obvious—particularly if one is concerned with measuring equilibrium concentrations of minor polymers of the system being studied. It is the usual case that substances will tend to associate in free-jet expansions, whether or not supersaturation is achieved. With high temperature species, the heats of vaporization are often so large that even a few hundred degrees' cooling, such as will occur in the vicinity of the Mach-1 condition at the sampling orifice, will lead to enormous supersaturations. Mitigating against extensive nucleation in such systems will be the very large condensation energy, relative to kT, released by such systems. Possible system parameters that can be varied to detect and correct for nucleation in sampling are discussed by Greene and Milne (1966).

(c) Along with changes in the molecular composition and the translational energy state brought about by the free-jet expansion will go changes, or relaxation, of internal energy states of molecular systems. Although such relaxation affects the course of the free-jet expansion by determining the effective γ, and hence is coupled with nucleation and chemical quenching, the principal implications for mass spectrometric sampling studies are likely to be in terms of changing fragmentation patterns and the postulated phenomena of metastability.

The problem of the temperature dependence of fragmentation patterns is not a new one in high temperature mass spectrometry, although it is largely ignored. High pressure sampling is further complicated in that, even in cases where temperature dependencies are known, the partial relaxation of vibrational degrees of freedom will lead to uncertainties in the internal energy state and hence to uncertainties in the fragmentation pattern to be expected. In addition, the ability to sample relatively complex, weakly bound species, present in significant concentrations only at high pressures, will involve new classes of molecules for which past experience may serve as a poor guide to expected cross section and fragmentation behavior. Some of the advantages of internal energy relaxation are discussed in a later section.

A related problem involves the postulated metastability of certain clusters that may be present in high pressure gases or be formed through nucleation. Leckenby and Robbins (1966) and Milne and Greene (1967a) have observed anomalously low concentrations of clusters in diatomic gases such as N_2 and O_2, Leckenby and Robbins argue that because of selective internal energy relaxation, one may obtain a very weakly bound cluster such as O_2—O_2 which is made up of two oxygen molecules that are internally each rather hot in terms of unrelaxed rotational energy (in high temperature species, vibrational energy as well). Such clusters can unimolecularly dissociate by exchange of rotational energy of the individual molecules with the weak van der Waals bond holding the oxygen molecules together. Such a phenomena can thus complicate the interpretation of sampled concentrations of weakly bound species.

(d) A very practical problem involved in the sampling of high temperature systems pertains to the mutual interaction of the system and the probe. It is seldom possible for the system being studied to be in both thermal and thermodynamic equilibrium with the probe. In fact, even in the circumstance of sampling a pure inert gas through a knife-edged orifice at the same temperature, there is a possibility of heat exchange from the orifice edge to the expansion-cooled, approximately Mach-1, gas passing by. This could perhaps be better phrased by asking to what extent the core flow of gas through an orifice, which ultimately makes up the molecular beam, feels the influence of the orifice wall and hence is nonisentropic. Data already cited indicate that even for orifices as small as a few mils in diameter, one-atmosphere expansions from isothermal systems proceed as though they were adiabatically expanding through an orifice of very nearly the geometrical diameter, i.e., boundary layers are quite small and heat exchange is negligible. In the many cases of interest to this review, where the sampled gas will be much hotter than the orifice walls, less is known. The success in achieving high energy beams from shock tubes (Skinner, 1968) and arcs (Winicur and Knuth, 1967) argues that nearly adiabatic sampling can be achieved with cold orifices. This conclusion is supported by the studies of Milne and Greene (1967d), who have sampled the burnt-gas region of hot, one-atmosphere flames.

In systems such as flames, there may be an aerodynamic perturbation of the system in addition to heat loss effects. Some observations on reaction-zone sampling with low pressure flames are recorded by Milne (1965a). A particularly serious case of probe interaction occurs when cold probes are used with systems containing condensable species. Physical plugging is unavoidable (Milne and Greene, 1967d), and it appears that the approach must be to obtain information rapidly, through optimization of detection signal-to-noise and the use of time-resolved spectroscopy, prior to major aerodynamic and heat-loss effects due to the buildup of condensate on the probe tip. A novel self-cleaning skimmer has been described (Hundhausen and Harrison, 1967), but it would be more difficult to achieve cleaning of the sampling orifice itself.

If one is sampling ions as well as neutrals, then one must contend with new effects involving the buildup of charge on the orifices and phenomena in the boundary layer near the orifice. Sampling from supersonic flows likewise can introduce problems for both ions and neutrals, since there is a possibility of detached shock structure in front of the probe. These phenomena are referred to under the appropriate applications below.

A collection of abstracts dealing with various aspects of direct sampling has been compiled from a conference on high pressure sampling (Milne, 1965b).

2. *Sampling Applications*

The potential applications of direct molecular beam sampling from high pressures are as broad as the interests in understanding the nature and reactions of either gaseous systems or systems forming gaseous products. In the following discussion we select those areas akin to the interests of high temperature chemists, where applications are either already being made or could readily be made. We presumably need the most liberal of past definitions of high temperature chemistry to justify some of the inclusions. Operationally our choice is limited to the use of high pressure molecular beams with mass spectrometric detection of reactive, condensable, or unstable neutral and ionic species, regardless of temperature. Of course, other detection techniques, such as EPR (Westenberg and Fristrom, 1965), could be used.

a. Transpiration and Knudsen Cell Vaporization. High temperature chemists have been properly wary of Knudsen cell measurements in which Knudsen numbers approached unity. With the detailed understanding being generated about high pressure expansions, this limit need no longer apply. With an awareness of the problems to be expected as continuum flow ensues, one can hope to make both third- and second-law interpretation of species intensity data. The usefulness of such measurements has been argued by Dimiduk (1967). In addition the problem of making valid measurements of noncondensable species from Knudsen cells has likewise been of concern. With conventional beam shuttering one may incur substantial errors in interpreting the relative intensities of condensable versus noncondensable species, as discussed by Foner and Hudson (1953) and Milne and Greene (1968). Use of modulation and phase-sensitive detection largely overcomes this problem, as discussed in an earlier section.

With the use of high pressure beam techniques and beam modulation, driving pressures of gas in transpiration systems such as $BeO+H_2O$ (Hildenbrand *et al.*, 1965), Li_2O+H_2O and HCl (Schoonmaker and Porter, 1960), $BaO+H_2O$ (Stafford and Berkowitz, 1964) precious metal$+O_2$ systems (Norman *et al.*, 1968) and carbon and boron$+H_2$ (Stafford *et al.*, 1968) can be extended well beyond the presently employed values of a few Torr or less. The coupling of conventional transpiration experiments, involving gaseous driving pressures of one atmosphere or more, with direct mass spectrometric sampling, should be fruitful. Greene *et al.* (1968b) have undertaken such studies with the graphite system. Corrosion phenomena involving the gaseous transport of material in high pressure gas could presumably be directly observed, such as the transport of oxides in steam.

b. Free Evaporation—Laser Heated Surfaces. If evaporation or decomposition rates exceed conditions for collision-free vaporization, then the phenomena involved in high pressure beam formation enter. Thus, recent observations of species generated by pulsed laser heating of metals and carbon in vacuum (Berkowitz and Chupka, 1964) may be complicated by the presence of nucleation and mass separation. Through the use of time-of-flight velocity analysis, variation of the evaporating area, the study of the angular variation in intensity of the products, and the application of a model of continuum expansion similar to that for flow through orifices (Edwards and Collins, 1968), one can hope to provide better interpretation of such results (Greene *et al.*, 1968b). If decomposition to noncondensable species occurs, then differential pumping may have to be incorporated to permit valid sampling (Pellett, 1968; Lincoln, 1965).

c. Flames, Flow Reactors, Gas Kinetics. The desire to observe reactive species in flames has historically represented one of the chief motivations for developing molecular beam sampling techniques (Foner and Hudson, 1953; Milne and Greene, 1965, 1967d). Wagner and Homann have used direct mass spectrometric sampling for several years to examine both neutral gas molecules and particles from low pressure flames (Homann, 1967, 1968; Homann *et al.*, 1965). Vriens *et al.* (1965) have also sampled high pressure flames. Milne and Greene (1966) demonstrated the ability to quantitatively sample free radicals from the burnt-gas region of one-atmosphere flames, while Milne (1965a) indicated some of the problems and possibilities for determining species profiles through flame reaction zones in low pressure flames. Initial attempts to observe highly condensable species in one-atmosphere flames failed (Milne and Greene, 1967d). A number of laboratories are planning direct flame sampling studies (Williams *et al.*, 1968; King, 1968; Pertel, 1968; Sawyer, 1968; Balwanz, 1968).

Ions have been sampled from high pressure flames, where continuum flow was clearly involved (Hayhurst and Sugden, 1966; Hayhurst and Telford, 1966; Knewstubb *et al.*, 1966; King and Scheurich, 1966; Feugier and Van Tiggelen, 1967; Peeters *et al.*, 1968; Hurle *et al.*, 1968). In lower pressure flames, Miller and Calcote have reported results with ion sampling (Miller, 1967).

Direct sampling of flow reactors has been applied by Wong and Potter (1966), Niki and Weinstock (1967), Talrose *et al.* (1966), and Westenberg and Fristrom (1965). Other discussions of direct sampling of reactive species are given by Bentley *et al.* (1961), Barber *et al.* (1963), and Mori and Takezaki (1968). It is reasonable to project that a whole range of gas kinetics experiments can be profitably instrumented for direct sampling (Foner and Hudson, 1968).

d. Shock Tubes—Flash Photolysis. The use of shock tubes with beam formation systems has been pioneered by Skinner (1968), who was seeking a means of generating neutral beam energies above 1 eV. Oman *et al.* (1966–1967), Peng and Liquornik (1967), and Jones and Byrne (1968) are also developing such systems. A number of workers have coupled shock tubes to mass spectrometers to study chemical reactions. (Kistiakowsky and Michael 1964; Dove and Moulton, 1965; Modica, 1967; Ryason, 1967; Rubin, 1968; Wang, 1968; Blauer, 1968; Marsters *et al.*, 1965; Diesen, 1966; Gay and Kern, 1968).

Flash photolysis–mass spectrometer coupling was initiated by Kistiakowsky and Kydd (1957), who pioneered in the development of the Bendix time-of-flight instrument for obtaining very rapid time resolved mass spectra. Meyer (1967) and Gutman *et al.* (1966) are also using this technique. The trade-off between determining complete spectra at 10 to 100 μsec intervals in a single flash, using the low duty cycle Bendix (Kistiakowsky and Kydd, 1957), versus following continuously one species at a time in successive flashes (Gutman *et al.*, 1966) is not immediately clear.

e. Ablation and Beam–Surface Reaction. As already mentioned, one of the motivations for developing supersonic molecular beams is to permit gas–surface interactions to be studied, including chemical reactions. (The many studies currently under way on recombination, angular reflection, and energy exchange are not covered.) Recently, mass spectrometry has been used to study gas reactions with hot surfaces, employing direct sampling of products (McKinley, 1964, $Ni+Cl_2$; Schissell and Trulson, 1965, and Berkowitz-Mattuch *et al.*, 1963, W with O_2 ; Nutt and Carter, 1968, CH_3I on W ; and Thorn and Holt, 1968, $NiCl_2$ on Ni). In principle, it requires only the incorporation of a high pressure sampling device to extend such measurements to higher pressures, where different species may be involved. Ablation processes are a related phenomena to which direct sampling should be applicable. Plans for such experiments have been discussed by Lincoln (1966) and Pope and Parker (1968). Conceivably, primary ablation products could be directly observed in both laboratory simulation situations and in large scale wind tunnel tests.

f. Propellant Burning. Molecular beam sampling is being applied to various aspects of propellant burning. Summers (1965) made early attempts at direct sampling of a small liquid rocket motor, using a direct sampler designed by Netusil *et al.* (1965). Sawyer *et al.* (1968) discuss results obtained with this sampler, while Kahrs (1968) analyzes its performance in detail and suggests improvements. Pellett and Saunders (1968) are using both high and low pressure sampling to determine species coming from laser-assisted solid propellant burning. Williams *et al.* (1968) plan direct sampling at high pressure for a variety of propellant related combustion studies. Salser *et al.* (1968) report transitory species in propellant reactions. Direct sampling, though beset with its own problems, is intended to remove some of the ambiguities involved in indirect sampling, e.g., the products of ammonium perchlorate decomposition are still subject to doubt (Verneker and Maycock, 1967). In combustion systems not involving highly condensible species such as occur with metallized propellants, the prospects for immediate success in sampling at pressures up to one atmosphere are great.

g. Otto Engines, Incinerators, Air Pollution. As in the cases above, the combustion processes of interest in air pollution studies can be more thoroughly characterized by direct sampling. In past studies of sulfur oxide formation [for example, Merryman and Levy (1967)], the species SO had to be inferred from mass balance arguments. Species in the N–O and S–O systems as well as free radicals, halogen compounds, O_3, and similar species can be individually studied by molecular beam sampling techniques (Milne and Greene, 1966). In addition to the study of laboratory flames already mentioned, there are the possibilities of direct sampling of the dynamic processes in an internal combustion engine cylinder. Sawyer (1968) is beginning studies with this goal in mind. Incinerator burning processes as well as effluents from various chemical processes can be monitored, perhaps with special advantage, by molecular beam sampling and detection techniques.

h. Plasmas, Afterglows, Radiolysis Sources, Discharges, and Arcs. Activity in the direct mass spectrometric sampling of ions has been greater than for neutrals—presumably because fewer alternatives exist for measuring ions than for neutrals. Most of this sampling has been from systems at pressures of a few Torr or less where conditions were supposed to be close enough to effusive to avoid sampling perturbations. A more serious concern, as mentioned above, has been that of the electrical effects existing at and near the probes and slits used in forming and collimating the beam. One must be cautious in assessing the likelihood of reaction or nucleation effects with ions, based on experience with neutrals, since cross sections for collision and reaction between ions and neutrals can be expected to be significantly larger than between neutrals.

A few attempts at sampling one-atmosphere plasmas have been made. Leigh (1960) tried a cryogenically pumped sampler design, but no results were achieved. Carrigan *et al.* (1960) and O'Halloran *et al.* (1964) used a Bendix mass spectrometer to sample ions and neutrals from plasma jets. In the latter work it was not definitely established that relatively undisturbed free-jet expansion was achieved. Many workers have reported studies on ion sampling from plasmas and discharges at lower pressures (Bohme and Goodings, 1966; Mosharrafa and Oskam, 1966; Hayhurst and Padley, 1967; Studniarz and Franklin, 1968; Brink *et al.*, 1968; Miller, 1968a; Sperling and Engh, 1968; Gaur and Chanin, 1968; Shahin, 1968). The sampling from arc-heated wind tunnels has already been mentioned (Knuth *et al.*, 1967; Winicur and Knuth, 1967; Kessler and Koglin, 1967).

An increase in the use of mass spectrometry in the sampling of stationary (Lineberger and Puckett, 1968) and flowing afterglows (Ferguson *et al.*, 1968; Dunkin *et al.*, 1968) has been motivated by a desire to understand the ion chemistry of the upper atmosphere. McAfee *et al.* (1967) and Miller *et al.* (1968) have used a mass spectrometer in conjunction with ion drift tubes. Kebarle *et al.* (1968) report an interesting series of sampling studies of a high pressure radiolysis source, where ionic equilibria at pressures up to several hundred Torr are measured. At lower pressure Janik and Conway (1967) report on simple dimerization reactions.

i. Upper Atmospheric Composition. For a number of years rocket and satellite borne mass spectrometers have been used to gather neutral and ionic composition data at altitudes of about 50 km and greater [see, for example, Von Zahn (1967), Ghosh *et al.* (1968), Narcisi *et al.* (1968)]. Sampling has generally been done under molecular flow conditions, although the gas is invariably allowed to randomize and possibly react in the ion source during detection. This is currently the subject of much concern in the detection of O atoms (Von Zahn, 1967) and other free radicals. The possibility of using modulated molecular beam inlets with phase-sensitive detection does not seem to have been discussed. Sampling effects and gas loads definitely limit the lower altitude of present flyable packages. For very low altitudes, or for a very dense planetary atmosphere such as that of Venus, a high pressure, differentially pumped, beam sampling system may be the only practical way to obtain concentrations of reactive species. Miller and Burke are studying experimentally the possible effects of shock waves and other perturbations at the probe tip on sampled ion compositions (Burke, 1968). Sonin (1967) and Frankenthal (1968) have explored such problems theoretically.

j. Weakly Bound Species. The use of direct high pressure sampling to characterize and study the thermodynamic properties of weakly bound clusters is not distinguishable, in principle, from the Knudsen cell and transpiration studies mentioned in Section IV,A,2,*a*, but it deserves special emphasis. It is now recognized that stable, well-defined molecules exist between any combination of atoms or molecules by virtue of the universal van der Waals attractive forces. Though involving very low energies of formation, these molecules can have appreciable concentrations at high pressures. For example, Milne and Greene (1967c) have observed about 0.1% of Ar_2 in equilibrium with gaseous argon at 1 atm and 300°K. Meaningful second- and third-law studies (Milne and Greene, 1967b) are thought possible for such complexes, although the problem of "metastability," if real (Leckenby and Robbins, 1966; Milne and Greene, 1967a), appears quite serious for some species. More strongly bound complexes, such as those involving hydrogen bonds (Greene *et al.*, 1968a) and charge transfer bonding (Passchier and Gregory, 1968) should be even more amenable to molecular beam observation and quantitative study.

k. Particle Sampling. The sampling of particles or droplets from high temperature systems involving high pressure expansion is of interest in following many processes. Micron-sized particles will achieve considerable acceleration from an expanding gas in a nozzle-beam sampling system and will continue into a collector device in the vacuum system (Israel, 1967). Bonne *et al.* (1965) have sampled and studied carbon particles formed in sooty flames. Thomann (1966) collected ice crystals formed in supersonic expansions. References to particle gas flow are given by Rudinger (1968) and Jarvinen and Draper (1967).

l. Future Sampling Possibilities. The following sampling applications have not, to our knowledge, been put into practice, but are representative of many useful possibilities involving high pressure beam formation.

The features of very rapid response and broad applicability to reactive species makes direct molecular beam sampling of potential usefulness in controlling chemical processes. The stability and absence of contamination and memory effects achieved when modulated molecular beams are used could recommend this type of sampling to automated control systems. Response times of the order of milliseconds are achievable, should such be required in controlling critical reactions.

Another sampling application of great potential involves the use of free-jet expansions to introduce high molecular weight species into the mass spectrometer for subsequent characterization or monitoring. Already, mass separation effects have been utilized to concentrate heavy species in light carrier gases (Ryhage, 1967; Dole *et al.*, 1968). Of potentially greater significance would be the use of free jets to form molecular beams from very high pressure gas used to transport measurable quantities of complex, high molecular weight species. Giddings *et al.* (1968) have discussed such transport in terms of chromatographic separations. Milne (1969) elaborates on this idea, which points to one possible way of obtaining, in the gaseous state, molecules that are too unstable to be directly volatilized from the condensed phase.

Among unusual high temperature processes that could be approached through direct sampling are included the formation of fallout nuclei in the vicinity of fireballs, the behavior of chemicals such as Ba released from high pressure in the upper atmosphere, and the fate of chemical or biological agents disseminated by explosive devices.

B. Studies Using High Pressure Beam Properties

Studies that take advantage of the phenomena accompanying high pressure beam formation are as important as the passive applications just presented. The ability of free jet expansions to achieve drastic translational cooling, to accelerate heavy solutes to the velocity of light solvents, to effect appreciable internal energy relaxation, to generate unusual clusters, and to produce intense, modulated, collimated beams of reactive and condensable species leads to a number of prospects for new approaches to the study of high temperature behavior.

1. *Nucleation Studies*

The fact that gases tend to nucleate during the isentropic expansion from high pressures, while probably representing the chief limitation to faithful sampling studies, provides a unique situation in which to study the condensation process itself. Mass spectroscopic observation of nucleation has been reported by Henkes (1961) (CO_2), Henkes (1962) (H_2), Bentley (1961) (CO_2), Turnbull (1962) (N_2O, N_2, CO_2, SO_2), Greene and Milne (1963) (Ne, Ar, CO_2, O_2, N_2O, N_2, NH_3, H_2O), Leckenby *et al.* (1964) (CO_2, Ar, N_2, O_2, N_2O, $SF_6 \cdot C_2H_6$, C_3H_8), Cuthbert *et al.* (1965) (CO_2, SO_2, N_2O), Robbins and Leckenby (1965) (CO_2, N_2O, N_2, O_2, Ar, Xe), Leckenby and Robbins (1966) (O_2, N_2, Ar, Xe), Milne and Greene (1967c) (Ar, CO_2, N_2, O_2, H_2O), Milne and Greene (1967d) (Hg, CsCl, CH_3OH), Milne and Greene (1967a) (NO, N_2, O_2, Ar, CO_2), Beck and Mogenstern (1967) (K), Buchheit and Henkes (1968) (H_2), Good *et al.* (1968) (NO), Golomb and Brown (1968) (NO, Ar), Knof and Maiwald (1968) (acetone), and Milne *et al.* (1969b)

(n-butane). Discussions of the extent of nucleation in different gases have been given by Bier and Hagena (1965) and by Hagena (1968). Anderson *et al.* (1965b) and Andres (1965) apply a simple freezing model to explain the orifice and pressure dependence of argon dimerization. Milne *et al.* (1969a) give a detailed kinetic interpretation of argon dimerization, based on a temperature-dependent, third order, recombination kinetic model.

The above observations of nucleation as a function of pressure, temperature, orifice size, and dilution with other gases provide a virtually untapped source for kinetic interpretation. Nucleation of higher temperature species is a logical extension of interest for sampling and for a basic understanding of condensation. There appear to be no insoluble problems, although the importance of cluster fragmentation on electron impact and the possibility of metastability of neutral species have not been determined. The great advantage of the free-jet environment is that extremely fast reactions, approaching collision frequencies, can be followed by virtue of the extremely short time scale involved and the limited number of collisions per molecule.

Free-jet nucleation studies so far have involved observations of the final extent of nucleation, or cluster growth, after the entire free-jet expansion is over, and collisions have ceased. This causes ambiguity in interpretation, considering the enormous range of temperature covered. It would be distinctly preferable if one could probe through the continuum flow field with the skimmer, having demonstrated that collisions effectively cease at the skimmer, and thus follow the growth of clusters as a function of time. The recent probing experiments by Bossel *et al.* (1968) indicate that such an approach may well succeed. In this connection, one should be able, with suitably designed probes so as to avoid detached shocks, to study the nature and changes in species in wind tunnel (Daum and Gyarmathy, 1968) or nozzle expansions. For example, one might be able to bridge the gap between monomers and the smallest particles that can be seen by light scattering in nozzle experiments such as those reported by Wegener (1967).

The study of nucleation takes on great importance in terms of some of the potential uses of nozzle beams or free jets for other studies, such as chemical reactions or surface interactions. For example, the study of nitric oxide nucleation (Milne and Greene, 1967a) was prompted by the postulate of Fontijn and Rosner (1967) of the crucial role that very small amounts of dimers or higher clusters might play in the reaction of nitric oxide with O atoms. As nozzle beams are enlisted in forming beams of high temperature molecules for reaction studies, great care will have to be exercised. For example, Robbins *et al.* (1967) observe extensive cluster formation in Na beams.

2. *High-Energy Molecules*

The fact that light and heavy gases in mixtures achieve virtually the same velocity in free jets, first demonstrated by Becker and Henkes (1956) was recognized by Anderson *et al.* (1966) as a means of achieving neutral molecular beams with heavy components having translational energies substantially greater than 1 eV (Klingelhöfer and Lohse, 1964; Anderson, 1967a). Thus a small percentage of O_2 expanding in helium at 2000°K achieves a kinetic energy in the resultant beam equivalent to a source temperature of 16,000°K. A large number of mixtures have been examined by Fenn and co-workers (Fenn, 1967; Abuaf *et al.*, 1967a) using velocity distribution measurements based on a time-of-flight scheme. The effect has been observed in mass spectrometric studies of mixtures by Milne and Greene (1967d), Greene *et al.* (1968b) (using a high resolution time-of-flight scheme suggested by Greene), and Miller and Andres (1968).

Given these very energetic molecules, one can simulate very high temperature reaction behavior in beam experiments involving either crossed beams or beam–surface interactions. A good simulation of satellite-velocity surface–gas reaction is achieved by obtaining beams with very high kinetic energy, yet low temperature internal energy.

3. *Properties of Clustered Neutrals and Ions*

The clusters generated by free-jet expansions provide an opportunity for some interesting studies on these unusual states of matter. Virtually any desired complex could conceivably be generated for study. Burghoff and Gspann (1967) and Gspann and Krieg (1968) have characterized nucleation in free jets using scattering measurements with potassium beams. Bauchert and Hagena (1965) have used retarding field measurements on clustered ions to deduce their mass distribution. The observation of angular distributions and survival of beams of clustered H_2 impinging on metal surfaces have recently been reported by Becker *et al.* (1968). An exciting prospect for characterizing these intermediate states between simple gases and liquids or solids has been demonstrated by the very recent application of electron diffraction techniques to the determination of interatomic distances and order in clusters of Xe generated in free jets (Audit and Rouault, 1967a; Audit, 1968). There is no reason to believe that spectroscopic techniques will not be brought to bear on these systems. Recent observation of absorption of vacuum ultraviolet radiation is pertinent (Shardanand, 1968; Verkhovtseva *et al.*, 1968). Tanaka (1968), in fact, has unambiguously identified Ar_2 and determined its molecular parameters.

The ionization behavior of clusters furnishes another opportunity for interesting experimental studies. For example, it appears that when neutral water clusters are ionized, they lose OH, forming the series $H^+ \cdot (H_2O)_n$. Since the ions of this series are the same ones that figure prominently in the upper atmosphere, in flames and discharges, and in radiolysis systems, it is of interest to determine their energetics of formation by electron impact and photoionization studies (Milne *et al.*, 1968). Precise determination of the energy of ionization processes in other clusters can provide information about ion binding energies. Processes such as $(H_2)_2+e \rightarrow H_3^+ + H + 2e$ or $HgAr+e \rightarrow HgAr^+ + 2e$ are examples. Robbins *et al.* (1967) have reported the photoionization energies of sodium clusters. Hagena and Henkes (1965) discuss effective ionization cross sections in condensed beams. Hydrocarbon clusters have been found to show interesting ionization behavior. Milne *et al.* (1969b) have observed quite different fragmentation ratios of mass $58^+/43^+$ from the *n*-butane dimer compared to the normal monomer ratio. In addition, parent dimer ions of the unusual formula $C_8H_{20}^+$ are formed in *n*-butane expanded in argon. As with neutrals, almost any desired ion can probably be generated and involved in subsequent reactions.

It is worth mentioning also that the use of modulated molecular beams from heated effusive sources represents an advantageous way to investigate the temperature dependence of fragmentation patterns. The temperature of the gas can be well characterized under effusive conditions, and the patterns are uninfluenced by regions of temperature variation in the ion source, by the presence of a hot filament, or by reaction of gas with the ion source. The onset of chemical reaction in the gas source can be detected by a number of diagnostic techniques, including time-of-flight velocity analysis. Insofar as the fragmentation behavior of a molecule is a function of the internal vibrational energy state, the observation of fragmentation ratios as a function of free-jet expansion conditions, for a given source temperature, can yield information about vibrational relaxation in a fairly direct manner. Milne *et al.* (1969b) report preliminary results for *n*-butane to 500°C, indicating that substantial vibrational cooling occurs even from expansion at 1 atm pressure through a 0.003 in. diameter orifice.

4. *Free-Jet Isolation Spectroscopy*

The ability of a free jet both to reduce the internal energy of its constituents and, by expansion to essentially collisionless flow, to effectively isolate them from interaction with other matter, leads to the fourth and potentially most significant application of supersonic nozzle beams to high temperature chemistry. Spectroscopy of high temperature species, particularly that in the infrared and microwave regions, has been limited by several problems arising from the enormous number of rotational and vibrational levels populated. Because so many levels are populated, the concentration of species in a particular level is very small and transitions from that level are difficult to detect. Spectra are complicated, and rotation–vibration bands are broad and overlapping. The matrix isolation technique has proven to be a very useful method of reducing the internal temperature of high temperature species, but many studies are limited or made uncertain by the presence of the matrix. The free-jet expansion provides an alternative of potentially universal applicability to high temperature molecules. Kantrowitz and Grey (1951) pointed out in their original paper on nozzle beams the reduction of the number of internal energy states of molecules. Subsequent workers have repeated this observation (Anderson *et al.*, 1965a, b). The specific application to spectroscopy appears to have first been appreciated by Greene and Milne (1969).

All that is required is to expand the species of interest as a free jet from a high pressure, which may be provided by a carrier gas. As pointed out earlier, translational cooling will be extreme. Direct evidence exists that appreciable rotational relaxation occurs under reasonable expansion conditions (Robben and Talbot, 1966; Campargue, 1966; Marrone, 1967; Anderson *et al.* 1968; Miller and Andres, 1967; Ashkenas, 1967; Lefkowitz and Knuth, 1968). For example, N_2 free jets have achieved rotational temperatures as low as 10°K (Robben and Talbot, 1966). Vibrational temperatures can also be significantly lowered, even though vibrational relaxation is much slower than rotational (Hagena and Henkes, 1960; Russo, 1967; Harbour, 1968; Milne *et al.*, 1969b). One can therefore obtain species generated at high temperatures but with cryogenic rotational temperatures, and in some cases at substantially reduced vibrational temperatures, without matrix interactions and with reduced Doppler and pressure broadening.

Spectroscopic studies could be carried out in either a free jet or in a portion of the jet collimated into a molecular beam. The direct use of the free jet has the advantage of relative simplicity, of allowing studies at high densities without possible skimmer interaction, and of readily permitting a combination of large cross section and long path length. Collimation of the jet into a molecular beam will, on the other hand, give exceptionally low Doppler broadening and facilitate modulation of the beam to improve the signal-to-noise ratio. As a compromise some collimation of the jet will probably be desirable in almost all cases.

The free-jet isolation technique has not yet been implemented, although approximate calculations indicate that it should be feasible in at least some cases (Greene and Milne, 1969). The problems are those of (1) obtaining sufficient relaxation of the internal energy states without obtaining excessive cluster formation, while (2) obtaining sufficient density and path length. In principle, and in most cases in practice, both problems reduce to a matter of having sufficient pumping speed, since cluster formation can be reduced to any desired limit by reducing the concentration of species, and any desired path length can be obtained by using either a long slit as an expansion orifice, or an array of jets.

5. *Collisions at Low Temperatures*

It might be pointed out, finally, that within the free jet, molecules interact with extremely small relative kinetic energies, so that one can hope to investigate collision phenomena and intermolecular forces at effectively very low temperatures (Knuth and Fisher, 1968). Such measurements may not be possible by other techniques due to the delicate balance between condensation and low energy interactions.

C. Electric and Magnetic Deflection Experiments

The deflection experiments to be considered here involve the measurement of the physical displacement of a species in a molecular beam due to the interaction of an inhomogeneous field with an electric or magnetic moment. Resonance experiments, which are fundamentally spectroscopic in nature and use deflecting fields to reduce the number of states detected, are not included here. It should be noted that the resonance experiments, when they are applicable and can be carried out and interpreted, can produce more precise and detailed data than the simple deflection experiments. However, the electric resonance technique is of no value for nonpolar species and is presently limited for complex species at high temperatures by the large number of energy levels involved and the small population in any given level. The advantages of the simple deflection experiments are that they can be carried out and the results interpreted, within the limitations of the experiment, for any species that can be produced in a molecular beam. As discussed below, significant and often unique data can be obtained from the simple deflection experiments.

The same general principles apply to both electric and magnetic experiments. If a molecule with an appropriate moment is placed in an inhomogeneous field, there will be forces on the molecule given by

$$f = \mu \nabla F$$

where f is the force produced by a field gradient ∇F on an effective moment μ.

If the field is one-dimensional, and a molecular beam is introduced perpendicular to the direction of the field, the beam profile will change according to the distribution of moments and velocities and the geometry of the system. That the velocity distribution be known is extremely important, since the deflection of an individual molecule is proportional to the time it spends in the field region and hence to its velocity. Most of the measurements of alkali metal polarizabilities were apparently in error due to a skewing of the velocity distribution as the result of small-angle scattering of the beam by residual gases in the system (Hall and Zorn, 1967; Hall, 1968). Other perturbations in the velocity distribution in effusive beams may be caused by too high pressures and by nonideal slits (McFee and Marcus, 1960; Miller and Kusch, 1956). In order to obtain quantitative results, therefore, it is necessary to either measure the velocity distribution or to use nearly ideal slits, low source pressures, and sufficiently low system pressures that small-angle scattering is unimportant. The achievement of the latter condition can be verified, in the absence of velocity measurements, by measuring the effective moment as a function of system pressure and extrapolating to zero pressure if necessary.

1. *Electric Deflection Measurements*

The electric deflection measurements are probably the more generally useful of the two types of experiments. Every atom or molecule has a significant (from the point of view of producing a readily measurable effect) induced electric moment and many have permanent moments (dipole and higher), whereas the number of species that have significant magnetic moments is more limited.

The simplest case for an electric deflection experiment is that of an atom or nonpolar molecule. The induced electric moment μ, which is always in the direction of the field E, is given by

$$\mu = \alpha E$$

where α is the polarizability. To be completely accurate, the polarizability should be expressed as a tensor. However, the value for the polarizability measured in this and most other experiments is an average over all directions and is essentially independent of the rotational and vibrational states. Different electronic states often have measurably different polarizabilities. This will not usually be a complication, because most electronic states will either decay before reaching the field or will not be populated sufficiently to be detected. Only in the case of either a high effective source temperature or a low-lying level, combined with a very long lifetime state, will levels other than the ground state persist into the field.

The case of polar molecules is more complicated, since there will be contributions from both the polarizability and the permanent moment or moments to the effective electric moment. The induced contribution will be the same as for a nonpolar molecule. The effective electric moments due to dipole moments have been investigated in connection with the Stark effect and are discussed by Herzberg (1950, 1966), Townes and Schawlow (1955), Gordy *et al.* (1953), and Ramsey (1956). These moments depend both on the value of the dipole moment or moments and on the rotational state of the molecule. For linear and asymmetric top molecules, which have only a second order Stark effect, the contribution of the dipole moments to the effective moment will also depend on the field strength and may be either positive or negative depending on the rotational quantum numbers for the particular rotational state. Symmetric tops and nonsigma electronic states, which show a first order Stark effect, will have an effective moment that is essentially independent of field strength. It is therefore possible to separate the contributions of polarizabilities and dipole moments by an analysis of the changes in the beam profile and their dependence on field strength (Greene and Milne, 1967).

The electric deflection technique has been applied to both dipole and polarizability measurements (Ramsey, 1956; Fraser, 1931; Bedersen and Robinson, 1966), but the range of applications has been limited by the detectors employed. With the addition of a universal mass spectrometric detector (Greene and Milne, 1968b), there are a number of applications of electric deflection measurements. The quantitative measurement of molecular dipole moments and polarizability should aid in understanding chemical bonding, particularly in some of the unusual species encountered in high temperature systems. The value of the dipole moment often has very strong structural implications, and may allow a definite choice between two or more chemically plausible structures. Some examples are discussed in conjunction with the quadrupole lens technique described below. It is also of interest that the polarizabilities of most of the elements have never been directly measured. The electric deflection method appears to be the only currently feasible method for measuring this basic property of the less volatile elements. A good discussion of electric deflection measurements of polarizabilities is given by Bederson and Robinson (1966).

In addition to explicit electric moment measurements, there are other potentially useful applications of the deflection technique. It can be used to measure rotational temperatures in a molecular beam (Herm and Hersbach,

1965). Since the distribution of electric moments of a polar molecule contains the temperature as a parameter, the " deflection " of the beam may be used to determine the rotational temperature in cases in which the dipole moments are known. This could be particularly important when studying the kinetics of vaporization or supersonic nozzle beams. It has been used as a rotational state selector in similar molecular beam studies. Since the translational temperature is also a parameter, it could be determined in some cases. However, time-of-flight techniques appear to be more useful. Electric deflection measurements can also be used in mass spectrometry to interpret fragmentation patterns and help identify the neutral precursors of the ions observed. For example, Greene and Milne (1968b) have easily detected a greater contribution of Na_2Cl_2 to Na^+ than to $NaCl^+$. In the case of beams from effusive sources this type of information is similar to that obtained from velocity distribution measurements. In supersonic beams, however, where all species have the same mean velocity and only the velocity distributions differ, the electric deflection techniques appear to have a significant advantage.

The above discussion has assumed that the beam is deflected by a field of known strength and gradient. Extremely valuable results can also be obtained using inhomogeneous fields of effectively unknown properties. Although there are several types of electric field that could be used for qualitative experiments, that produced by a quadrupole lens has proven to be particularly useful in conjunction with high temperature mass spectrometry. In these experiments a quadrupole lens is placed between a Knudsen cell and the collimating slit preceding the electron gun on a mass spectrometer. The beam strength is first measured with the field off. A stop wire is then inserted in the beam path, blocking the detector, and the field turned on. Molecules entering the quadrupole off-axis and having appropriate electric moments, entrance angles, and velocities, will be refocused through the mass spectrometer slit. If refocused molecules are detected, this proves the presence of a polar molecule. For qualitative data the quadrupole lens has several advantages over the classical electric deflection technique. These include relative simplicity, the ability to use wide beams with a smaller source-to-detector distance for better signal-to-noise, and the partial elimination of the need to detect small changes in a large signal. The principal disadvantage is the inability to make quantitative moment measurements. It also may be less useful in the detection of small dipole moments in the presence of large induced moments, which tend to defocus the beam.

The quadrupole refocusing technique has been used in a number of studies to prove the presence or absence of a dipole moment. Recent studies of this type include those of Buchler *et al.* (1963; 1964; 1967), Kaufman *et al.* (1967), Falconer *et al.* (1968), and Kaiser *et al.* (1968). This technique has also been applied to the determination of asymmetric electronic states (Berg *et al.*, 1965).

2. *Magnetic Deflection Experiments*

Magnetic deflection experiments are very similar to the electric deflection experiments discussed above. Neglecting nuclear moments, induced moments, and those associated with the rotation of the entire molecule, all of which are relatively small, a magnetic moment will be present if there is resultant electronic angular momentum. As in the electric case, the value of the effective magnetic moment depends on the rotational state as well as on the coupling between orbital and spin angular momentum. Detailed discussions of magnetic moments are given by Townes and Schawlow (1955), Herzberg (1950; 1966) and Ramsey (1956).

Magnetic deflection experiments, coupled with mass spectrometric detection, would appear to have their principal value to high temperature chemistry in the determination of molecular ground states, in obtaining structural information, and in the separation of species with different magnetic moments. Code *et al.* (1967) have used magnetic deflection experiments to disprove a hypothesized electronic state in XeF_6, Berkowitz and Goodman (1964) have used magnetic deflection experiments with a high temperature mass spectrometer in the determination of the structures of sulfur and other species. A number of investigators, classical and modern, have used magnetic deflection to separate atomic and molecular species (Ramsey, 1956; Gordon *et al.*, 1968). Like electric deflection, this is of potential value in sorting fragmentation effects. A magnetic deflection apparatus for use with a mass spectrometer has been reported by Blue and Meschi (1966).

D. Miscellaneous Applications of Molecular Beams

In addition to the topics included in the previous sections, there are a number of important applications of molecular beams to high temperature chemistry that can only be mentioned here. Beside the beam–beam and beam–surface experiments alluded to in the introduction, these include several kinds of spectroscopic experiments, ionization cross-section measurements, and high temperature electron diffraction studies.

There are several reviews of molecular beam spectroscopy (Kusch and Hughes, 1959; Trischka, 1962; Smith and Unsworth, 1965) as well as some very pertinent discussion in a short review of high temperature spectroscopy by Brewer (1967). The first three reviews emphasize resonance techniques, which have been applied to a number of molecules of high temperature interest [see, for example, Wharton *et al.* (1962)]. Molecular beams have also been used in a fluorescence technique to establish whether an observed absorption spectrum results from a transition from the ground state (Brewer and Green, 1969; Brewer and Walsh, 1965). Another spectroscopic application of molecular beams is in microwave spectroscopy, where molecular beams are used to isolate unstable or high temperature species and to obtain reduced pressure and Doppler broadening (Rusk and Gordy, 1962; Barrett and Curl, 1967). And finally, optical transition probabilities have been measured by following the intensity of a given transition as a function of distance along the beam path [see, for example, Bell *et al.* (1958)]. It appears that all of these spectroscopic techniques could be substantially improved by the use of high pressure beam techniques.

There are a number of molecular beam methods that have been used to determine ionization cross sections (Kieffer and Dunn, 1966). A novel technique that allows the absolute determination of electron bombardment ionization cross sections without requiring a knowledge of the density of the species of interest (Crawford and Wang, 1967; Crawford, 1968) appears to be of considerable potential interest in high temperature mass spectrometry.

Molecular beams are also used in high temperature electron diffraction (Vilkov *et al.*, 1967; Akishin, *et al.* 1967). Here again, high pressure beam techniques would appear to be useful in at least some cases, and have been used by Audit (1968) for the study of van der Waal clusters, as already discussed.

ACKNOWLEDGMENTS

The authors wish to acknowledge the Advanced Research Projects Agency, the Office of Naval Research, the Office of Saline Water, the Air Force Office of Scientific Research, the Defense Atomic Support Agency, and the Air Force Cambridge Research Laboratories for their support of the authors' research mentioned herein, and their partial support of the writing of this review.

REFERENCES

Abuaf, N. A., Anderson, J. B., Andres, R. P., Fenn, J. B., and Marsden, D. G. (1967a). *Science* **155**, 997.

Abuaf, N. A., Anderson, J. B., Andres, R. P., Fenn, J. B., and Miller, D. R. (1967b). *Proc. Intern. Symp. Rarefied Gas Dyn. 5th, Oxford*, 1966, **2**, p. 1317. Academic Press, New York.

Akishin, P. A., Rambidi, N. G., and Spiridonov, V. P. (1967). *In* " Characterization of High Temperature Vapors " (J. L. Margrave, ed.). Wiley, New York.

Amend, W. E., and Hurlbut, F. C. (1968). *Proc. Intern. Symp. Rarefied Gas Dyn. 6th*, 1969, p. 1205. M.I.T., Cambridge, Massachusetts.

Anderson, J. B. (1967a). *Amer. Inst. Chem. Eng. J.* **13**, 1188.
Anderson, J. B. (1967b). *Entropie* **18**, 33.
Anderson, J. B., and Cauley, J. M. (1968). Paper presented at 6th *Rarefied Gas Dyn. Symp.* M.I.T., Cambridge, Massachusetts.
Anderson, J. B., and Fenn, J. B. (1965). *Phys. Fluids* **8**, 780–787.
Anderson, J. B., Andres, R. P., and Fenn, J. B. (1965a). *Advan. At. Mol. Phys.* **1**, 345.
Anderson, J. B., Andres, R. P., Fenn, J. B., and Maise, G. (1965b). *Proc. Intern. Symp. Rarefied Gas Dyn., 4th, Toronto*, 1964, **2**,p. 106. Academic Press, New York.
Anderson, J. B., Andres, R. P., and Fenn, J. B. (1966). *Advan. Chem. Phys.* **10**, 275.
Anderson, J. B., Fenn, J. B., Moonan, J. F., and Tang, S. (1968). Proj. SQUID Semiann. Progr. Rept., October, Office of Naval Res. Contract N00014-67-A-0226-0005, NR-098-038.
Andres, R. P. (1965). *Ind. Eng. Chem.* **57**, 24.
Andresen, S. G. (1966). *Rev. Sci. Instr.* **37**, 974.
Angenmaier, P. (1966). *Z. Angew. Phys.* **20**, 184.
Ashkenas, H. (1967). *Phys. Fluids* **10**, 2509.
Ashkenas, H., and Sherman, F. S. (1965). *Proc. Intern. Symp. Rarefied Gas Dyn., 4th, Toronto*, 1964, **2**, p. 84. Academic Press, New York.
Audit, P. (1968). *Proc. Intern. Symp. Rarefied Gas Dyn., 6th*, 1969, p. 1703. M.I.T., Cambridge, Massachusetts.
Audit, P., and Rouault, M. (1967a). *Compt. Rend.* **265**, 1100.
Audit, P., and Rouault, M. (1967b). *Compt. Rend. Ser. A* **B264B**, 373.
Balwanz, W. W. (1968). Private communication.
Barber, M., Farren, J., and Linnett, J. W. (1963). *Proc. Roy. Soc.* **A274**, 285, 293.
Barrett, A. H., and Curl, R. F. (1967). *In* " Characterization of High Temperature Vapors " (J. L. Margrave, ed.). Wiley, New York.
Bauchert, J., and Hagena, O. (1965). *Z. Naturforsch.* **20a**, 9.
Beck, D., and Morgenstern, R. (1967). Diplomarbeit, Freiberg Univ., Freiberg, West Germany.
Becker, G. (1961). *Z. Physik* **162**, 250.
Becker, E. W., and Henkes, W. (1956). *Z. Physik* **146**, 320.
Becker, E. W., Bier, K., and Burghoff, H. (1955). *Z. Naturforsch.* **10a**, 565.
Becker, E. W., Bier, K., and Henkes, W. (1956). *Z. Physik* **146**, 333.
Becker, E. W., Klingelhofer, R., and Mayer, H. (1968). *Z. Naturforsch.* **23a**, 274.
Bederson, B., and Robinson, E. J. (1966). *In* " Molecular Beams " (J. Ross, ed.). Wiley (Interscience), New York.
Bell, G. D., Doris, M. H., King, R. B., and Routly, P. M. (1958). *Astrophys. J.* **127**, 775.
Bennewitz, H. G., and Wedemeyer, R. (1963). *Z. Physik* **172**, 1–18.
Bentley, P. G. (1961). *Nature* **190**, 432.
Bentley, P. G., Bishop, J., and Leece, J. (1961). *Instr. Meas. Chem. Anal. Elec. Quant. Nucl. Process Control Proc. Intern. Conf., Stockholm*, 1960, **1**, pp. 253–259. Academic Press, New York.
Berg, R. A., Wharton, L., Klemperer, W., Buchler, A., and Stauffer, J. L. (1965). *J. Chem. Phys.* **43**, 2416.
Berkowitz, J., and Chupka, W. A. (1964). *J. Chem. Phys.* **40**, 2431.
Berkowitz, J., and Goodman, L. S. (1964). Stern-Gerlach experiments using magnetic detection. Paper presented at the *Ann. Conf. Mass Spectry and Allied Topics, 12th, Montreal, Canada*, 1964.

Berkowitz, J., and Walter, T. A. (1968). *J. Chem. Phys.* **49**, 1184.

Berkowitz-Mattuck, J. B., Buchler, A., Engelke, J. L., and Goldstein, S. N. (1963). *J. Chem. Phys.* **39**, 2722.

Bernstein, R. B., and Muckermann, J. T. (1967). Intermolecular forces. *Advan. Chem. Phys.* **12**, 389.

Bier, K., and Hagena, O. (1965). *Proc. Intern. Symp. Rarefied Gas Dyn.*, 4*th*, *Toronto*, 1964, **2**, 260. Academic Press, New York.

Blauer, J. A. (1968). Private communication.

Blue, G. D., and Meschi, D. J. (1966). Inhomogeneous field deflection for determination of electronic and geometric structure of high temperature molecules observed with a mass spectrometer. Paper presented at the *Ann. Conf. Mass Spectry. and Allied Topics*, 14*th*, *Dallas, Texas, 1966*.

Bohme, D. K., and Goodings, J. M. (1966). *J. Appl. Phys.* **37**, 4261.

Bonne, V., Homann, K. H., and Wagner, H. G. (1965). *Symp. Combust.*, 10*th*, *Univ. Cambridge*, *Cambridge*, *Engl.*, 1964. Combustion Inst., Pittsburgh, Pennsylvania.

Bossel, U., Hurlbut, F. C., and Sherman, F. S. (1968). *Proc. Intern. Symp. Rarefied Gas Dyn.*, 6*th*, 1969, p. 945. M.I.T., Cambridge, Massachusetts.

Boyer, M. H., Murad, E., Inami, Y. H., and Hildenbrand, D. L. (1968). *Rev. Sci. Instr.* **39**, 26.

Brewer, L. (1950). Chemistry and metallurgy of miscellaneous materials: thermodynamics. *Natl. Nucl. Energy Ser. Div. IV* **19B**, 261.

Brewer, L. (1967). High temperature chemistry. *Natl. Acad. Sci. Natl. Res. Council Publ.* **1470**.

Brewer, L., and Walsh, R. M. (1965), *J. Chem. Phys.* **42**, 4055.

Brewer, L., and Green, D. W. (1969). *J. High Temp. Sci.* To be published.

Brink, G. O. (1966). *Rev. Sci. Instr.* **37**, 857.

Brink, G. O., Fluegge, R. A., and Hull, R. J. (1968). *Rev. Sci. Instr.* **39**, 1171.

Brown, R. F., and Heald, Jr., J. H. (1967). *J. Spacecraft Rockets* **4**, 476.

Brundin, C. L., ed. (1967). *Proc. Intern. Symp. Rarefied Gas Dyn.*. 5*th*, *Oxford*, 1966. Academic Press, New York. (Suppl. *Advan. Appl. Mech.*).

Buchheit, K., and Henkes, W. (1968). *Z. Angew. Phys* **24**, 191.

Buchler, A., Stauffer, J. L., Klemperer, W., and Wharton, L. (1963). *J. Chem. Phys.* **39**, 2299.

Buchler, A., Stauffer, J. L., and Klemperer, W. (1964). *J. Chem. Phys.* **40**, 3471; *J. Am. Chem. Soc.* **86**, 4544.

Buchler, A., Stauffer, J. L., and Klemperer, W., (1967). *J. Chem. Phys.* **46**, 605.

Burghoff, H., and Gspann, J. (1967). *Z. Naturforsch.* **22a**, 684.

Burke, R. R. (1968). Ion sampling from supersonic gas streams. Paper presented at the *Ann. Conf. Mass Spectry. and Allied Topics*, 16*th*, Pittsburgh, 1968.

Busol, F. I., Skibenko, E. I., and Yuferov, V. B. (1967). *Soviet Phys. Tech. Phys.* (*English. Transl.*) **11**, 1609.

Campargue, R. (1966). *Chem. Soc.* (*London*) *Spec. Publ.* **20**, 287.

Campargue, R. (1967). *Entropie* **18**, 55.

Campargue, R. (1968). *Proc. Intern. Symp. Rarefied Gas Dyn.*, 6*th*, 1969, p. 1003, M.I.T., Cambridge, Massachusetts.

Carrigan, T., Raisen, E., and Schmude, K. (1960). WADD TR 60-359 AD262-364.

Code, R. F., Falconer, W. E., Klemperer, W., and Ozier, I. (1967). *J. Chem. Phys.* **47**, 4955.

Crawford, C. K. (1968). *J. Vacuum Sci. Technol.* **5**, 131.

Crawford, C. K., and Wang, K. I. (1967). *J. Chem. Phys.* **47**, 4667.
Cuthbert, J., Preece, E. R., and Turnbull, A. H. (1965). *Appl. Spectry.* **19**, 18.
Datz, S., and Taylor, E. H. (1959). *In* " Recent Research in Molecular Beams " (I. Estermann, ed.), p. 157. Academic Press, New York.
Datz, S., Minturn, R. E., and Taylor, E. H. (1961). *Rev. Sci. Instr.* **32**, 210.
Daum, F. L., and Gyarmathy, G. (1968). *AIAA J.* **6**, 458.
de Leeuw, J. H., ed. (1965). *Proc. Intern. Symp. Rarefied Gas Dyn., 4th, Toronto*, 1964. Academic Press, New York (*Suppl. Advan. Appl. Mech.*).
Diesen, R. N. (1966). *J. Chem. Phys.* **44**, 3662.
Dimiduk, P. W. (1967). Importance of the high pressure mass spectrometer sampling system to materials technology. AFML-TR-67-97, April.
Dole, M., Mack, L. I., Hines, R. C., Mobley, R. C., Ferguson, L. D., and Alice, M. B. (1968). *J. Chem. Phys.* **49**, 2240.
Dove, J. E., and Moulton, D. M. L. (1965). *Proc. Roy. Soc.* **A283**, 216.
Drowart, J., and Goldfinger, P. (1967). *Angew. Chem.* **6**, 581.
Dunkin, D. B., Fehsenfeld, F. C., Schmeltekopf, A. L., and Ferguson, E. E. (1968). *J. Chem. Phys.* **49**, 1365.
Edwards, R. H., and Collins, R. L. (1968). *Proc. Intern. Symp. Rarefied Gas Dyn., 6th*, 1969, p. 1489. M.I.T., Cambridge, Massachusetts.
Estermann, I. (1946). *Rev. Mod. Phys.* **18**, 300.
Estermann, I., ed. (1959). " Recent Research in Molecular Beams," Academic Press, New York.
Falconer, W. E., Buchler, A., Stauffer, J. L., and Klemperer, W. (1968). *J. Chem. Phys.* **48**, 312.
Fenn, J. B. (1967). *Entropie* **18**, 11.
Fenn, J. B., and Anderson, J. B. (1965). *Proc. Intern. Symp. Rarefied Gas Dyn., 4th, Toronto*, 1964, **2**, p. 311. Academic Press, New York.
Ferguson, E. E., Fehsenfeld, F. C., and Schmeltekopf, A. L. (1968). Flowing afterglow measurements of ion-neutral reactions. *Advan. At. Mol. Phys.* **5**. To be published.
Feugier, A., and Van Tiggelen, A. (1967). *Combust. Flame* **11**, 234.
Fisher, S. S., and Knuth, E. L. (1968). Private communication.
Fite, W. L., and Datz, S. (1963). *Ann. Rev. Phys. Chem.* **14**, 61.
Fite, W. L., and Brackmann, R. T. (1958). *Phys. Rev.* **112**, 1141.
Foner, S. N., and Hudson, R. L. (1953). *J. Chem. Phys.* **21**, 1374.
Foner, S. N., and Hudson, R. L. (1968). *J. Chem. Phys.* **49**, 3724.
Fontijn, A., and Rosner, D. E. (1967). *J. Chem. Phys.* **46**, 3275.
Frankenthal, S. (1968). *U.S. Govt. Res. Rept.* **68**, 12; AD666-450.
Fraser, R. G. J. (1931). " Molecular Rays," Cambridge Univ. Press, London and New York.
Fraser, R. G. J. (1937). " Molecular Beams," Methuen, London.
Freeman, R. D., and Edwards, J. G. (1967). " The Characterization of High Temperature Vapors," p. 508. Wiley, New York.
French, J. B. (1965). *AIAA J.* **3**, 993.
French, J. B. (1966). " Molecular Beams for Rarefied Gas dynamic Research," AGARD-ograph 112. Inst. for Aerospace Studies, Univ. of Toronto, Toronto.
French, J. B., and Locke, J. W. (1967). *Proc. Intern. Symp. Rarefied Gas Dyn., 5th, Oxford*, 1966, **2**, p. 1461. Academic Press, New York.
Freund, R. S., and Klemperer, W. (1967). *J. Chem. Phys.* **47**, 2897.

Gaur, J. P., and Chanin, L. M. (1968). *J. Appl. Phys.* **40**, 256 (1969).

Gay, I. D., and Kern, R. D. (1968). Mass spectrometry in inorganic chemistry. *Advan. Chem. Ser.* **72**, 83.

Gazzola, C., Scoles, G., and Torello, F. (1968). *Proc. Intern. Symp. Rarefied Gas Dyn., 6th*, 1969, p. 977. M.I.T., Cambridge, Massachusetts.

Gersing, E.. Hundhausen, E., and Pauly, H. (1963). *Z. Physik* **171**, 349.

Ghosh, S. N., Hinten, B. B., Jones, L. M., Leite, R. J., Mason, C. J., Schaefer, E. J., and Walters, M. (1968). *J. Geo. Res. Space Phys.* **73**, 4425.

Giddings, J. C., Myers, M. N., McLaren, L., and Keller, R. A. (1968). *Science* **162**, 67.

Giordmaine, J. A., and Wang, T. C. (1960). *J. Appl. Phys.* **31**, 463.

Golomb, D., and Brown, R. F. (1968). Private communication.

Good, R. E., Golomb, D., Del Greco, F. P., Hall, D. W., and Whitfield, D. L. (1968). *Proc. Intern. Symp. Rarefied Gas Dyn., 6th*, 1969, p. 1449. M.I.T., Cambridge, Massachusetts.

Gordon, R. J., Herm, R. R., and Herschbach, D. R. (1968). *J. Chem. Phys.* **49**, 2684.

Gordy, W., Smith, W. V., and Trambarulo, R. F. (1953). " Microwave Spectroscopy." Wiley, New York.

Govers, T. R., LeRoy, R. L., and Deckers, J. M. (1968). *Proc. Intern. Symp. Rarefied Gas Dyn., 6th*, 1969, p. 985. M.I.T., Cambridge, Massachusetts.

Greene, F. T. (1968). A method for making precise velocity distribution measurements with a time-of-flight mass spectrometer. Paper presented at the *Ann. Conf. Mass Spectry. and Allied Topics, 16th, Pittsburgh*, 1968.

Greene, F. T., and Milne, T. A. (1963). *J. Chem. Phys.* **39**, 3150.

Greene, F. T., and Milne, T. A. (1966). *Advan. Mass Spectrometry*, **3**, 841.

Greene, F. T., and Milne, T. A. (1967). A study of the electric deflection of molecular beams. Contract No. AF49(638)-1724. Midwest Res. Inst., Kansas City, Missouri.

Greene, F. T., and Milne, T. A. (1968a). *Advan. Mass Spectrometry*, **4**, 301.

Greene, F. T., and Milne, T. A. (1969). " Free-Jet Isolation Spectroscopy—A New Approach to the Study of High-Temperature Species," To be published.

Greene, F. T., and Milne, T. A. (1968b). A study of the electric deflection of molecular beams. Contract No. AF49(638)-1724. Midwest Res. Inst., Kansas City, Missouri.

Greene, F. T., Brewer, J., and Milne, T. A. (1964). *J. Chem. Phys.* **40**, 1488.

Greene, F. T., Vandegrift, A. E., and Milne, T. A. (1968a). An experimental study of the structure, thermodynamic and kinetic behavior water. OSW Contract No. 14-01-0001-1479. Midwest Res. Inst., Kansas City, Missouri.

Greene, F. T., Beachey, J., Bennett, S., and Milne, T. A. (1968b). Mass spectrometric studies of graphite vaporization at high temperatures. Quart. Repts. Contract No. F33615-68-C-1709. Midwest Research Inst., Kansas City, Missouri.

Grosser, A. E. (1966). *Bull. Am. Phys. Soc.* **11**, 889.

Grosser, A. E., Iczkowski, R P., and Margrave, J. L. (1963). *Rev. Sci. Instr.* **34**, 117.

Gspann, J., and Krieg, G. (1968). *Z. Naturforsch.* **23a**, 726.

Gutman, D., Hay, A. J., and Belford, R. L. (1966). *J. Phys. Chem.* **70**, 1786.

Hagena, O. (1968). *Proc. Intern. Symp. Rarefied Gas Dyn., 6th*, 1969, p. 1465. M.I.T., Cambridge, Massachusetts.

Hagena, O., and Henkes, W. (1960). *Z. Naturforsch.* **15a**, 851.

Hagena, O., and Henkes, W. (1965). *Z. Naturforsch.* **20a**, 1344.

Hagena, O., and Morton, H. S. (1967). *Proc. Intern. Symp. Rarefied Gas Dyn., 5th*, Oxford, 1966. **2**, p. 1396. Academic Press, New York.

Hagena, O., and Varma, A. K. (1968). *Rev. Sci. Instr.* **39**, 47.

Hall, W. D. (1968). Thesis, Dept. of Phys., Univ. of Michigan, Ann Arbor, Michigan.

Hall, W. D., and Zorn, J. C. (1967). *Bull. Am. Phys. Soc.* **12**, 131.

Hall, W. D., Hoeberling, R. F., and Zorn, J. C. (1968). *Bull. Am. Phys. Soc.* **13**, 21.

Hamel, B. B. (1968). Proj. SQUID, Semiann. Progr. Rept., October, Office of Naval Res., Contract N00014-67-A-0226-005, NR-098-038.

Hamel, B. B., and Willis, D. R. (1966). *Phys. Fluids* **9**, 829.

Harbour, P. J. (1968). *Proc. Intern. Symp. Rarefied Gas Dyn.*, *6th*, 1969, p. 1713. M.I.T., Cambridge, Massachusetts.

Harrison, H., Hummer, D., and Fite, W. L. (1964). *J. Chem. Phys.* **41**, 2567.

Hayhurst, A. N., and Sugden, T. M. (1966). *Proc. Roy. Soc.* **A293**, 36.

Hayhurst, A. N., and Telford, N. R. (1966). *Nature* **212**, 813.

Hayhurst, A. N., and Padley, P. J. (1967). *Trans. Faraday Soc.* **63**, 1620.

Henkes, W. (1961). *Z. Naturforsch.* **16a**, 842.

Henkes, W. (1962). *Z. Naturforsch.* **17a**, 786.

Herm, R. R., and Herschbach, D. R. (1965). *J. Chem. Phys.* **43**, 2139.

Herschbach, D. R. (1966). Molecular beams. *Advan. Chem. Phys.* **10**, 319.

Herzberg, G. (1950). "Molecular Spectra and Molecular Structure. I. Spectra of Diatomic Molecules," Van Nostrand, Princeton, New Jersey.

Herzberg, G. (1966). " Molecular Spectra and Molecular Structure. III. Electronic Spectra and Electronic Structure of Polyatomic Molecules," Van Nostrand, Princeton, New Jersey.

Hildenbrand, D. L., Theard, L. P., Murad, E., and Ju, F. (1965). AFRPL-TR-65-95.

Homann, K. H. (1967). *Combust. Flame* **11**, 265.

Homann, K. H. (1968). Investigation of reacting gases with the mass spectrometer. *AGARDograph.* To be published.

Homann, K. H., MacLean, D. I., and Wagner, H. G. (1965). *Naturwissenschaften* **52**, 12.

Hanes, G. R. (1960). *J. Appl. Phys.* **31**, 2171–2175.

Hostettler, H. V., and Bernstein, R. B. (1960). *Rev. Sci. Instr.* **31**, 872–877.

Hundhausen, E., and Harrison, H. (1967). *Rev. Sci. Instr.* **38**, 131.

Hurle, I. R., Sugden, T. M., and Nutt, G. B. (1968). *Symp. Combust.*, *12th*, *Poitiers*, *July* 1968. The Combustion Inst. To be published.

Inghram, M. G., and Drowart, J. (1960). *Proc. Intern. Symp. High Temp. Technol.*, *Asilomar Conf. Grounds*, *Calif.*, 1959. p. 219. McGraw-Hill, New York.

Israel, G. W. (1967). *J. Colloid Interface Sci.* **24**, 330.

Janik, G. S., and Conway, D. C. (1967). *J. Phys. Chem.* **71**, 823.

Jarvinen, P. O., and Draper, J. S. (1967). *AIAA J.* **5**, 824.

Johnston, W. D., and King, J. G. (1966a). *Rev. Sci. Instr.* **37**, 475.

Johnston, W. D., and King, J. G. (1966b). *Phys. Rev. Letters* **16**, 1191.

Jones, T. V., and Byrne, M. A. (1968). *Proc. Intern. Symp. Rarefied Gas Dyn.*, *6th*, 1969. p. 1311. M.I.T., Cambridge, Massachusetts.

Kahrs, J. (1968). Combustion species sampling final report. AFRPL-TR-68-120 June. Thiokol Chem. Corp.

Kaiser, K. W., Muenter, J. S., and Klemperer, W. (1968). *J. Chem. Phys.* **48**, 3339.

Kantrowitz, A., and Grey, J. (1951). *Rev. Sci. Instr.* **22**, 328.

Kaufman, M. (1964). Thesis, Dept. of Chem., Harvard Univ., Cambridge, Massachusetts.

Kaufman, M., Muenter, J., and Klemperer, W. (1967). *J. Chem. Phys.* **47**, 3365.

Kebarle, P., Arshadi, M., and Scarborough, J. (1968). *J. Chem. Phys.* **49**, 817.

Kessler, R., and Koglin, B. (1967). *Z. Angew. Phys.* **23**, 316.
Kieffer, L. J., and Dunn, G. H. (1966). *Rev. Mod. Phys.* **38**, 1.
King, I. R. (1968). Private communication.
King, I. R., and Scheurich, J. T. (1966). *Rev. Sci. Instr.* **37**, 1219.
Kinsey, J. L. (1966). *Rev. Sci. Instr.* **37**, 61.
Kistiakowsky, G. B., and Kydd, P. H. (1957). *J. Am. Chem. Soc.* **79**, 4825.
Kistiakowsky, G. B., and Michael, J. V. (1964). *J. Chem. Phys.* **40**, 1447.
Klingelhöfer, R., and Lohse, P. (1964). *Phys Fluids* **7**, 379.
Knewstubb, O. F., Hayhurst, A. N., and Telford, N. R. (1966). *Nature* **212**, 504.
Knof, H., and Maiwald, B. (1968). *Z. Naturforsch.* **23a**, 279.
Knuth, E. L. (1964). *Appl. Mech. Rev.* **17**, 751.
Knuth, E. L., and Fisher, S. S. (1968). *J. Chem. Phys.* **48**, 1674.
Knuth, E. L., Kuluva, N. M., and Callinan, J. P. (1967). *Entropie* **18**, 38.
Kusch, P., and Hughes, V. W. (1959). *In* " Handbuch der Physik " (S. Flugge, ed.). Springer, Berlin.
Laurmann, J. A., ed. (1963). *Proc. Intern. Symp. Rarefied Gas Dyn., 3rd, Paris,* 1962. Academic Press, New York (*Supple. Advan. Appl. Phys.*).
Leckenby, R. E., and Robbins, E. J. (1966). *Proc. Roy. Soc.* **A291**, 389.
Leckenby, R. E., Robbins, E. J., and Trevalion, P. A. (1964). *Proc. Roy. Soc.* **A280**, 409.
Lefkowitz, B., and Knuth, E. L. (1968). *Proc. Intern. Symp. Rarefied Gas Dyn., 6th,* 1969, p. 1421. M.I.T., Cambridge, Massachusetts.
Leigh, C. H. (1960). Exploratory research on the vaporization of solids. Status Rept. RAD-SR-18-60-120 AD249-086, Office of Naval Res., December.
Leonas, V. B. (1964). *Soviet Phys. Usp.* (*English Transl.*) **7**, 121.
LeRoy, R. L., Gowers, J. R., and Deckers, J. M. (1968). Private communication.
Levethal, M., and Robiscoe, R. T. (1967). *Phys. Rev.* **158**, 49.
Lincoln, K. A. (1965). *Anal. Chem.* **37**, 541.
Lincoln, K. A. (1966). On the feasibility of analyzing ablator blow-off gases with a time-of-flight mass spectrometer. USNRDL-TR-1014.
Lineberger, W. C., and Puckett, L. J. (1968). Reactions of NO^+ Ions. *Symp. Phys. Chem. Upper Atmosphere, Waltham, Massachusetts, June* 1968.
McAfee, Jr., K. B., Sipler, D., and Edelson, D. (1967). *Phys. Rev.* **160**, 130.
McElligott, P. E., Roberts, R. W., and Jernakoff, G. (1963). *Vacuum* **13**, 419.
McFee, J., and Marcus, P. M. (1960). Velocity distributions for direct and reflected atomic beams. Techn. Rept. No. 1, January. Carnagie Inst. Technol., Schenley Park, Pittsburgh, Pennsylvania.
McKinley, J. D. (1964). *J. Chem. Phys.* **40**, 120.
McMichael, G. E., and French, J. B. (1966). *Phys. Fluids* **9**, 1419.
Marrone, P. V. (1967). *Phys. Fluids* **10**, 521.
Marsters, G. F., Bauer, S. H., and Resler, F. L. (1965). Optimized geometry for coupling a mass spectrometer to a shock tube. TID-23214.
Merryman, E. L., and Levy, A. (1967). *J. Air Pollution Control Assoc.* **17**, 800–806.
Meyer, R. T. (1967). *J. Chem. Phys.* **46**, 967.
Miller, D. R. (1968a). Private communication.
Miller, D. R. (1968b). Development of a supersonic nozzle molecular beam facility. IPAPS-67/68-212. Univ. California, San Diego, California.
Miller, D. R., and Andres, R. P. (1967). *J. Chem. Phys.* **46**, 3418.
Miller, D. R., and Andres, R. P. (1968). *Proc. Intern. Symp. Rarefied Gas Dyn., 6th,* 1969. M.I.T., Cambridge, Massachusetts.

Miller, R. C., and Kusch, P. (1955). *Phys. Rev.* **99**, 1314–1321.

Miller, R. C., and Kusch, P. (1956). *J. Chem. Phys.* **25**, 860.

Miller, W. J. (1967). *Symp. Combustion, 11th*, p. 311. The Combustion Institute, Pittsburgh, Pennsylvania.

Miller, T. M., Moseley, J. T., Martin, D. W., and McDaniel, E. W. (1968). *Phys. Rev.* **173**, 115.

Milne, T. A. (1965a). A study of the kinetics in low pressure flames by the direct mass spectrometric sampling of unstable intermediates. Institute Rept. No. FS-127-P, December Midwest Res. Inst., Kansas City, Missouri.

Milne, T. A., ed. (1965b). High-pressure sampling conf. rept. July. Midwest Res. Inst., Kansas City, Missouri and Office of Naval Res.

Milne, T. A. (1969). "Proposed use of Nozzle-beam sampling of super-critical dense gases in the mass spectrometry of non-volatile compounds." To appear in *Intern. J. Mass Spectry. Ion Phys.*

Milne, T. A., and Greene, F. T. (1965). *Symp. Combustion, 10th*, p. 153. The Combustion Institute, Pittsburgh, Pennsylvania.

Milne, T. A., and Greene, F. T. (1966). *J. Chem. Phys.* **44**, 2444.

Milne, T. A., and Greene, F. T. (1967a). *J. Chem. Phys.* **47**, 3668.

Milne, T. A., and Greene, F. T. (1967b). *J. Chem. Phys.* **47**, 3684.

Milne, T. A., and Greene, F. T. (1967c). *J. Chem. Phys.* **47**, 4095.

Milne, T. A., and Greene, F. T. (1967d). Mass spectrometric study of metal-containing flames. Final Tech. Summ. Rept., 1961–1967, Contract Nonr-3599100.

Milne, T. A., and Greene, F. T. (1967e). Mass spectrometric study of homogeneous nucleation in free jets. Paper presented at the *Conf. Mass Spectry. and Allied Topics, 15th*, Denver 1967.

Milne, T. A., and Greene, F. T. (1968). Mass spectrometry in inorganic chemistry. (J. L. Margrave, ed.). *Advan. Chem. Ser.* **72**, 68.

Milne, T. A., Vandegrift, A. E., and Greene, F. T. (1969a). " Mass Spectrometric Observations of Argon Clusters in Nozzle Beam. II. The Kinetics of Dimer Growth." To be published.

Milne, T. A., Beachey, J., and Greene, F. T. (1968). A direct mass spectrometric study of the formation and reactions of water cluster ions. Quarterly Reports, Contract No. F19629–68–C–0276, Midwest Research Inst., Kansas City, Missouri.

Milne, T. A., Beachey, J., and Greene, F. T. (1969b). " Molecular Beam Studies of the Temperature Dependence of Fragmentation of n-Butane, and its Vibrational Relaxation in Free Jets." To be published.

Minten, A., and Osberghaus, O. (1958). *Z. Physik* **150**, 74.

Modica, A. B. (1967). *J. Chem. Phys.* **46**, 3663.

Moran, J. P., Wachman, H. Y., and Trilling, L. (1967). Fundamentals of Gas-Surface Interactions (H. S. Saltsburg, Ed.). Academic Press, New York. To be published.

Mori, S., and Takezaki, Y. (1968). *Ber. Bunsenges. Physik Chem.* **72**, 157.

Mosharrafa, M., and Oskam, H. J. (1966). *Physica* **32**, 1759.

Muntz, E. P. (1967). *Proc. Intern. Symp. Rarefied Gas Dyn., 5th, Oxford*, 1966, **2**, p. 1257. Academic Press, New York.

Narcisi, R. S., Bailey, A. O., and Della Lucca, L. (1968). Negative ion composition measurement in the lower ionosphere. *Symp. Phys. Chem. Upper Atmosphere, Waltham, Massachusetts, June* 1968.

Netusil, V. F., Proffit, R. L., and Enloe, J. D. (1965). Development of a system for the identification of rocket exhause products. Final Rept. AFRPL-TR-65-70 AD461-444.

Niki, H., and Weinstock, B. (1967). *J. Chem. Phys.* **47**, 3249.
Norman, J. H., Staley, H. G., and Bell, W. E. (1968). Mass spectrometry in inorganic chemistry. *Advan. Chem. Ser.* **72**, 101.
Nutt, C. W., and Carter, A. J. (1968). *Trans. Faraday Soc.* **64**, 771.
Nutt, C. W., Penmore, G. W., and Biddlestone, A. J. (1959). *Trans. Faraday Soc.* **55**, 1516.
O'Halloran, G. J., Fluegge, R. A., Betts, J. F., and Everett, W. L. (1964). ASD TDR 62-644, Pts. I and II, AD 600-996 and AD 608 867.
Oman, R. A., Calia, V. S., and Weiser, C. H. (1966–1967). Research on Gas-Surface Interactions. Pt. I. Project. Summary. Grumman Res. Dept. Rept. RE-310, December 1967.
Omrod, G., and Patterson, W. (1967). *J. Sci. Instr.* **44**, 639.
Passchier, A. A., and Gregory, N. W. (1968). *J. Phys. Chem.* **72**, 2697.
Pauly, H., and Toennies, J. P. (1965). *Advan. At. Mol. Phys.* **1**, 201.
Peeters, J., Vinckier, C., and Van Tiggelien, A. (1968). " Oxydation and Combustion Reviews." Elsevier, Amsterdam.
Pellett, G. W. (1968). Private communication.
Pellett, G. L., and Saunders, A. R. (1968). *Aerospace Sci. Meeting*, 6*th*, *New York*, *January* 1968, AIAA Preprint No. 68-149.
Peng, T. C., and Liquornik, D. L. (1967). *Rev. Sci. Instr.* **30**, 989.
Pertel, R. (1968). Private communication.
Phipps, J. A., Griffith, D. R., and Scott, J. E. (1963). *Bull. Am. Phys. Soc.* **8**, 550.
Polak, M., and Trilling, L. (1968). *Proc. Intern. Symp. Rarefied Gas Dyn.*, 6*th*, 1969, p. 1213. M.I.T., Cambridge, Massachusetts.
Pope, R. B., and Parker, J. A. (1968). Analysis of ablation products and boundary-layer chemistry of ablating materials with a mass spectrometer. Preprint. NASA Ames Res. Center.
Ramsey, N. F. (1956). " Molecular Beams." Oxford Univ. Press, London and New York.
Rebrov, A. K., and Sharafutdinov, R. G. (1968). *Proc. Intern. Symp. Rarefied Gas Dyn.*, 6*th*, 1969, p. 965. M.I.T., Cambridge, Massachusetts.
Robben, F., and Talbot, L. (1966). *Phys. Fluids* **9**, 644.
Robbins, E. J., and Leckenby, R. E. (1965). *Nature* **206**, 1253.
Robbins, E. J., Leckenby, R. E., and Willis, P. (1967). *Advan. Phys.* **16**, 739.
Roberts, R. W., and McElligott, P. E. (1963). *Trans. Natl. Symp. Am. Vacuum Soc.*, 10*th*, 1911. Macmillan, New York.
Rosenblatt, G. M., and Lee, P. K. (1968). *J. Chem. Phys.* **49**, 2995.
Ross, J. (1966). Molecular Beams. *Advan. Chem. Phys.* **10**.
Rothe, D. E. (1966). *Phys. Fluids* **9**, 1643.
Rubin, L. (1968). Private communications.
Rudinger, G. (1968). Proj. SQUID, Semiann. Rept., October, Office of Naval Res. Contract N00014-67-A-0226-055, NR-098-038.
Rusk, J. R., and Gordy, W. (1962). *Phys. Rev.* **127**, 817.
Russo, A. L. (1967). *J. Chem. Phys.* **47**, 5201.
Rutgaizer, Yu. S. (1967). *Zh. Tekhn. Fiz.* **37**, 562.
Ryason, P. R. (1967). *Rev. Sci. Instr.* **38**, 607.
Ryhage, R. (1967). *Arkiv Kemi.* **26**, 305.
Salser, G. E., Dauermann, L., Ray, A. B., and Koehler, G. (1968). *AIAA J.* **6**, 2186.
Sawyer, R. F. (1968). Private communication.
Sawyer, R. F., McMullen, E. T., and Purgalis, P. (1968). *AIAA J.* **6**, 2111.
Schissel, P. O., and Trulson, O. C. (1965). *J. Chem. Phys.* **43**, 737.

Schoonmaker, R. C., and Porter, R. F. (1960). *J. Phys. Chem.* **64**, 457.
Schügerl, K. (1968). *Proc. Intern. Symp. Rarefied Gas Dyn.*, *6th*, 1969, p. 909. M.I.T., Cambridge, Massachusetts.
Schwarz, J. A., and Madip, R. J. (1968). *Rev. Sci. Instr.* **39**, 1366.
Scoles, G. (1965). *Physica* **31**, 233.
Scoles, G., and Torello, F. (1968). *Meccanica* **3**, 20.
Scott, J. E., Jr., and Phipps, J. A. (1967). *Proc. Intern. Symp. Rarefied Gas Dyn.*, *5th*, *Oxford*, 1966, **2**, p. 1337. Academic Press, New York.
Scott, P. B. (1965). "Molecular Beam Velocity Distribution Measurements." M.I.T. Report, Cambridge, Massachusetts.
Scott, P. B., Bauer, P. H., Wachman, H. Y., and Trilling, L. (1967). *Proc. Intern. Symp. Rarefied Gas Dyn.*, *5th*, *Oxford*, 1966, **2**, p. 1353. Academic Press, New York.
Sebacher, D. I. (1968). *AIAA J.* **6**, 51.
Sebacher, D. I., Guy, R. W., and Lee, L. P. (1968). *Proc. Intern. Symp. Rarefied Gas Dyn.*, *6th*, 1969, p. 931. M.I.T., Cambridge, Massachusetts.
Shahin, M. M. (1968). Nature of charge carriers in negative coronas. *Appl. Opt.* To be published.
Shardanand, (1968). *J. Quant. Spectry. Radiative Transfer* **8**, 1533.
Sherman, F. S. (1963). Rept. 6-90-63-61, May. Lockheed Missiles and Space Co. Palo Alto, California.
Sherman, F. S. (1965). *Phys. Fluids* **8**, 773.
Skinner, G. T. (1968). *Proc. Intern. Symp. Rarefied Gas Dyn.*, *6th*, 1969, p. 1325. M.I.T., Cambridge, Massachusetts.
Skofronick, J. G. (1967). *Rev. Sci. Instr.* **38**, 1628.
Skofronick, J. G., and McArdle, R. T. (1968). *Rev. Sci. Instr.* **39**, 1965.
Smetana, F. O., Sherrill, II, W. A., and Schort, Jr., D. R. (1967). *Proc. Intern. Symp. Rarefied Gas Dyn.*, *5th*, *Oxford*, 1966. **2**, p. 1243. Academic Press, New York.
Smith, K. F. (1955). " Molecular Beams." Methuen, London.
Smith, K. F., and Unsworth, P. J. (1965). Molecular bean spectroscopy. *Sci. Progr.* (*London*) **53**, 45.
Sonin, A. A. (1967). *J. Geophys. Res.* **72**, 4547.
Sperling, R. L., and Engh, R. O. (1968). Paper presented at the *Ann. Conf. Mass Spectry. and Allied Topics*, *16th*, *Pittsburgh*, 1968.
Stafford, F. E., and Berkowitz, J. (1964). *J. Chem. Phys.* **40**, 2963.
Stafford, F. E., Pressley, Jr., G. A., and Baylis, A. B. (1968). Mass spectrometry in inorganic chemistry. *Advan. Chem. Ser.* **72**, 137.
Stair, A. T. (1968). OAR *Res. Rev.* **7**, No. 4.
Stern, S. A., Waterman, P. C., and Sinclair, T. F. (1960). *J. Chem. Phys.* **33**, 805.
Stickney, R. E. (1968). Low-energy molecular scattering as a tool for studying gas-solid interaction potentials: Comparison of theory and experiment. *Proc. Intern. Materials Symp.*, *4th*, *Berkeley*, *June* 1968. To be published.
Studniarz, S. A., and Franklin, J. L. (1968). *J. Chem. Phys.* **49**, 2652.
Stuart, R. V., Brower, K., and Mayer, W. (1963). *Rev. Sci. Instr.* **34**, 425.
Summers, W. H. (1965). Applied rocket combustion and thermodynamics investigation program. *Res. Technol. Briefs.* **3**, No. 4.
Talbot, L. (1961). *Proc. Intern. Symp. Rarefied Gas Dyn.* *2nd*, *Berkeley*, 1960. Academic Press, New York. (Suppl. *Advan. Appl. Mech.*).
Talrose, V. L., Strunin, V. P., Dodonov, A. F., and Lavrovskaya, G. K. (1966). *Advan. Mass Spectrometry*, **3**, 993.

Tanaka, Y. (1968). Absorption spectra of Ar_2 in the vacuum-uv region. Paper presented at the *Symp. Mol. Structure and Spectry. 23rd, Columbus, Ohio, 1968.*
Taylor, W. J. (1963). *J. Chem. Phys.* **38**, 779.
Thomann, H. (1966). *Phys. Fluids* **9**, 896.
Thorn, R. J., and Holt, B. (1968). Private communication.
Townes, C. H., and Schawlow, A. L. (1955). " Microwave Spectroscopy." McGraw-Hill, New York.
Trilling, L., and Wachman, H. Y., eds. (1969). *Proc. Intern. Symp. Rarefied Gas Dynamics, 6th, Cambridge, 1968.* Academic Press, New York, 1969.
Trischka, J. W. (1962). *Methods Exptl. Phys.* **3**, 589.
Turnbull, A. H. (1962). Paper presented at the *Ann. Conf. Mass Spectry. and Allied Topics, 10th. New Orleans, 1962.*
Verkhovtseva, E. T., Fogel, Ya. M., and Osyka, V. S. (1968). *Opt. i Spektroskopiya* **25**, 440.
Verneker, V. R. P., and Maycock, J. N. (1967). *J. Chem. Phys.* **47**, 3618.
Vidal, R. J., Skinner, G. T., and Bartz, J. A. (1967). *Proc. Intern. Symp. Rarefied Gas Dyn., 5th, Oxford*, 1966, **2**, p. 1287. Academic Press, New York.
Vilkov, L. V., Rambidi, N. G., and Spiridonov, V. P. (1967). *Zh. Strukt. Khim.* **8**, 786.
Von Zahn, U. (1967). *J. Geophys. Res.* **72**, 5933.
Vriens, L., Boers, A. L., and Smity, J. A. (1965). *Appl. Sci. Res.* Sect. B, **12**, 65.
Vyse, R., Heggie, J. C., and Craddock, M. K. (1968). *Proc. Intern. Symp. Rarefied Gas Dyn., 6th*, 1969, p. 939. M.I.T., Cambridge, Massachusetts.
Wang, C. P. (1968). *Phys. Fluids* **11**, 1865.
Wang, J. C. F., and Bauer, P. H. (1968). *Proc. Intern. Symp. Rarefied Gas Dyn., 6th*, 1969, p. 1009. M.I.T., Cambridge, Massachusetts.
Wang, K. C., and Wahlbeck, P. G. (1968). *J. Chem. Phys.* **49**, 1617.
Ward, J. W., Mulford, R. N. R., and Bivans, R. L. (1967). *J. Chem. Phys.* **47**, 1718.
Wegener, P. P. (1967). Light scattering measurements of condensation in supersonic flow. *Symp. Nucleation Phenomena Gas Dynamics, U.S. Naval Acad. December* 1967. A. A. Pouring, Chairman.
Weiss, R. (1961). *Rev. Sci. Instr.* **32**, 397.
Westenberg, A. A., and Favin, S. (1963). *Symp. Combustion, 9th*, p. 785. The Combustion Inst.
Westenberg, A. A., and Fristrom, R. M. (1965). *Symp. Combustion, 10th*, 1965, p. 473. The Combustion Inst. Pittsburgh, Pennsylvania.
Wharton, L., Kaufman, M., and Klemperer, W. (1962). *J. Chem. Phys.* **37**, 622.
Williams, G. J., Pearson, G. S., and Heath, G. A. (1968). Private communication.
Winicur, D. H., and Knuth, E. L. (1967). *J. Chem. Phys.* **46**, 4318.
Winterbottom, W. L. (1968). *J. Chem. Phys.* **49**, 106.
Wong, E. L., and Potter, A. E. (1966). Mass spectrometric investigation of reaction of oxygen atoms with methane. NASA TND-3371.
Woods, R., and Fenn, J. B. (1966). *Rev. Sci. Instr.* **37**, 917.
Yamamoto, S., and Stickney, R. E. (1967). *J. Chem. Phys.* **47**, 1091–1099.
Yealland, R. M., LeRoy, R. L., and Deckers, J. M. (1967). *Can. J. Chem.* **45**, 2651.
Zapata, R. N., Ballard, R., and Cabrera, N. (1968). *Proc. Intern. Symp. Rarefied Gas Dyn., 6th*, 1969, p. 739. M.I.T., Cambridge, Massachusetts.
Zmbov, K. F., Ficalora, P., and Margrave, J. L. (1968). *J. Inorg. Nucl. Chem.* **30**, 2059.
Zorn, J. C. (1964). *Am. J. Phys.* **32**, 721.

Chemical Research in Streaming Thermal Plasmas

M. P. Freeman

CENTRAL RESEARCH DIVISION, AMERICAN CYANAMID COMPANY, STAMFORD, CONNECTICUT

I. Introduction 151
A. Streaming Thermal Plasmas 152
B. Current Industrial Applications 170
II. Recent Research 175
A. Product Oriented Research 176
B. Mechanism Studies 183
C. Diagnostics 191

I. Introduction

The concept of directly expending electrical energy for the production or modification of bulk chemicals is rather daring when one considers the high cost of industrial electricity, presently on the order of one cent per kilowatt hour (about 861 kcal or 3410 BTU). The projected large scale nuclear generators promise to depress the price to less than one third, but this is still expensive compared to the price of heat from the combustion of fossil fuels, currently about one cent for 20,000 BTU. Even the projected lower price of electricity is about ten times that extant when the pre-World War 1 nitrogen fixation arc processes were in operation. How, then, can one account for the ever increasing chemical research activity in streaming thermal plasmas? To some extent the impetus must be ascribed to the development, for various aerospace applications, of inexpensive, well-packaged and easy-to-operate plasma jet units. Because of their intrinsic high throughputs and because the exigencies of the confinement and control of streaming thermal plasmas make the reactors particularly resistant to attack from corrosive reagents and products, the principal motivation for the chemical industry probably lies in the expectation that the saving in plant capital investment will offset the high price of the energy consumed. Then too, energy alone does not tell the whole story. The high temperature with which a streaming thermal plasma starts far exceeds that attainable by combustion, and it can and does have a significant effect on subsequent reaction history, making in isolated instances unique products not otherwise easily formed or not otherwise formed in that physical modification.

As Reed (1967) has pointed out, the bulk of work in this field is being done in or for industrial concerns; because of their need for secrecy, one cannot get a clear picture of the total research activity from the literature alone. Those in the field see equipment and hear rumors and allegations from their various professional contacts, but these can scarcely be explicitly cited in a work of this kind. Nonetheless, they have instructive value and so undoubtedly influence to some extent the published work cited here, hopefully without betraying any confidences.

This chapter is not intended to be merely a review, but rather an analysis of current published results and apparent significant advances which, hopefully, will stimulate even more activity and help newer investigators channel their efforts into the more rewarding areas. For this reason, many current contributions are neglected simply because they are difficult to interpret, while to gain perspective some of the newer developments are placed in the context of older, and in some cases, well-known work. The writer must apologize in advance for the certainty of having omitted major developments published in recent months.

Section I,A is a discussion of the nature of streaming thermal plasmas (abbreviated STP), how they are generated, and their properties. Industrial applications of the present and very near future are covered in Section I,B. Section II reviews current published results in three major categories: in II,A, product oriented chemical research, in II,B, studies of mechanism, and finally, in II,C, research in diagnostic techniques.

A. Streaming Thermal Plasmas

In a recent review article (Freeman, 1968a) the concept of the streaming thermal plasma (STP) was defined so as to emphasize the similarities in the problems that must be considered when undertaking chemical research in these environments, however they are formed. That they are plasmas, indeed very dense plasmas in comparison to those produced in most terrestrial laboratories, is at the present time somewhat incidental. To be sure, their intrinsic streaming nature is a consequence of their plasma properties (Section I,A,1), and the plasma properties are used in certain diagnostic techniques (Section II,C). However, for the chemist they have been of little concern, despite the possibility that in some instances ion–molecule reactions may very well govern the chemistry (Section II,B,2), and that the possibility does exist of physically manipulating the stream with electric and magnetic fields. Of immediate concern to the chemist is the high heat content concomitant to the ionization process of the plasmas thus produced, out of all proportion to the " average temperature " of the plasma. Furthermore, the facts that these plasmas are thermal in nature and are committed to high velocity flow (streaming) are of the utmost importance and must be seriously considered when planning and performing experiments.

Because the plasmas are thermal, no compound unstable at high temperatures is likely to be produced in significant quantities unless it forms spontaneously from " precursors " present in the quenched gas. Also, confinement problems are inevitably severe, and means must be found for removing the sensible heat of the reaction products. In fact, it often must be removed at a controlled rate to ensure the formation of the correct product in the desired physical form. Further, the high heat flow to the necessarily cold confining walls makes the plasma extremely nonisothermal, both radially and axially, although the sheer magnitude of this heat flow apparently works to make the flow radial profile both predictable and reproducible (Skrivan and Von Jaskowsky, 1965; Sherman and Grey, 1964; Grey and Jacobs, 1964). On the plus side, the short time required for elementary reactions to occur at these high temperatures implies the possibility that, at least for pressures near atmospheric, it may be feasible in some instances to predict chemistry on the basis of thermodynamic considerations (see discussion in Section II,B,1,*c*) and thus to predict the outcome of experiments on a sounder basis than the rather difficult and often unnecessarily discouraging " cut and try " method.

Finally, the streaming of these plasmas is an ever present factor that must be seriously considered for effective work. It ensures short residence times in the reactor and, indeed, gives the experimenter sophisticated control over the temperature history of the reactants. More subtly, it affects mixing properties and the maximum rate of quenching product streams. Regrettably, the dynamic properties of these fluids are as yet virtually unknown, but occasional studies are reported and it seems likely that an increasing number of useful correlations will continue to become available.

1. *Generation*

Streaming thermal plasmas are inevitably formed in some sort of high intensity arc device. Unless one of the electrode materials is to be used as a reactant, the electrodes are well-cooled and the constriction necessary to stabilize the arc and to attain the high temperatures desired (Finkelnburg and Maecker, 1956) is achieved by surrounding the arc by cold walls capable of withstanding a high heat flux, and/or surrounding the arc with a cold liquid or gas which is forced to flow into the arc (or is convected in by natural processes) and is converted to the thermal plasma. Now the current flowing in the arc column causes an inwardly directed " magnetic pressure " from the Lorentz ($J \times B$) forces acting when the current interacts with the strong solenoidal magnetic field it produces. This is sometimes known as the Maecker effect. Should the arc diameter be anywhere constricted more than the average (as it always is at the point of attachment to a cold cathode, for example), then the current density and consequently the magnetic pressure will be higher at that point. Even though the pressure differential is only on the order of one inch of water or less, a high velocity streaming sets in along the arc axis away from the points of greater constriction. Velocities due to this magnetohydrodynamic pumping are typically on the order of 100 m/sec. The pumping action is so strong that, due to the cold gas drawn in by aspiration, high intensity arcs are to some extent self constricting and are thus sometimes run without cold confining walls, particularly when one of the electrodes is to be vaporized as a chemical reagent. Whether or not there is a superimposed gas flow, this magnetohydrodynamic (MHD) streaming would appear to control the flow (Freeman *et al.*, 1966).

2. *Devices*

Several STP generating devices are shown in Figs. 1–5.

a. Plasma Jet. In the plasma jet, Fig. 1, the Maecker effect streaming velocity is increased typically by a factor of 3 by forced throughput of the constricting fluid. Temperatures in the core attain about 13,000°K (23,400°R) with, say, argon (Freeman, 1968a; Ahlborn, 1965), while with liquid water constriction they may go as high as 50,000°K (Burhorn *et al.*, 1951). Throughputs on laboratory scale devices (less than 40 kW dissipation) are typically 0.1–1.0 ft^3/min. Single units may readily be scaled up from laboratory scale by factors of 10^2 or 10^3, so that transitions in scale from research to pilot plant to full scale chemical production unit are relatively easy.

The plasma jet has been around long enough to warrant serious study (Freeman, 1969b). In recent years there has been some effort to develop a comprehensive theory for these devices (Tiknomerov and Varlamov, 1964; Ahlborn, 1965), and one must suppose that since the advent of the segmented-anode torch (Fig. 1b) (Dean and Runstadler, 1966; marketed by Hypertherm Corporation, a subsidiary of CREARE Inc., Hanover, N. H.), which is particularly well-adapted to the study of the arc process itself, the plasma jet will soon become one of the best characterized arc devices. We shall say more about the plasma jet below (Section I,A,3) when we discuss its chemically significant characteristics in greater detail; however, it seems appropriate to say a word about hardware. While some outstanding chemical investigators have successfully developed their own plasma jet hardware [e.g., Leutner (1963), Harnish *et al.* (1963)], most successful work has been done with the commercially available, highly evolved laboratory-scale units available from a number of firms. The factors to consider when purchasing a unit are its electrical efficiency, its ease of starting, whether it produces a clean, uniform jet uncontaminated by electrode materials, whether it can be readily adapted to completely sealed or even vacuum operation, whether the parts most likely to be damaged by malfunction or misuse are inexpensive and easy to replace, and of course the reputation the firm has for lending a helping hand with initial start-up problems.

b. Consumable-Electrode Devices. Two consumable-electrode devices are schematically depicted in Fig. 2. Developed by Sheer and Korman (1952) in cooperation with the Vitro Corporation, the consumable-electrode device may be used whenever it is necessary to vaporize a refractory reagent (Gibson and Weidman, 1963). The many process variations developed by these workers and their colleagues are covered by subsequent patents too numerous to cite here. In this device, the refractory reagent (e.g., alumina, silica, various metal oxides) is incorporated in a carbon-containing electrode that is operated so as to overload and hence evaporate. The resulting high velocity jet of vaporized material, the so-called anode flame, may then be manipulated like any other streaming thermal plasma to recover product (Sections I,A,3,*b* and *c*). This type of arc may either operate on dc or, with three electrodes, on 3φ ac, as shown in Fig. 2b. Despite unquestioned commercial importance (Section I,B,3) and even greater potential (Section II,A,2) of this device, there are no off-the-shelf vendors of research equipment known to the writer. Presumably the investigator must build his own. This situation will probably change when the basic patents expire this year.

c. Transpiration Cooled Devices. The high intensity arc with transpiring electrode (Fig. 3a) (Sheer, 1965) is at present more interesting than useful. It is similar to the preceding device except that the anode is not consumed and the makeup feed for the streaming plasma comes from a gas transpiring through a sintered porous electrode. As a scientific enigma it is outstanding. There is an appreciable nonluminous gap between the electrode and the plasma that has repeatedly been shown to be quite cold, and yet, based on the voltage drop across the gap, it has better electrical conductivity than any other part of the arc column (Sheer and Korman, 1965).

A variation of the principle is shown in Fig. 3b. This is a plasma jet with transpiration-cooled front electrode studied by Pfender *et al.* (1969). They

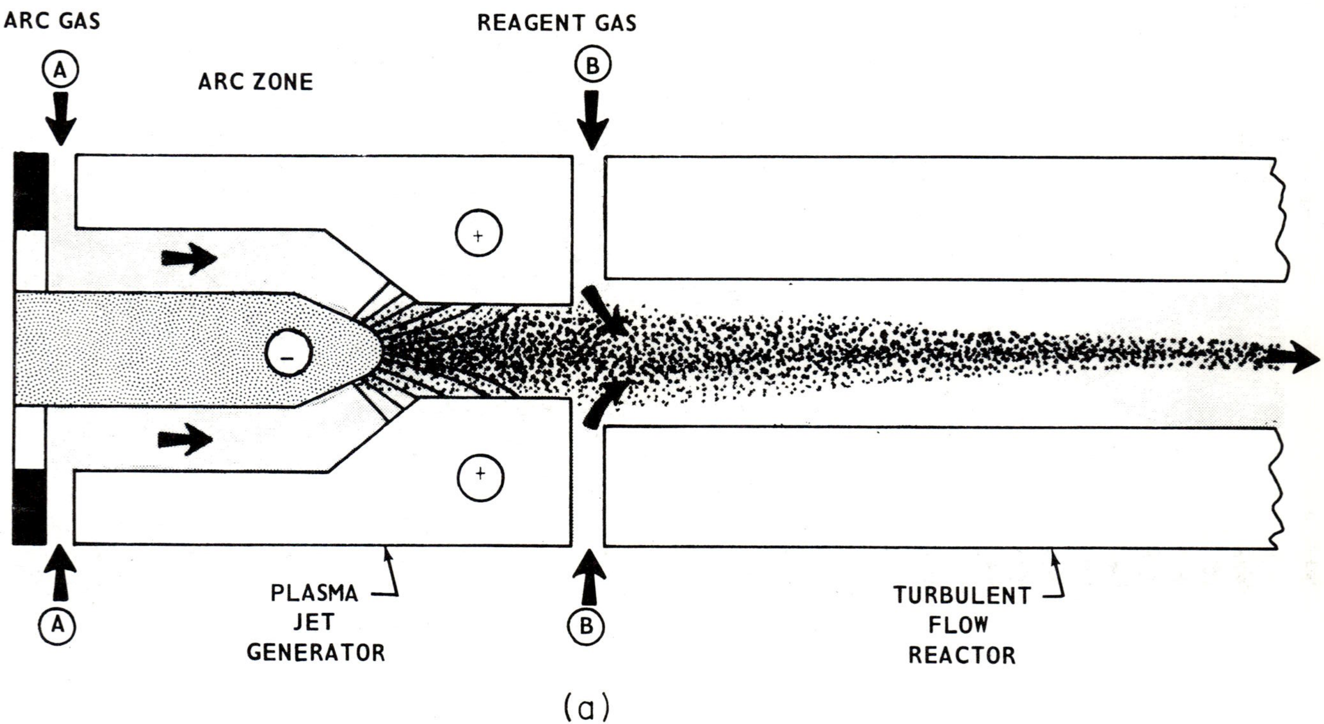

Fig. 1. (a) A typical plasma jet reactor configuration.

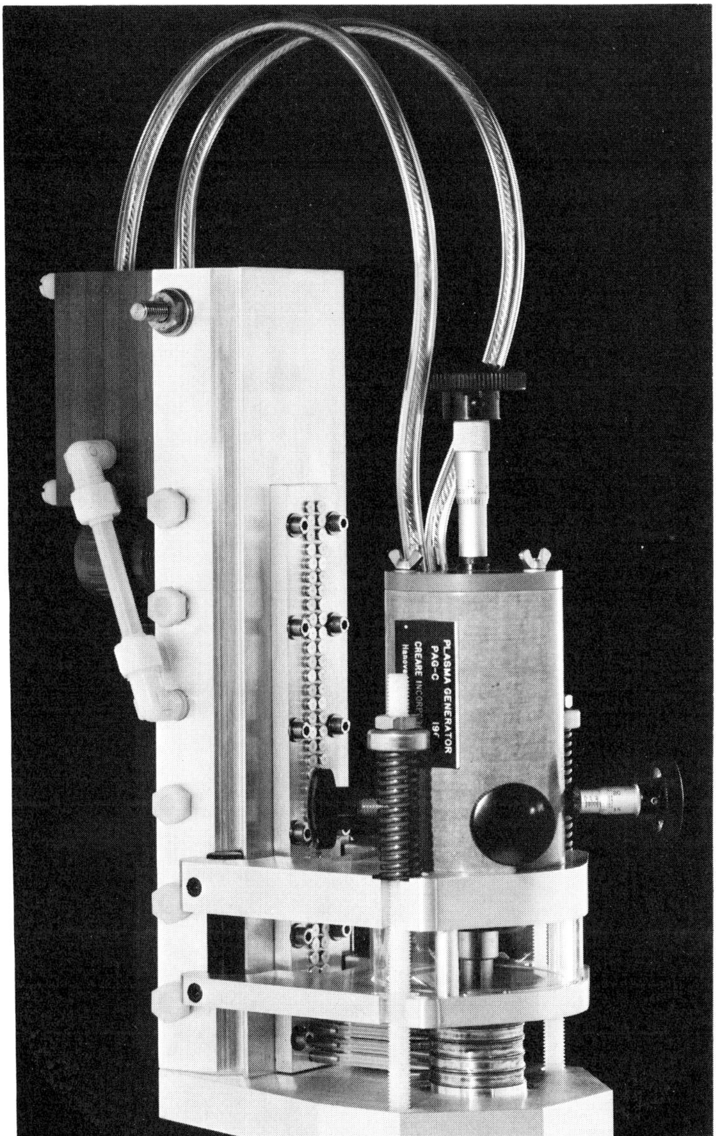

(b)

Fig. 1. (b) A segmented-anode torch (courtesy of the Hypertherm Corp., a subsidiary of CREARE Inc., Hanover, N. H.)

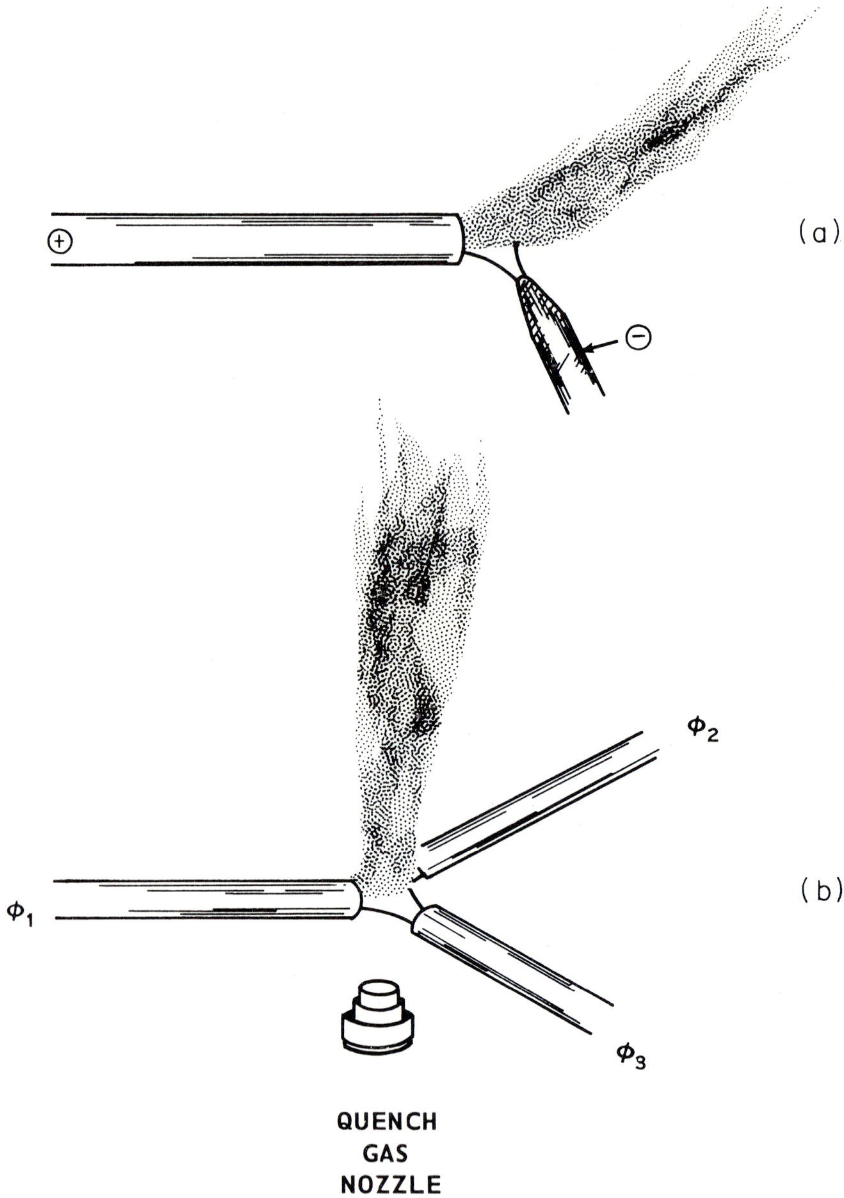

Fig. 2. Consumable-electrode devices: (a) dc configuration; (b) 3φ ac operation.

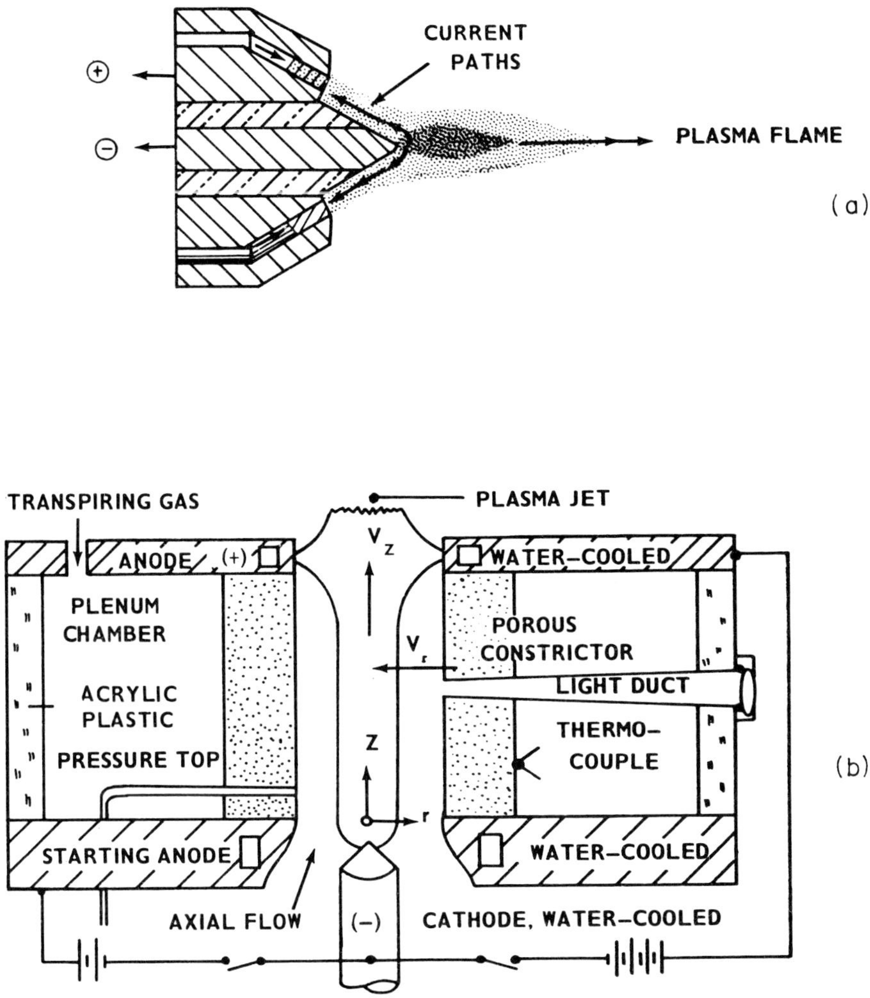

Fig. 3. Transpiration cooled devices: (a) high intensity arc; (b) plasma jet.

report temperatures far higher than those attainable in an ordinary plasma jet or even in a welding arc, and this device may very well represent *the* plasma jet configuration of tomorrow.

d. Welding Arc. The welding arc is very well-understood. Under highly stylized conditions in an argon atmosphere it is known as the Olsen arc, after the investigator who has spent many years investigating its radiative properties (Olsen, 1963a,b). The completely consistent heat flow and velocity profiles are also known for this device, so it serves as a prototype of what we mean by a well-understood thermal plasma device. The characteristic bell shape of such an arc is due again to the high velocity streaming caused by the Maecker effect as the arc impinges on the anode plate. This accounts for the enormous heat flux transferred to the anode plate directly under the electrode (whether it is conducting there or not). Such a device is used in the cold-hearth melting and crystallization of refractory substances (Reed, 1969) and in the cutting and of course the welding of metals, where the MHD pumping causes droplets of the melting cathode to impinge on the workpiece with high velocity.

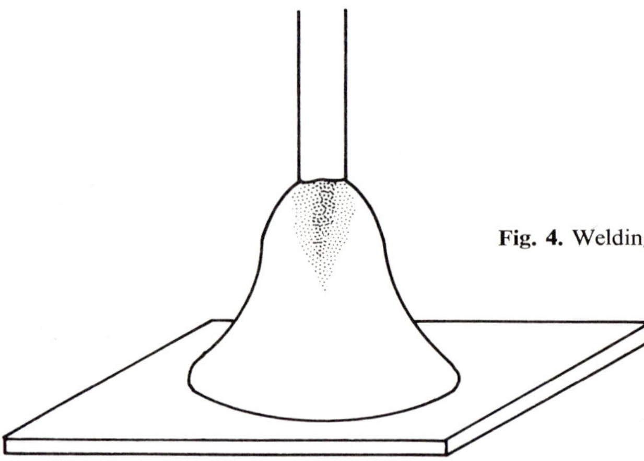

Fig. 4. Welding (Olsen) arc.

e. Large Volume Arcs. Two types of large volume and hence intrinsically low temperature arc devices are shown in Fig. 5. Through use of magnetic induction and cold-wall constriction it is possible to maintain an electrodeless high intensity arc, the so-called radio frequency plasma generator [Reed (1961), for further references see Reed (1967)]. Many of its operating principles are surprisingly similar to those of the high intensity arc (Freeman and Chase, 1968; Eckert, 1968), but from a pragmatic viewpoint it is substantially different and should properly be reviewed separately, perhaps together with the rotating arc (Fig. 5b). Thus, except for its commercial aspects (Sections I,B and II,A), it will be largely ignored in this chapter. In the few years since the first simple devices were fabricated, there has been amazing progress in hardware development (principally by the TAFA division of Humphreys Corp., Concord, N. H.), so that plasmas of nearly any substance can be produced at nearly any average enthalpy and,

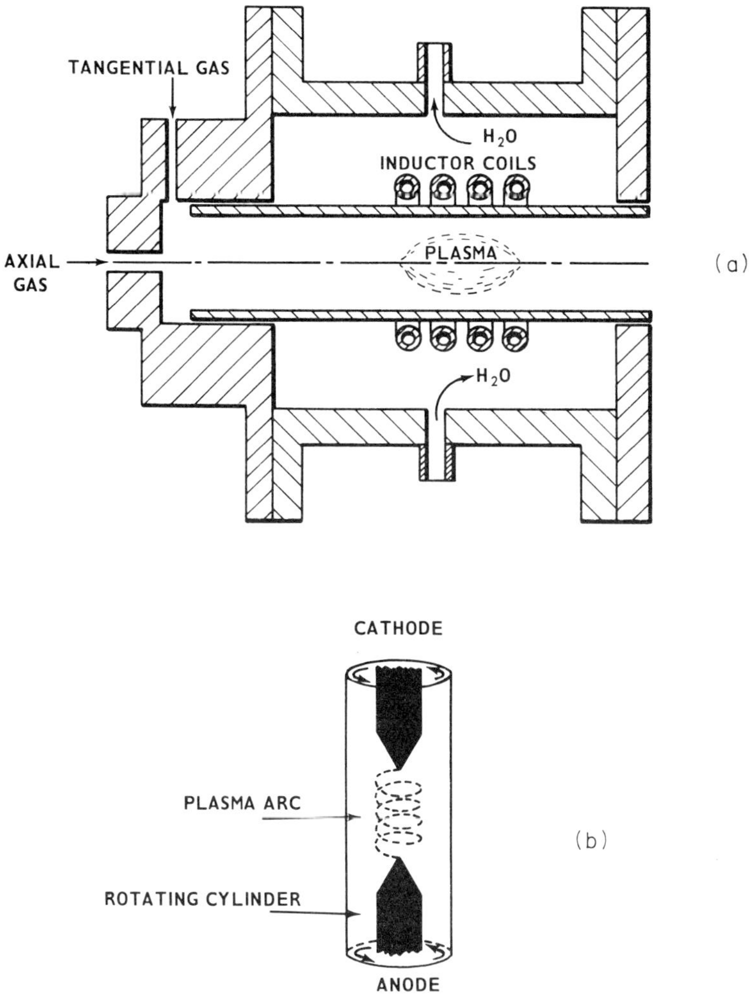

Fig. 5. Large volume arcs: (a) radio frequency induction arc; (b) rotating arc.

despite the traditionally high cost of radio frequency power, at reasonable cost. Note that contrary to what was initially " obvious," it has recently been established that the Maecker effect produces streaming in this device too (Chase, 1969), but now in such a way as to hinder the experimentalist who fails to take it into account. That is, the streaming is from the center of the fireball out both ends, so that at the upstream end it is counter-current to the impressed flow. Attempts to run solid reagents through the fireball thus can be somewhat frustrating, because unless the flow system is designed to override the streaming, the particles of solid will appear to simply bounce off the fireball as though it were a solid object.

When electrodes can be tolerated in the discharge, the rotating-arc device would seem to produce a rather similar expanded plasma. That is, a very long, high intensity arc has some constriction due to aspiration by the Maecker effect at the electrodes, but is mainly stabilized by thermal convection currents that keep the arc from growing indefinitely (Finkelnburg and Maecker, 1956). Von Engel and Steenbeck (1931) have long since shown that if these convection currents can be arrested (e.g., in free fall) the arc will expand to fill its container. In particular, rotating the arc was observed to have the same effect. This effect has recently been exploited to develop an arc reactor (Whyman, 1967) that looks quite promising for certain chemical purposes (Aust *et al.*, in press) in that it generates an extended field of plasma at temperatures as low as 7000°K, lower even than the rf plasma. Unlike the rf plasma, the Maecker effect now promotes long residence time by causing streaming from both ends toward the center. An attempt is currently being made to market hardware for research in this country (Plasma Physics Corporation, Hicksville, L. I., N. Y.), but it is too early to tell with what success.

f. Other Devices. The configurations of possible devices with potential for the chemist are by no means exhausted. The rotating-arc configuration (Fig. 11) currently being employed in coal research (Section II,A,1,*b*) is one case in point. One must also include the high intensity arc devices where the consumable anode is a fluid (Goldberger, 1964) or packed (AVCO, 1968) bed of conducting particles. Another configuration that has never been used by a chemist and yet would seem ideal for the most sophisticated sort of research is the doubly constricted arc (Freeman, 1968a). One must suppose that as the more established laboratories continue to refine their research, a whole new generation of devices will emerge. Recent research on electric arcs has raised the permissible operating pressures to 200 atm, and 1000 atm devices are under construction (Bauder, 1968). When these techniques have been refined for routine operation, a whole new vista of possibilities should open for the chemist.

3. *Properties*

In Section I,A, we outlined the general character of streaming thermal plasmas. Here, without becoming involved in the details of how reliably, or even how they were determined (See Section II,C), we digest the latest, most credible results to more clearly define the STP problem. Plasma jets are emphasized because most of the better work has been done with them and because the writer is on most familiar ground in this area, but the extension of the concepts to other STP devices should be obvious.

In practice, only a limited number of interesting reagent fluids can be passed indefinitely through a plasma jet device without some deterioration of the device. The chemist must either be content with short, dirty runs or else form the plasma of some " safe " material (A, N_2, H_2, NH_3, CO, etc.) and add further reagents after the plasma has been removed from the arc process. Thus we see that a plasma reactor divides naturally into three parts involving different kinds of considerations. They are (a) the arc process itself, where a relatively inert reactant or spectator fluid is converted to plasma by the arc process; (b) an intermediate section, where additional reagents are mixed with the plasma thus produced and the space- and hence time-resolved removal of heat from the plasma mixture is controlled (within limits) as necessary; and finally (c) the quenching section, where residual heat is removed and chemical products recovered. Each of these parts will be discussed in turn.

a. Arc Process. In its usual configuration the research scale plasma jet arc occurs between a point cathode and a cylindrical anode millimeters away. It is a high intensity arc with the concomitant flow due to Maecker-effect streaming passing through the hole in the anode to make a field-free jet of plasma. This would occur even were it not for the make-up gas, which constricts the arc further on the one hand and dilutes it on the other. The end result is a plasma jet with a velocity about three times that of the Olsen arc (Section I,A,2,*d*), but somewhat colder. The Olsen arc can easily exceed 16,000°K (100% ionization), while the core of an argon plasma jet can only be made to exceed 13,000°K (about 20% ionization) with difficulty. A number of luminosity fluctuations travel with the jet, some at jet speed, but most at jet speed plus the local speed of sound (Freeman *et al.*, 1962). The sonic fluctuations are discussed below; however, careful observation of the slow fluctuations shows that even in this device the velocity of the jet is still controlled by the current flow and hence by the Maecker effect (Freeman *et al.*, 1966).

A complete theoretical model of the arc process would involve a description of the cathode attachment, the distribution of current into the anode, and the flow of constricting fluid into the arc. Given this information and the material properties of the plasma fluid (electrical and thermal conductivities, density and heat content versus temperature), it should in principle then be possible to predict temperatures and axial velocity profiles, the rate of heat loss to the anode, and other quantities of interest. Using a drastically simplified arc channel model [cf. Maecker (1959)], Tiknomerov and Varlamov (1964) have in fact succeeded at reproducing in first approximation some plasma jet measurables; however, it is more convenient simply to say that the physical principles governing the arc are individually well-known, but in toto are too complex to go into here. Instead we describe arc operation and the resulting jet phenomenologically.

Most plasma jet arcs are in the falling or flat V vs I characteristic appropriate to a moderate intensity arc. Of the common jet materials, only argon at currents on the order of 200 A and higher enters the rising regime characteristic of the high intensity arc. Clearly, the power supply must have a sufficiently " falling " characteristic (a series resistor or its equivalent in reactance in the ac feed lines) to make operation stable according to the Kaufmann criterion, which states that the negative slope of the power supply characteristic must exceed that of the arc characteristic at the point of intersection for stable operation to occur. The average electrical conductivity of the plasma produced, about 2 to 20 mho/cm (Freeman, 1962), is not very different from that of a strong electrolyte solution, and strong currents may readily be drawn by inserting a not-too-well-cooled rod anywhere in the field-free plasma jet and returning it electrically to ground. (The anode fall provides the emf for the current.) Curiously, all else equal, the current through the arc process anode will exactly decrease to compensate for current drawn to the probe, so that cathode current remains constant.

The jet itself is said to be turbulent or laminar, depending on the noise level and on its mixing properties when discharged into stagnant cold gas. When discharged into the laboratory the so-called turbulent jet will " hiss," " howl," or " shriek," depending on the observer and the arc fluid used, and form a short cone of intense luminosity. The laminar jet, on the other hand, emerges with a most soothing hum. Although somewhat unstable in direction, its slowly fading luminosity will persist for several inches, but the most brilliant part disappears in even less length than in the turbulent case. Given the proper electrodes, one can operate the same arc unit in either mode simply by changing the gas flow an order of magnitude. Incropera and Lippert (1966) have suggested a Reynolds number criterion for predicting the transition, but as is always the case they encounter some confusion when defining their dimensionless number because of the nonisothermal situation. (The radial temperature and velocity profiles of argon jets leaving the arc process are described below.) The noise of the turbulent jet is far from " white," carrying a dominant component usually between 2 and 10 kHz.

This component, which under certain conditions of unconfined operation with molecular gases can assume magnitudes dangerous to personnel, is apparently the result of the fluctuating anode attachment process studied so exhaustively by Pfender and Eckert (1965) and their associates. Indications are that for certain gases these fluctuations are unavoidable and, indeed, one might suppose that by distributing the average heat load this rapid fluctuation decreases electrode deterioration to the low levels observed.

b. Intermediate Section. The plasma jet as it emerges from the arc device varies reproducibly in temperature from 13,000°K in the center to about 600°K at the walls. Thus far these temperatures are known exactly only for ambient pressure argon (Freeman, 1968b; Ahlborn, 1965), but indications are that other common plasma materials may not be much different. Now, this is a flow situation that is just not covered by standard fluid dynamic analyses. Investigators who attempt to define conventional dimensionless groups based on " average " plasma properties find the flow behavior to be highly unpredictable. It is small wonder, because it has been shown, for example, that even though the argon jet exhibits local thermodynamic equilibrium (LTE) throughout, the rates of production of ions and metastable atoms are as much as four orders of magnitude higher than those predicted on the basis of average jet enthalpy (Fig. 6) (Freeman, 1968b).

A jet model that does correlate both heat loss and species production data is a two-step model (Fig. 7) (Freeman and Longfield, 1968). Without mention of the substantial body of experimental data that went into the model, it is worth discussing here to clarify the function of the intermediate section. The assumptions are:

(1) Mass flux is constant over the jet cross section.

(2) The flow is considered to be in LTE throughout.

(3) Virtually all heat in excess of ambient flows in a central core of uniform temperature T_c, the only adjustable parameter in the model. Typically about 12,000°K, T_c must be determined by experiment for a particular jet device at a particular throughput.

(4) Heat loss from the hot core to the walls occurs by normal conduction processes, and the temperature profile in the sheath surrounding the hot core is determined by the heat flux and the boundary condition at the wall. For most purposes the heat flow in the outer sheath is thus negligible.

(5) As heat is lost from the plasma, T_c remains constant, but the diameter of the hot core shrinks accordingly.

Thermal development of the flow appears to occur instantaneously so that one should think of a flow of wall-confined thermal plasma as an isothermal

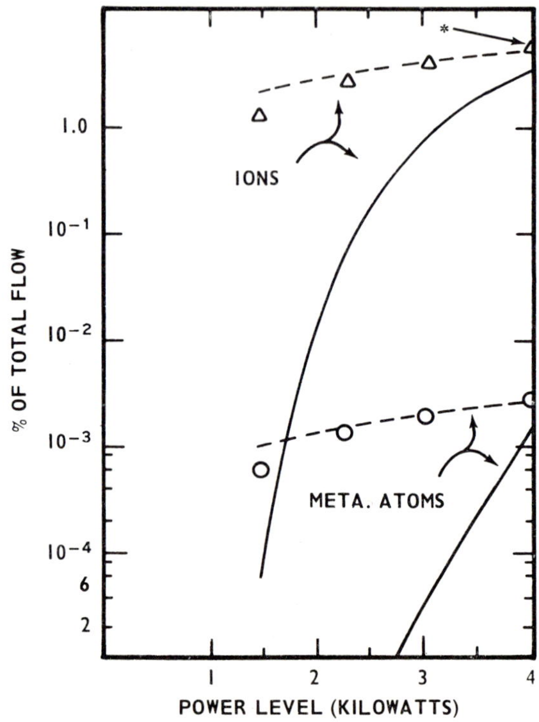

Fig. 6. Measured rate of conversion of argon to argon ions and metastable atoms in an atmospheric pressure argon plasma jet (molar flow rate of argon=0.016 g-mole/sec) as a function of net power flowing in plasma. Solid lines are equilibrium values based on average jet enthalpy predicted by a one-step model. Dashed lines show values predicted by the two-step model. Asterisk indicates point used for normalization.

hot core continuously diminishing in area as heat is lost to the walls. This hot core is surrounded by a somewhat nonisothermal but in any case low-temperature flow involving usually more than half the total mass flow. The core has very high viscosity and low density compared to the sheath and moves on the average with 30 to 40 times the velocity (Fig. 7).

i. Heat flow. For confined flow the heat loss of a jet to the wall as measured by W, the ratio of net power (and hence core cross-sectional area) flowing in the hot core at point z to that flowing at the entrance of the intermediate section W_0, is given by (Skrivan and Von Jaskowsky, 1965)

$$W/W_0 = (s/(z+s))^{2/3} \tag{1}$$

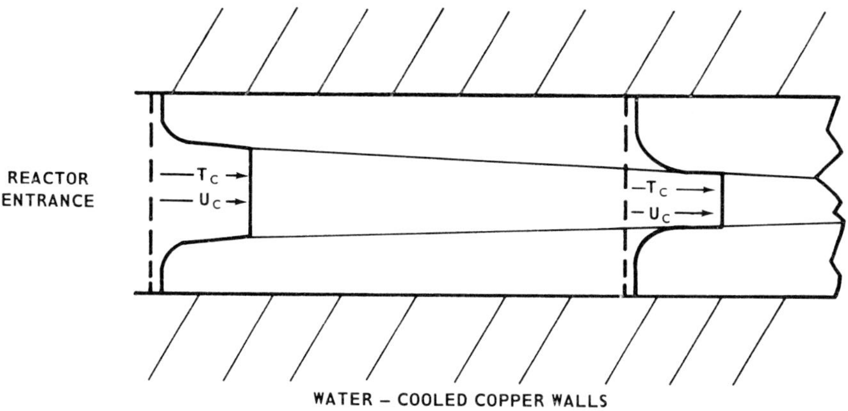

Fig. 7. The two-step model showing temperature (and velocity) profiles, assumed to be tapering but isothermal. Mass flux is assumed everywhere constant.

[The s is an adjustable parameter reflecting jet prehistory but is sometimes just the length of the anode channel in the arc unit (Freeman and Skrivan, 1962).] This uniformly confined flow represents the slowest possible cooling of the plasma. (Note however that if the walls are allowed to become very hot, the parameter s becomes correspondingly larger and the rate of heat loss diminishes further.) If the hot core is forced to impinge on a cold wall, for example, when using a plenum chamber with a constricting orifice, then the heat loss becomes much greater.

Very little is known about heat loss from plasma flow through discontinuous diameters, but in one study an interesting and illuminating transition was observed (Eubank and Johnson, 1967). At low powers the jet, emerging into a larger pipe about four times the diameter, turbulently entrained the cold surrounding gas to make a low temperature mixture and to effect a high heat transfer rate near the junction. At high power, on the other hand, the hot plasma apparently disengaged from the surrounding cold gas and simply adjusted its profile to conform to the new diameter. This was reflected in an order-of-magnitude decrease in Nusselt number. The transition was not sudden but occurred over a substantial range of heat flows.

Galyaev *et al.* (1965) have shown that one might expect to obtain significant additional control by adding additional heat to the intermediate section downstream of the arc process. This novel idea should certainly be a powerful tool in the hands of the sophisticated investigator, but to the writer's knowledge has not yet been experimentally implemented.

ii. Mixing. The mixing of a reagent introduced through a slot or orifice around the periphery of the jet with the jet itself is clearly of paramount

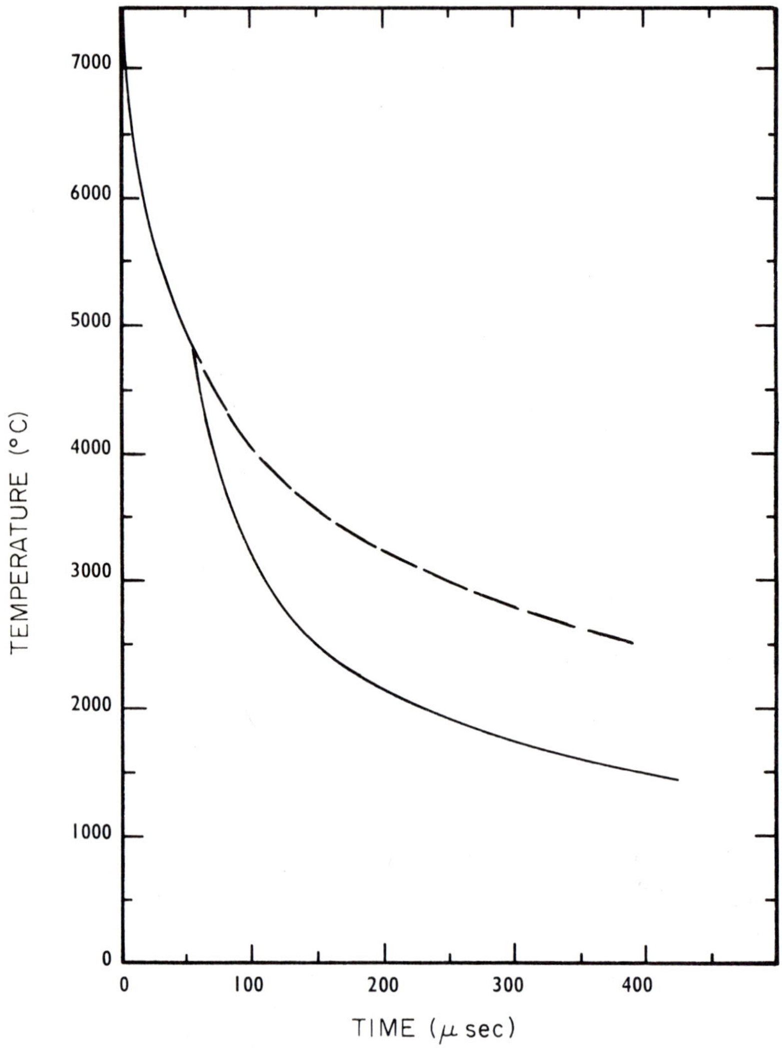

Fig. 8. Typical " average temperature " versus time history for argon flowing in a ¼-in. intermediate section. Lower line shows effect of adding one third the amount of ammonia after 1 cm of jet travel (60 μsec). Mixing is assumed to be diffusion controlled.

importance to the investigator (Fig. 8), particularly considering the very short residence times intrinsic to the apparatus. The classic work of Grey reported by his co-workers (Jacobs, 1962; Sherman, 1964) has done much to clarify this process and verify early indications that mixing by diffusion into laminar jets and by eddy diffusivity into turbulent jets is very rapid (Freeman and Skrivan, 1962). Recently developed mixing-kinetic computer programs (Cohen, 1966) may ultimately put quantitative control of this variable on a firm footing.

Presumably the mixing of gases and easily volatilized liquids proceeds by two steps. The cold added reagent fluid will start out in a film along the walls of the reactor or as isolated pockets largely undisturbed by the ineffectual turbulent shear of the very low density plasma. The rate limiting step must then be the diffusion of the cold reagent into the plasma. Subsequent mixing would be expected to be very fast.

Nonvolatile liquids and solids simply become entrained in the jet and emerge relatively unscathed because of the short residence times intrinsic to this situation. There is always some evidence of surface melting of a solid, and if chemical reaction is possible the outer part of the particles will show evidence of the reaction, but the whole particle is seldom converted (Stokes, 1967). As mentioned above, a much preferred device for working with refractory reagents is the consumable-electrode arc (Section I,A,2,*b*), unless of course spheroidizing or surface treatment of particles is the desired objective.

iii. Supersonic jet. The particulars thus far have dealt with subsonic jets at nearly atmospheric pressure. Most chemical research has been performed under these circumstances; however, it is easy to expand such jets through more or less designed nozzles and achieve supersonic flow. Under these conditions the kinetic degrees of freedom of the gas are suddenly cooled and one can arrest salient chemical reaction farther from equilibrium than is possible with a subsonic jet (Bray, 1959). Unfortunately, such an expansion has not yet proved practical for chemical synthesis because the gas must undergo the reverse transition to subsonic flow. If the pressure is high enough to make the recovery of chemical products attractive, the recovery shocks are so strong that the metastable chemistry is generally undone.

c. Quenching Section. In the terminal section of the reactor it is usually desirable to remove the residual sensible heat of the reaction stream as rapidly as possible, particularly when the desired product is not thermodynamically favored at low temperatures. The best " quench " is unquestionably the injection of a large excess of a cold inert fluid that will itself react with substantial reaction endotherm to make further desirable products. This procedure, time honored in the production of acetylene, has recently been theoretically analyzed by Khait (1965). In the case of an electrode flame, a cross jet of a quenching fluid fulfills the same purpose. For research-scale equipment where the product is not decomposed by further exposure to heat, a plasma jet may be effectively quenched simply by discharging it into a large water-cooled vessel. Turbulent entrainment of the cold reaction products then forms a most effective quench.

Sometimes, for purposes of research, it is desirable to sample a portion of the still hot stream with a water-cooled probe (Hirayama and Maniero, 1967; Bronfin and Hazlett, 1966). The probe walls then provide the quenching function that for real production one would have to provide some other way. Care must be taken, however, because if the probe is too large, a boundary layer will form and the heat loss will more nearly resemble that of an intermediate section. To compare with the quenching obtainable by turbulent entrainment, the probe diameter should not be over 0.050 in. i.d. For a theoretical treatment of probe quenching see Timmins and Ammann (1967).

Another superficially attractive method of quenching is to impinge the jet directly onto a well-cooled surface. Although Harnish *et al.* (1963) have reported excellent results using a water-cooled rotating copper cylinder, the writer has generally found stationary quenching objects of all shapes to be uniformly bad. They should not be categorically dismissed, however, for should they ever prove useful they would offer the simplest method for recovering high temperature heat from the quenching process, with consequent economic advantages.

B. Current Industrial Applications

As indicated in Section II,A, the acetylene-from-hydrocarbon synthesis is something of a red herring, and this is just as true when the commercial aspects of STP devices are considered as it is for current research. The skeptic will say that acetylene is the only bulk chemical produced in these devices and that the only operational plant, Chemische Werke Hüls, was built thirty years ago (Kirk and Othmer, 1963). He will probably also cite the failure of the recently constructed DuPont plant in Montague, Michigan (DuPont, 1964) to produce acetylene from heavy hydrocarbons economically (for which unusually high electrical costs are blamed) and may even hint at the rumor prevalent that the Hüls plant has not long to survive—a rumor, by the way, that is completely without foundation as far as the writer has been able to determine (Gladish, 1967). But if one disregards this one large, long-standing commercial success and instead views the developments of the past two decades, one sees that the growth is far from discouraging. The easy availability of the devices bears testimony to their utility in welding, cutting, and materials testing. They are also finding increasing use in the spray coating of objects with refractory substances (Section I,B,2) and especially the radio frequency plasma is being used for the spheroidization of refractory particles. Furthermore, the ready availability of large scale STP equipment for sale and rent reflects its general utility in driving large scale " hyper-thermal " operations for aerospace wind tunnel and materials testing research applications.

Although these fields are not traditionally thought of as " chemical," and are certainly not carried out on the grand scale of a chemical plant, it should be clear from the discussion of product oriented research (Section II,A) that these uses are probably the true indicators of the coming large scale chemical and metallurgical applications of these devices. There is a general argument why this should be. As we have previously pointed out (Hulburt and Freeman, 1963), bulk chemicals tend to sell in the same range per unit weight, but heat is absorbed by molecular bonds, so that the reagent chemicals tend to require the same amount of heat per average pound atomic weight—heat derived from costly electricity. Thus, to improve the profit picture it is most desirable to employ substances involving atomic species of high atomic weight. This forces one to consider products involving the heavier elements, which in many cases tend to be refractory. Then too, the more refractory the reagent, the more use is made of the high temperature advantage of STP devices.

1. *Acetylene Synthesis*

Because the only existing industrial arc acetylene process is fully described elsewhere (Kirk and Othmer, 1963), it is not necessary to elaborate further here except to inject some personal observations. First, the question often arises whether this is a " plasma jet " or an " arc " process (Fig. 9). This largely semantic question can be easily resolved in favor of the former if one scales the unit down to laboratory size with similarity transformations (not yet reduced to hardware), using the necessary dual constraints of constant instantaneous current density and constant gas velocity (Table I). Although the hollow cathode through which one eighth of the arc gas flows is a somewhat unusual but nonetheless workable feature (Peters, 1964), by and large any investigator of thermal plasma would instantly recognize it as a perfectly good plasma jet device.

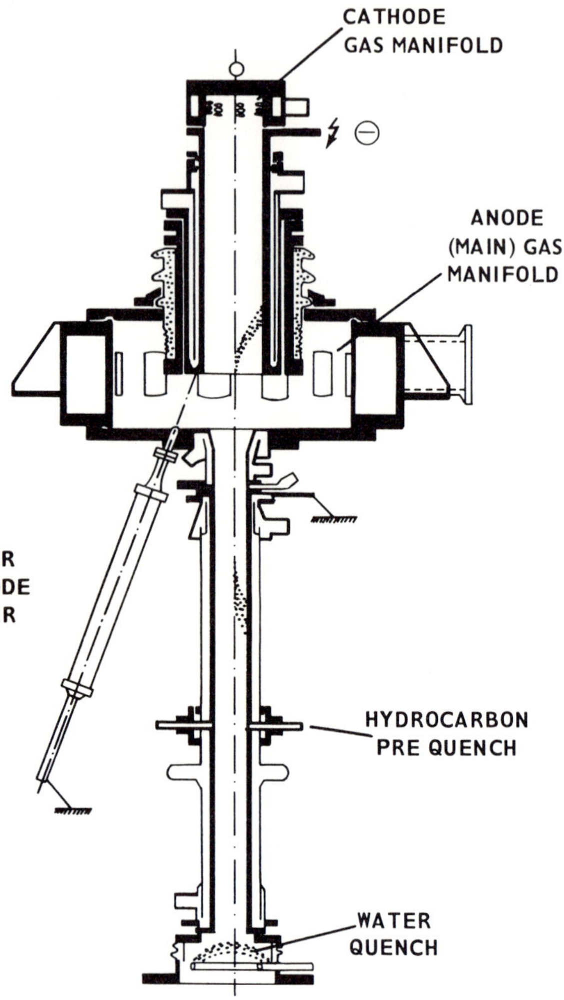

Fig. 9. Hüls acetylene arc jet unit. [Adapted from a drawing in Kirk and Othmer (1963).]

As a second comment, the writer would like to encourage anyone contemplating a production scale STP process to make a " pilgrimage " to Hüls to visit this well laid out and run 105 megawatt plant, if for no other reason than to inspect the modern instrumentation and control apparatus.

TABLE I. SCALING RELATIONS FOR STP DEVICES

Quantity	Factor
Enthalpy Pressure Velocity Instantaneous current density	Constraints
Volume	φ
Areas Throughput Net power into gas Heat lost to cooling water Net power into head	$\varphi^{2/3}$
Linear dimension Residence time Voltage Current Cathode striking area	$\varphi^{1/3}$
Field Density Efficiency Mean free path Temperature	φ°
Average current density	$\varphi^{-1/3}$

2. *Surface Modification*

It is difficult to assemble published information on the commercial utilization of plasmas for the modification of surfaces, especially spray coating. It is a well-established technique, and most of the technology has been borrowed directly from conventional flame spraying technology. Plasma spray coatings are substantially superior to oxyacetylene sprayed coatings in both density of the sprayed layer and adhesion. Commercial hand-held spray units are available (Fig. 10), and in general this procedure is used whenever the need for the superior properties of the coating justifies the somewhat higher equipment costs. Kirk and Othmer (1967) lists a few industrial metal spray-coating applications but asserts that the principal industrial application of the method lies in the application of nonmetallic refractories and intermetallic compounds. A short review of plasma spraying

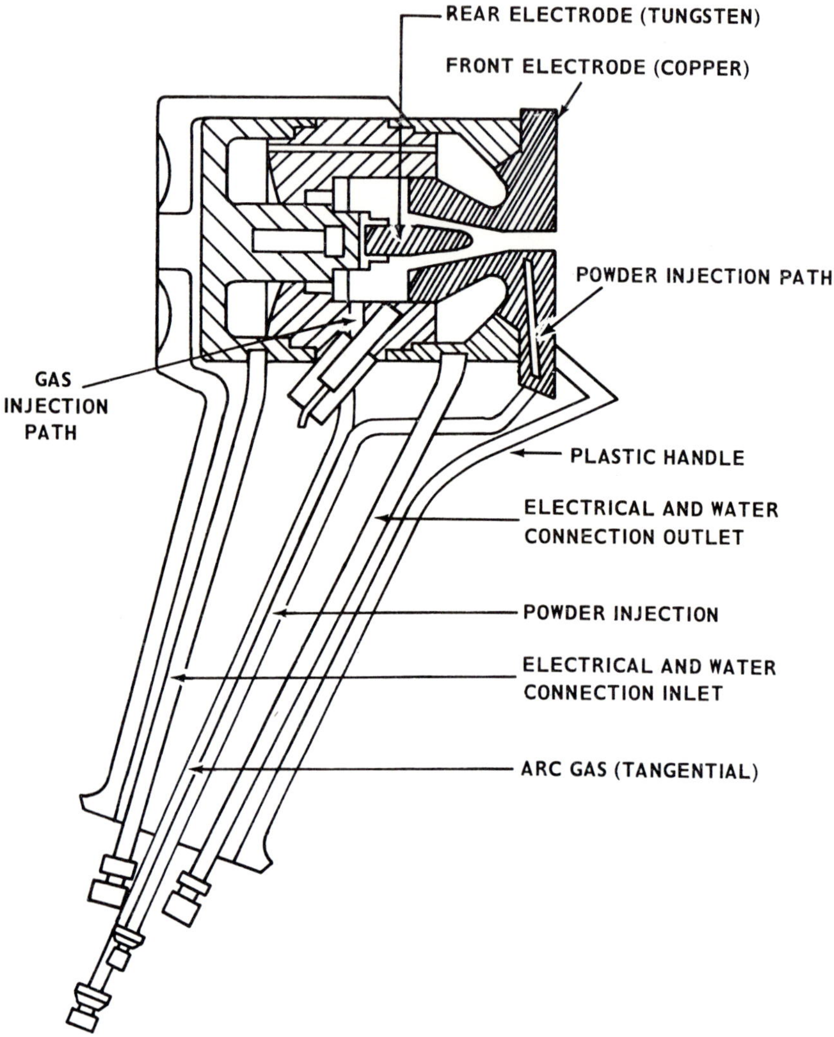

Fig. 10. A hand held plasma spray gun. [Adapted from a drawing in Kirk and Othmer (1967).]

technology has been published in Austria (Buettner, 1965), but of course it lacks many of the important developments from the American aerospace industry.

Another surface modification technique, as yet unreported in open literature but rumored to be in production use, involves the nitriding of metal surfaces by direct impingement with a nitrogen plasma jet.

3. *Spheroidizing*

The rendering of more or less refractory materials into smoothly flowing microspheres is a natural process for STP devices of all types. The only current application is in the somewhat limited production spheroidizing of refractory uranium- and plutonium-containing compounds for use in nuclear fuel elements (Strigl and Bildstein, 1964; Gibson and Weidman, 1963), but the writer has been given to understand (Thorpe, 1968) that large scale spheroidization processes will soon go on stream for less esoteric materials such as steel and titanium dioxide. Note that the cost of spheroidization is very much a function of the STP device used, but is most strongly influenced by the melting point of the materials and varies between $10 and $4000 per ton. The interested reader is referred to the TAFA (1966) literature for cost data relative to specific processes. Apparently the melting point does not tell the whole story, for Meyer (1964) has found in some instances that the catalytic activity for atomic recombination is of overriding importance.

The Vitro Corporation has for some time been in the business of synthesizing, on order, specialty chemicals with the consumable-anode arc (Section I,A,2,*b*). In some instances the electrode flame is quenched with cold oxygen so as to recover an oxide product, while in others the carbon is allowed to reduce the reactant to the desired product, be it element or carbide. Control of this so-called carbothermal reduction is achieved by varying the proportion of carbon to refractory in the electrode composition (Gibson and Weidman, 1963).

II. Recent Research

Reflecting the fact that most research on STP devices is done in industry, the bulk of research is definitely product oriented (Section II,A). In spite of this, some fairly sophisticated mechanism studies have been and are being made, mainly by industrial researchers (Section II,B). This probably reflects a growing awareness that to realize the full potential of these devices, " cut and try " methods have their limitations. Supplementing this work is the rather large continuing effort on diagnostic techniques (Section II,C), mostly done in academic and government laboratories, without which systematic investigation would be impossible.

A. Product Oriented Research

As Reed (1966) has pointed out, it is in a way regrettable that the formation of acetylene from hydrocarbon feedstocks proceeds so well in STP devices (indeed in almost any discharge device). Workers in the field will recognize the syndrome of the new investigator who starts out to make acetylene just to become oriented in STP research through repeating a well-known synthesis. In this laudable manner he is able to resolve and solve problems associated with plasma containment, reagent feeding, product quenching and recovery, waste heat disposal, instrumentation and control, and numerous other problems that come under the heading of " diapering and feeding " the plasma device. These details are all important because of the high throughputs of even laboratory-scale equipment. Because they are seldom discussed in the literature (due no doubt to the need for brevity on the one hand and an understandable disinclination to disclose know-how on the other), the new investigator will often find the patience of his sponsor nearly exhausted by the time he has developed his STP process to the incredibly high conversions and energetic efficiency cited in the literature [e.g., Anderson and Case (1962)]. He must then, usually in some haste, find an area of research somewhat less exploited, only to discover that unless he has taken great care, he has worked himself into a trap. He has, perhaps, committed himself to the feed and sometimes to the recovery of gaseous reagents, but worse he has adopted a way of thinking that ties him to the second row of the periodic table. Two difficulties arise. First, nearly all feasible products of these elements can be made quite economically by other routes, and second, none of the reactions approach closer than an order of magnitude to the thermal efficiency of the acetylene-forming reaction. Because molecular weights are low, this introduces a prohibitive electrical cost per pound for these low-cost materials. Few indeed are the laboratories that survive this phase and go on to the development of a commercially attractive process based on the deeper insight gained from the experience.

Thus it comes as no surprise that the bulk of published STP chemical research work in recent years, as always, concerns the formation of acetylene one way or another (Section II,A,1). What is new is the sophistication brought to bear on the problem and the use of unusual feedstocks such as low grade gasoline and coal. This unquestionably reflects the moral and monetary support of conservationists throughout the world, and we can only wish them every success. To put the problem in perspective, however, the total U.S. production of acetylene for all uses in 1966 was about 1.3 billion pounds (Stanford Res. Inst., 1966). In the same year 168 billion pounds of coke were produced, the total coal production being considerably higher, and in this country alone 750 billion pounds of crude petroleum were pumped out of the ground.

For reasons already cited (Section I.B), the synthesis of refractory materials, especially oxides, is more apt to reflect the immediate commercial future of STP devices. This field, which shows continuing activity, is of course somewhat difficult to separate from spheroidization discussed earlier (Section I,B,3). Considerable activity has been shown in the fields of metallurgy and metallurgical processing (Section II,A,3), while some surprisingly interesting developments have emerged in the modification of particulate surfaces (Section II,A,4). Finally, plasma phosphorus production and halocarbon chemistry have shown sufficient activity to warrant some comment (Section II,A,5).

For further information on chemical synthesis in plasmas we refer the reader to the excellent review by Reed (1967), the book edited by Baddour and Timmins (1967), and the more popular expositions of Goldberger (1966) and Korman (1964).

1. *Acetylene Synthesis*

a. Acetylene and Olefins from Methane. Work continues on the production of acetylene from methane injected into a hydrogen plasma jet. For a review of recent literature we refer the reader to Happel and Kramer (1967). One of the big surprises in recent years has been the announcement by Westinghouse scientists that their new high power STP device, utilizing a magnetically rotated ac arc between toroidal electrodes surrounding a pure methane flow, can produce acetylene at the unbelievably low power consumption of 2.7 kWh/lb (Maniero *et al.*, 1966). If we postulate a freezing temperature of 1150°C for the reaction (Freeman and Skrivan, 1962), the thermodynamically required endotherm for the reaction is 2.65 kWh/lb. Thus 98% of the energy input must go to reaction endotherm, which leaves little for heat loss and for the formation of their considerable by-products, mainly elemental carbon (Hirayama and Maniero, 1967).

For interested readers whose command of Russian exceeds the writer's, the book edited by Polak (1965) probably represents the most comprehensive and thorough study of this reaction available.

b. Upgrading of Coal. Although the upgrading of coal had been attempted before with only indifferent success, Bond and co-workers (1963, 1966) were the first to achieve satisfactory yields of interesting products from injecting small quantities of fluidized coal through a hole in the anode of an argon plasma jet. The limiting yields are obtained using coal particles of 50 μ and smaller. Adding 10% hydrogen to the plasma (equipment limitations precluded the use of more) they doubled the acetylene yield from 20 to 40%, while changing over to a nitrogen jet gave them HCN in good yield (based on carbon) for the principal product. Addition of hydrogen to a nitrogen jet optimized yields of both HCN (24.1%) and acetylene (7.4%). Similar experiments have since been reported by Kawa *et al.* (1966) and by Kawana *et al.* (1967). In every case the consumption of electricity, the expensive ingredient, was prohibitively high, but of course these experiments were in no way designed to optimize this factor. However, this work apparently stimulated federal support of considerable second generation work in several laboratories which, though not as yet all available for comment, seems to show promise. In particular, we may cite the work done under Krukonis (1968) at AVCO in which gas product streams with as much as 11% acetylene content at a power consumption of as little as 5 kWh/lb of acetylene have been achieved. Their reactors have evolved logically through a number of STP configurations to the latest magnetically rotated arc design shown in Fig. 11. This work is being vigorously pursued at the present time, for only small improvement would appear to make this process economically attractive. According to an independent engineering report included in the cited AVCO report, a 300 million pound per year plant with a product stream 12.8% in acetylene and a power consumption of 5 kWh/lb would produce acetylene competitively at a power cost of 5.25 mills per kWh.

c. Upgrading of Petroleum Fractions. For some reason, work on the conversion of intermediate weight hydrocarbons in STP devices to acetylene, ethylene, and other unsaturated low molecular weight substances has proceeded mainly in the Soviet Block countries. A rather complete review of the cracking of low grade gasolines is given by Vurzel and Polak (1965a). These same authors (1965b) present some fascinating results using (1) C_4 and lower hydrocarbons and (2) gasolines and b.p. 200°C fractions in a hydrogen plasma. Their yield of unsaturated products is always very high. Especially noteworthy is the datum that when the jet is very heavily loaded with the hydrocarbon (mixed temperature 900°C) the ratio of acetylene to ethylene to propylene is about unity. The high ethylene and propylene yields suggest that under these conditions the primary plasma cracking reaction is followed by an ordinary thermal cracking step. The economic implications of the product stream are at once obvious and exciting.

Fedoseev and Shteiner (1965) report on the production of acetylene and

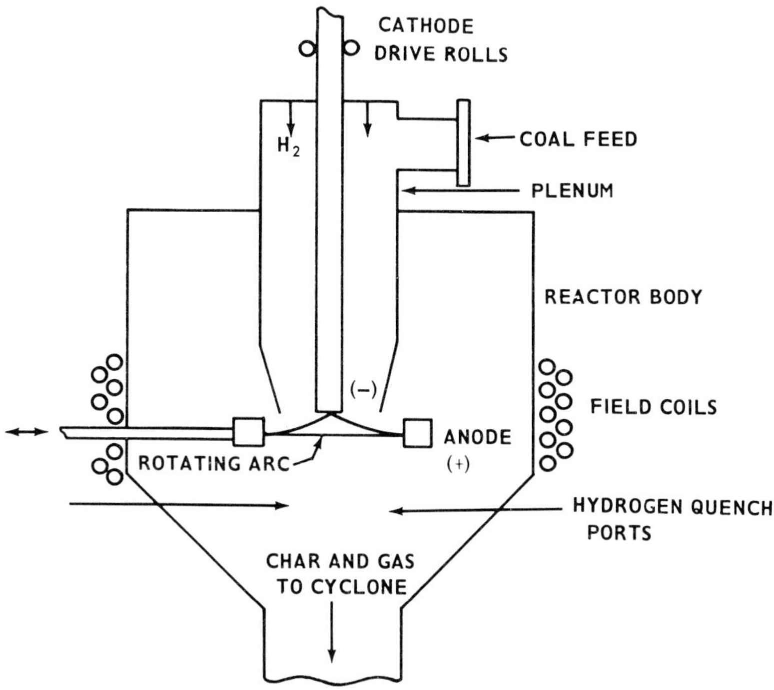

Fig. 11. Diagram of rotating-arc reactor.

ethylene from kerosene by directing an argon plasma jet at the surface of the liquid. The concentration of acetylene in the off gas is 29% by volume (ethylene 10%), which may well be the highest concentration of acetylene observed in any product stream. (As discussed in the above cited AVCO report (1968), this can do much to offset somewhat high electrical costs.) Singularly enough, they report no degradation of the liquid hydrocarbon.

2. *Synthesis of Oxide Materials*

The synthesis of oxide materials in an STP device is a " natural " reaction second only to the formation of acetylene. Presumably, the formation of nitric oxide, obsoleted by the Haber process in World War I, has been laid to rest for the past fifty years [see, however, Timmins and Ammann (1967)], but other oxide products represent products of the very near future, in many cases awaiting only the development of suitably large markets. Here, although one is competing with flame processes, the cost of electricity becomes small because of the high atomic weights involved (Section I,B), and the natural advantages of a plasma process, i.e., controlled atmosphere and temperature history for product formation, become important. With the consumable-anode device (Section I,A,2,*b*) it is possible and economical to completely evaporate an initially refractory material such as sand (Korman and Sheer, 1966) and to recover it as highly upgraded, finely divided (150 Å) silica using an oxygen cross flow for quenching the gaseous silica-containing electrode flame. Not enough work has been published on this technique to show just how much control one has over the product distribution resulting from it. Certainly for smaller or much larger particles the problem would be easier if the investigator had the much more dynamic control of thermal history available in an intermediate section of a plasma jet, say. Patents on the plasma jet synthesis of TiO_2 for pigmentary use ($\approx 0.25\ \mu$) on the one hand (Freeman and Fussman, 1966) and for fine particle titania ($\approx 0.01\ \mu$) on the other (Freeman, 1965) show the extent of control possible. However, although Stokes (1967) and his co-workers have had some success feeding solid reagents to a plasma jet, the extent of reaction has almost always been incomplete, so that for these plasma jet processes one must always start with a gaseous or easily volatilized compound (usually the halide) of the desired metal. This is a serious restriction. Furthermore, because the subsequent reaction with oxygen is exothermic, to effect the sudden quench necessary to get the very finely divided product, the residence time in the reactor at temperatures above the boiling or sublimation point of the product must be long enough for the reaction to go to completion. Because of the unavoidable heat loss, this entails a relatively high electrical consumption.

3. *Metallurgical Applications*

The use of STP devices in metallurgy is obvious, but it is a difficult field to review [see however, Warren and Shimizu, (1965)]. It is rife with rumors, but shows only a few patents and publications, no doubt reflecting the proprietary aspects of the field. One particularly promising process being developed by the Molybdenum Corporation of America (Kruesi, 1968) involves the direct carbo-thermal reduction of molybdenum and columbium oxides to the respective pure metals. This process, covered by one or more of the Sheer-Korman patents, has been known for some time, but many practical problems, especially the recovery of usable products, have only recently been solved.

Along the same lines, the centrifugal plasma furnace (Fig. 12) (Grosse *et al.*, 1965) is of considerable interest. Here the plasma serves as an intense source of heat, which is contained by a surrounding bed of the refractory material in question. Proper design can bring the temperature of the surface to virtually any value (radiation losses are of little importance in this geometry),

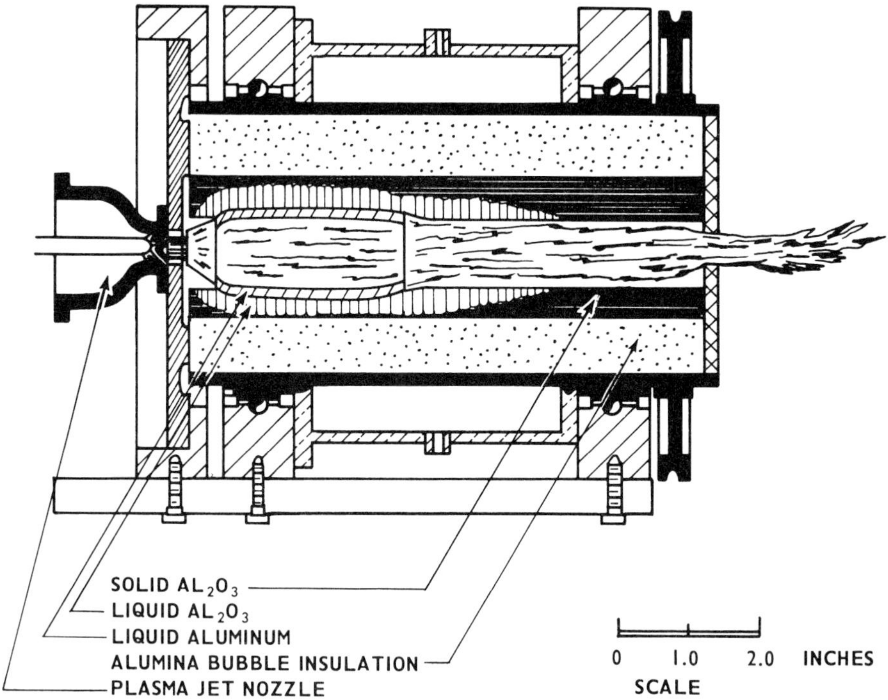

Fig. 12. The centrifugal plasma furnace. (Adapted from Fig. 2 of Grosse *et al.*, 1965, with the permission of the American Society for Testing Materials.)

so this would appear to be a most promising configuration for metallurgical operations. Product recovery should be much less a problem than for the carbothermal process. A patent (U.S. Patent, 1966b) has been issued using this device for the refining of vanadium.

Jones (1967) of M. Jones Associates, Newport Beach, Calif., reports the development of a single-stage iron-ore-to-steel plasma jet process, while W. C. Heraeus G.m.b.H. has patented a process for the purification of molten steel using an argon–hydrogen plasma (German Patent, 1966). An application for another patent has been filed in the Netherlands (1966) for the plasma jet case hardening of steel, while steel has been successfully and economically spheroidized in a transferred arc (TAFA, 1966).

Aside from their obvious applications in cutting and welding, STP devices have found little use in the forming of metals, except for the fabrication of complex shapes of highly refractory metals by means of spray coating suitable forms with very thick coats of the material and then dissolving the form. This procedure, referred to for tungsten and molybdenum by Kirk and Othmer (1967), has been extended to beryllium (Roseberry *et al.*, 1964).

Moss *et al.* (1964) report on an interesting phenomenon often observed but seldom published. They report that in the rapid quenching attendant on plasma jet spraying of metals, metastable phases sometimes form in normally two-phase regions of their phase diagram. A 60% molybdenum–ruthenium alloy formed in this way was observed to exhibit an anomalously high superconductivity transition.

4. *Surface Modification*

Quite apart from plasma spray coating and spheroidizing applications, there has been a limited amount of work on the modification of solids sent through or topically treated with a plasma. One such application is the formation of improved supported catalysts by spraying evaporated metals and refractories onto tumbled carrier pellets. In this way Mo has been supported on steel and chromia on alumina (U.S. Patent, 1966c). Another U.S. patent (1966a) claims to achieve the same end by placing the plasma jet beneath a fluidized bed of the particles.

A less obvious process and one that is closer to commercial realization is the plasma reduction of alluvial sands to unlock previously unrecoverable precious metals (Anon., 1968). This already large scale development work involves the (not very) high temperature partial decomposition in reactive atmospheres of otherwise intractable matrices so that the subsequent leaching operation becomes economically feasible. This is a process particularly well-suited to an rf plasma (Section I,A,2,*e*). The development work is being done for Ionarc Smelters, Ltd. by the TAFA division of Humphries Corp.

5. *Other Product Research*

Because calcium carbide and phosphorus are produced commercially in electric furnaces, it is always tempting to try to produce them more effectively, or at least with different capital equipment costs, using an STP device. The reactions have been made to proceed in various plasma devices (Goldberger, 1964) using elemental carbon as the reducing agent, and the Battelle Development Corporation is currently attempting to license yet another process that shows considerable promise for the production of phosphorus on an intermediate industrial scale.

Halogen chemistry in STP devices is already important because of the widespread use of volatile metal salts in syntheses, but research centered on the halogen-containing products is only in its infancy. The various chlorofluorocarbon products produced when chlorine- and fluorine-substituted methanes are decomposed in an rf argon plasma are reported by Beguin *et al.* (1967) to be mainly the benzenic derivatives, whereas cracking CCl_4 in a dc plasma, say, yields mainly C_2Cl_6 (Salvemini, 1966). Bronfin (1967) summarizes work on fluorocarbon decomposition and reactions rather exhaustively. Baddour and Bronfin (1965) report the carbothermal reduction of CF_4 to the important monomer C_2F_4 in very significant yields. Their quench was done by means of a fine probe, but even so, the survival of the unsaturated product is certainly surprising. (Note that numerous patents have been issued to DuPont for similar syntheses.) Several other reports on halogen chemistry are scattered through the literature, but they are adequately reviewed in the above cited literature.

Combustion processes for producing the very large amounts of high grade carbon black used in the tire industry are apparently limited in the quality of carbon they can produce by the temperature of the combustion process (Witt, 1967). It would be surprising if work were not currently in progress in the carbon black companies to supplement the combustion with additional heat from an STP device so as to bypass that problem. The Cabot corporation, for example, apparently already possesses the know-how, judging from their French Patent (1965) which deals with the production of fully amorphous carbon blacks by the decomposition of perhalocarbons in a plasma device.

Finally, with a total power dissipation of about 9 kWh/lb, Stokes and Streng (1965) have succeeded in producing ozone by quenching a helium plasma jet with liquid oxygen. This remarkable synthesis is a landmark, not so much because it is economically attractive, but because it stands as a warning to those who would try to justify STP chemistry on the basis of thermodynamic calculations. This widespread and ever more popular practice can only serve to discourage the imaginative investigator, while the writer, for one, has never been convinced the predictions thus made (actually often explanations after the fact) were more than coincidentally true. That is, the reactions of interest are almost inevitably favored by high temperature, so that if thermodynamic calculations are carried out for a sufficiently high temperature, the " predicted " yield can always be made to equal or exceed that observed. It is not clear that one has learned much in the process, however.

B. Mechanism Studies

Anyone starting out to do chemistry in an STP device is confronted with several questions. First, does the reaction go at all? Second, what reactor configuration is optimum for production of the desired product in its most useful or valuable form? Assuming he finds answers to these first two questions, he must then consider the problems attendant on scaling the process up, first to pilot plant and finally to full scale operation. To answer these questions he must have some understanding of the underlying principles of operation of these devices. This has given rise to several studies of greater or lesser sophistication that provide some guidance and hopefully in the near future will provide much more. The day is probably not far off when exploratory thermal plasma chemistry can safely be performed on a computer. In this it differs markedly from chemistry in other discharge devices, which are generally still very much in the " black box " category.

1. *Methane Decomposition*

There are three explanations for the observed yield of acetylene when methane is introduced into a hydrogen plasma. First, the ingredients are well mixed and form an equilibrium mixture, which is frozen in its high temperature composition in the quenching step (Melamed *et al.*, 1965). Second, the methane and hydrogen mix immediately, and the amount of acetylene formed is determined by the rate of decomposition of the methane and the reactor temperature history, terminated by a sudden quench (Anderson and Case, 1962). Third, the reaction is essentially instantaneous at the plasma temperatures, and the controlling factor is the rate of (spatially resolved) mixing of the methane with the plasma—that is, the degree of conversion depends on the fraction of methane introduced that mixes with the plasma before the mean temperature of the resulting mixture and unmixed methane falls below some freezing temperature (Freeman and Skrivan, 1962). It is clear that the truth of the matter is some mixture of these three extreme points of view, perhaps with other factors as well, but the question remains: Which of these interpretations can be used most effectively to model the reaction for practical purposes of scale-up and for the prediction of new chemical syntheses? Or does each viewpoint have its own range of validity in the type of reactor chosen for the study? Certainly Reed (1967) is inclined to this latter view, so the three models perhaps can be best accepted as guidelines for pragmatic use depending on the particular " boundary conditions " being considered.

a. Mixing Kinetic Model. The apparatus used in the mixing kinetic model study (Freeman and Skrivan, 1962) is schematically shown in Fig. 13. Under these well-defined conditions, with only very small amounts of methane added to an argon plasma, the sensible heat flowing in the plasma is well-known at every point (Section I,A,3,*b*,*i*) and the analysis is straightforward.

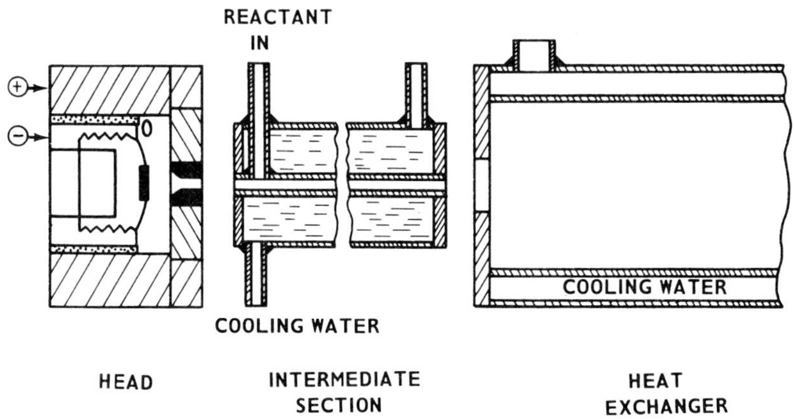

Fig. 13. Apparatus used in mixing kinetic model study of methane and ammonia decomposition.

Mixing is presumed to occur by diffusion from a concurrent cold sheath flow into the hot center, where reaction is assumed to occur instantaneously and is complete. This slow mixing process is assumed to continue until the heat available in some cross section, divided by the combined heat capacities in that cross section, falls below some freezing temperature, an adjustable parameter of the model. The conversion of methane stops at that point. The data correlate very well over the entire range of heat flows available. The freezing temperature for methane at 300 Torr total pressure is a very reasonable 1150°C.

There can be little doubt that for the experiment described, this is the correct model. The equilibrium hypothesis can be completely disregarded here, because exactly the same cracking behavior was observed for ammonia in the same reactor at two different pressures. The correct behavior is in fact more closely adhered to by the ammonia than by the methane (analytical problems are less severe). Ammonia is completely unstable at all attainable temperatures in the plasma jet, so that equilibrium simply cannot come into the picture. On equally sure grounds we can disregard the reaction kinetic model here. This was the first model the investigators tried, and the resulting activation energies, a parameter of the model, were an order of magnitude too low, as were the reaction rates. This, in fact, is what first suggested that the process was really diffusion controlled.

b. Reaction Kinetic Model. The quite different apparatus used for the reaction kinetic model study of Anderson and Case (1962) is shown in Fig. 14. We see significant differences from the preceding experiment in the

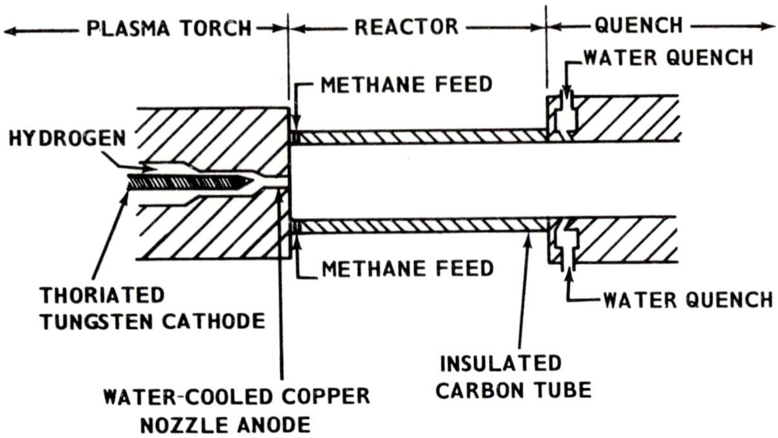

Fig. 14. Apparatus used in reaction kinetic model study of methane decomposition.

mismatch of diameters at the head–intermediate section interface, the hot-walled intermediate section with high volume to surface ratio, the hydrogen plasma, the much heavier loading of methane, and the water quench. Disregarding for the moment the rather higher conversions of methane than predicted by their model at low power levels, one can scarcely argue with their interpretation of the data.

As an added note of confusion, Ammann *et al.* (1968) have performed a careful experiment intermediate between the two reported above in the hope of resolving what appeared to be an unreconcilable conflict. Predictably they find that their data can be made to fit either model, but with substantial strain of the adjustable parameters in either case. Of the two, the data seem to fit the mixing kinetic model best, especially at high power levels, but require a somewhat high freezing temperature.

c. Equilibrium Model. Unfortunately the writer's command of Russian precludes a critical analysis of the equilibrium model (Melamed *et al.*, 1965). Although the studies cited above precluded this model on the basis of the product distribution observed (no elemental carbon or ethylene was formed), it is altogether possible that for a sufficiently large scale, long residence time reactor this could be the controlling factor. (Note in Table I that the residence time necessarily increases on scale-up.) Although their sampling procedures are suspect, the limited data reported by the Westinghouse engineers for their multimegawatt ac torch operating on pure methane indicate that this may be the case (Hirayama and Maniero, 1967).

2. HCN *Synthesis*

The synthesis of HCN from methane added to a nitrogen plasma jet captures the imagination, for it proceeds in good, if in somewhat economically unattractive yield, and at the same time this reaction is not known to proceed under any circumstances in conventional chemical reactors. The reagents are easy to handle, and at least the product HCN is easy to determine quantitatively. Furthermore, the dependence of the yield on operating parameters is simple and reproducible. It is clear that this reaction is a most attractive key for revealing the distinctive features of STP chemistry. One might even hope to further understand basic chemical mechanisms by studying this reaction, because of the hitherto unattainable high temperature regime and because of the fact that the chemical reactions, whatever they might be, are spatially resolved in the jet where characteristic times are on the order of a few microseconds per millimeter.

Although others have worked on this problem [cf. Leutner (1962, 1963)], probably the most substantial body of quantitative experimental data on this system has been assembled by the writer and his colleagues (Freeman, 1968a; Freeman and Mentzer, 1969) in reactor configurations schematically represented in Fig. 15. The variables studied thus far are methane-to-nitrogen ratio, net heat in plasma, intermediate section length (2–32 diameters), nitrogen flow rate, total pressure, and composition dependence when the nitrogen plasma is partly replaced by argon. The following facts are now known about the reaction:

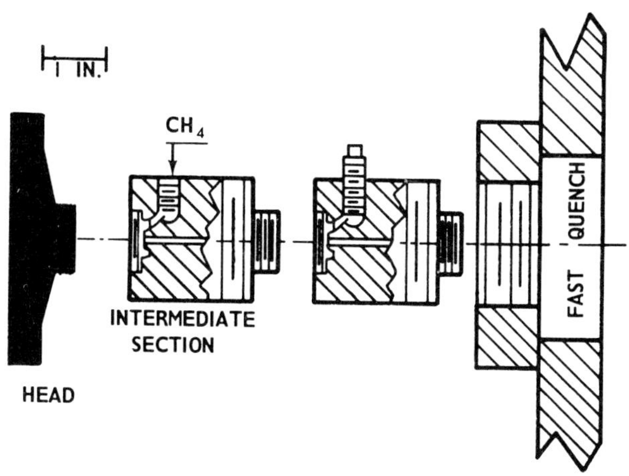

Fig. 15. Apparatus used in the production of HCN from methane in a nitrogen plasma jet.

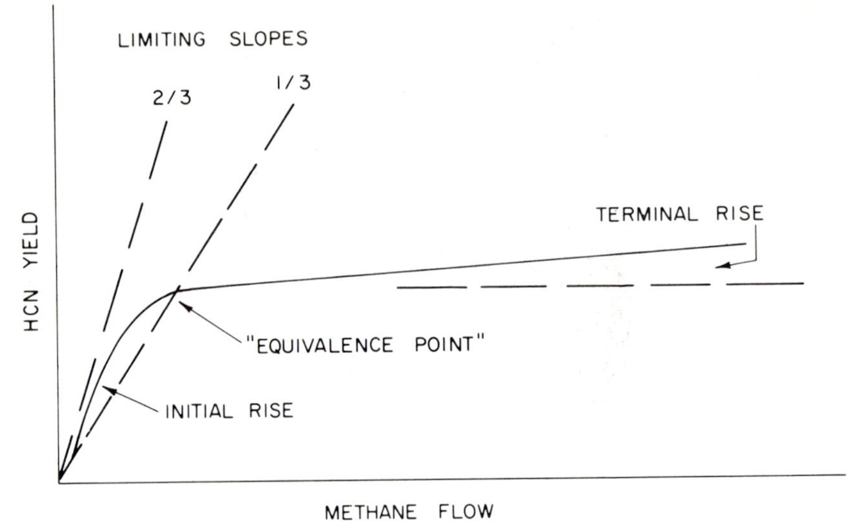

Fig. 16. Typical titration curve.

(1) A " titration curve " of the type shown in Fig. 16 is observed under all conditions, distinguished by an initial rise region, a break or " equivalence point" that always seems to occur on the line that denotes that one third of the available methane is being converted to HCN, and a plateau region with an indefinitely continuing low rate of rise. No decrease with increasing methane is ever observed.

(2) The amount of HCN produced in the plateau region (measured in every case at $CH_4 : N_2 = \frac{1}{2}$) is a simple linear function of net heat flow in the plasma (Fig. 17).

(3) The amount of HCN produced is independent of reactor length.

(4) The amount of HCN produced depends only on the heat flow at the point of mixing and not at all on where the methane is introduced (Fig. 18).

(5) As argon is substituted for nitrogen, the absolute production rate of HCN is, in first approximation, a constant (Fig. 19).

There are, unfortunately, two equally appealing explanations that would appear to explain the quantitative results. The first explanation rests on the assumption that the attainment of equilibrium is very slow in this reaction, so that only some of the reaction paths leading to the formation of HCN or its precursors are active, utilizing species present in trace amounts in a conventional or shock tube reactor. We are thus led to the concept of titrating such species present in the plasma prior to mixing. Utilizing the two-step

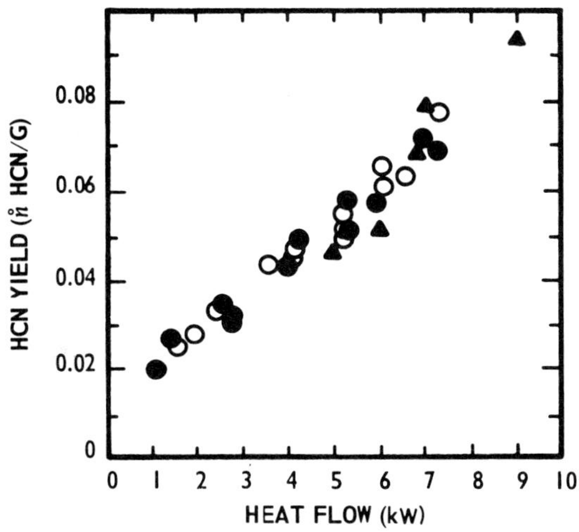

Fig. 17. Dependence of the plateau value of HCN production rate $\mathring{\eta}$ (normalized to 0.0171 g-moles/sec$=G$, the most common flow rate of nitrogen) on heat flow in the plasma. 2-in. reactor ●, 8-in. reactor ○, $\frac{1}{4}$-in. reactor ▲.

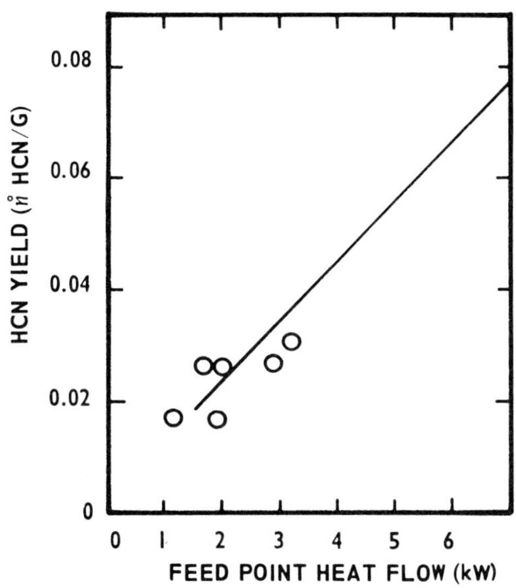

Fig. 18. Dependence of the plateau value of normalized HCN production on heat flow at the 6-in. point in the 8-in. reactor. Methane is introduced at the 6-in. point. The line is the best line through data using heat flow and methane introduction at the reactor entrance.

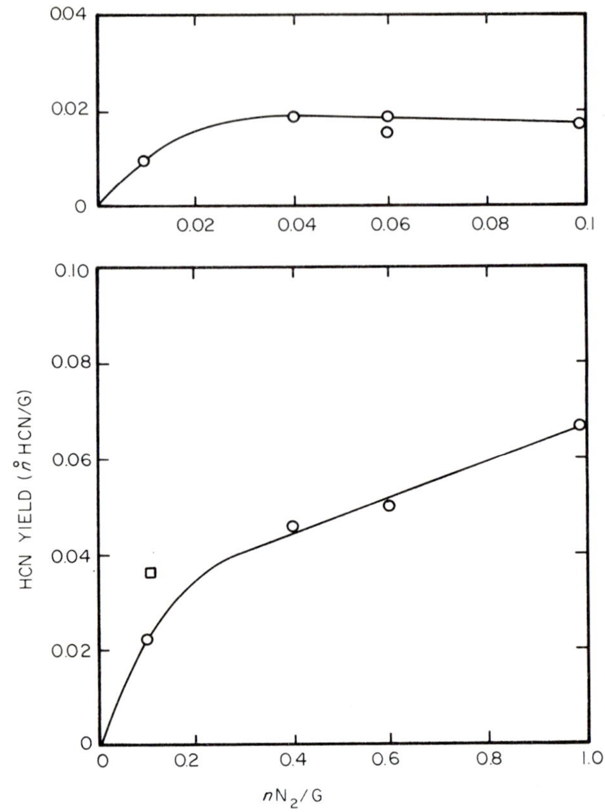

Fig. 19. HCN normalized plateau production rate versus mole fraction nitrogen in the nitrogen–argon jet at two net power levels; upper figure 1.5 kW, lower figure 6 kW.

model (Section I,A,3,*b*) (Freeman and Mentzer, 1969) in a rather long extrapolation from the atmospheric argon jet for which it was formulated (Freeman and Longfield, 1968), it is possible to quantitatively account for every observation above, but only if each nitrogen ion reacts with carbon fragments from the decomposition of the methane to make two molecules of HCN. This is quite reasonable because of the large excess of nitrogen atoms. One might postulate the following scheme:

$$N^+ + N + M = N_2^+ + M \quad \text{rapid}$$

or

$$N^+ + N_2 = N_2^+ + N$$

followed by

$$N_2^+ + C_2 = N—C{\equiv}C—N^+ \text{ rapid}$$
$$N—C{\equiv}C—N^+ + e^- = 2CN \text{ slow}$$

and finally in the quench

$$H + CN = HCN \text{ slow}$$

The second explanation (Bronfin and Cohen, 1969) is, from some viewpoints, more attractive because of its greater predictive value. It apparently accounts for the data almost as well and is based on a mixing kinetic model with local thermodynamic equilibrium throughout. The reactor is divided into thousands of zones, and difference equations are set up to calculate temperature and composition from thermodynamic and diffusion relations at every point. Eddy diffusivity enters as an adjustable parameter, as does a freezing temperature below which it is assumed that no more CN, the assumed precursor of HCN, can form.

Note that there is no possibility that these two explanations are merely two viewpoints of the same phenomenon, because the former is a frozen flow model requiring the " freezing in " of the ion concentration for long periods of time, while the latter assumes at all times complete local thermodynamic equilibrium (except that to conform to experiment, solids formation is suppressed). Rao *et al.* (1965, 1967) have measured reaction rates for this reaction in a shock tube and have found a very slow reaction rate indeed, as compared to jet residence times, which would tend to favor the titration or frozen flow model. But these workers are not at all satisfied with their results, and so one cannot decide on the basis of their work alone. It is clear, however, that this reaction is on the verge of being very well understood, awaiting only the answer to the one vital, well-defined, and answerable question of the reaction kinetics.

3. *Other Studies*

There have been no other studies in STP devices to compare with those cited above, although a study on the thin-probe quenching of nitric oxide formed in a cascade arc is worth noting (Ammann and Timmins, 1966; Timmins and Ammann, 1967). The cascade arc is not streaming, but the probe study is nonetheless interesting.

C. Diagnostics

It is difficult to intelligently design experiments of any sophistication in thermal plasmas, because of the high sensible heat flow and the fact that virtually nothing is known about flow properties in strongly nonisothermal situations. Because of the inescapably cold surroundings, the latter, especially, is a ubiquitous factor in research on all of these devices. The flow is complicated further by the arc process originating the plasma. Quite generally, the techniques used to monitor the chemically significant parameters in other situations are of little or no use in a plasma reactor.

The first information that the experimenter needs for both first order scientific work and for economic considerations is how much heat is flowing in the plasma, and maybe some measure of flow velocity (Section II,C,1). For refined work he will want to know something about radial heat flux, composition, and velocity profiles (Section II,C,2) so that he can determine total flows of interesting species and perhaps so that he can determine transport properties. With this information he can formulate mixing and quenching calculations. To the extent that the arc process itself interacts with the flow, special diagnostic techniques for the arc process are also mentioned (Section II,C,3), and finally, a few attempts have been made to directly determine the high temperature species that make up the plasma flow; these too merit attention (Section II,C,4).

1. *Average Properties*

The total mass of reagents and plasma flowing in a confined reactor, and consequently the average atomic composition, can of course be readily determined by conservation of the properly metered input flows. Total heat flow is nearly as easy, but requires some planning (Section II,C,1,*a*). If the composition of the plasma is sufficiently simple, given heat flow and atom flow one can calculate an average velocity, a quantity that is difficult to measure directly (Section II,C,1,*b*).

a. Heat Flow. Unless a plasma is loaded with solids, the amount of heat it radiates is usually inconsequential. (This may not always be true for a radio frequency plasma.) One may therefore get a measure of the heat flowing in the plasma itself from a knowledge of the easily measurable electrical energy expended in the arc process and the amount of heat carried away by the cooling water almost always employed to protect the arc device. (Where cooling water is unnecessary and/or it is desired to include the heat lost by radiation, the entire device may be surrounded by a water-cooled enclosure.) Thus it is only necessary to monitor current, voltage, phase angle when ac is used, and the flow rate and temperature rise of the cooling water. The difference is the total heat flowing in the streaming plasma itself. Measurement to 5% accuracy is easy. The writer has in fact monitored this information with a small homemade analog computer (Fig. 20) for some time. It gives a continuous output on a strip chart recorder of net heat

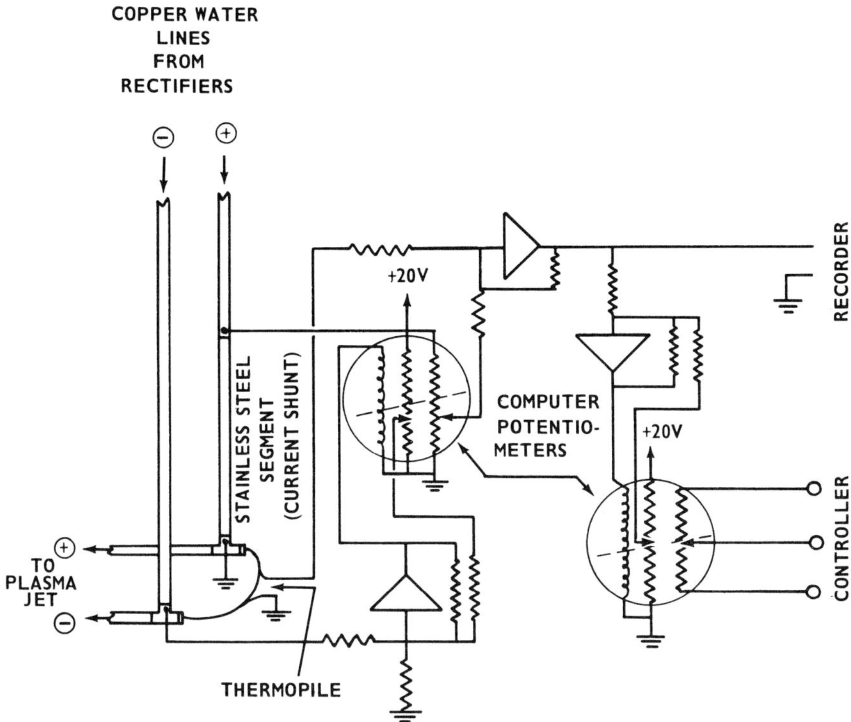

Fig. 20. Heat flow computer.

flowing in the plasma, and if desired it directly controls the rectifiers to keep this quantity precisely constant. Such a device is definitely to be recommended for careful work.

If the plasma flows down a water-cooled tube that is segmented and suitably instrumented, further heat balances can be carried out as a function of axial position and axial heat flow profiles can be constructed. Skrivan and Von Jaskowsky (1965) have done this for tubes of several diameters with several gases and have correlated their results (Section I,A,3,*b*,*i*).

Another way to measure heat flow that requires prior calibration is to measure the average electrical conductivity of a well-defined portion of the plasma stream. It can readily be done for a confined jet using ablating cylindrical probes for electrical contact (Freeman, 1962), while for an open jet plume, radio frequency (Sprouse, 1965) and microwave (Kelly, 1965) measurements show some promise. Note, however, that as yet none of these conductivity measurement techniques have been exploited by chemists.

A magnetic induction method that seems promising for routine diagnostics is being developed by Burkhardt and Schnell (1966). They hope to extend the method eventually to velocity profile measurement.

Because the atom flow is known, one approach to the measurement of average velocity is through thrust measurements (Bryson and Fröhlich, 1966). The technology for this is probably a bit too involved to be attractive to the chemical investigator, however.

In developed flow conditions the center-line velocity is related in a simple reproducible way to the average velocity, so methods of measuring center-line velocity should be considered in the same category. Freeman *et al.* (1962, 1966) and Cabannes *et al.* (1965) have simply used separated photomultiplier stations to measure the time required for naturally occurring luminosity fluctuations in the jet center to travel between stations. Ahlborn (1965) has achieved the same end by momentarily short-circuiting the arc supply with a capacitor and measuring the time required for the resultant dark region to travel between photomultiplier stations. A similar somewhat simpler method that, however, requires a substantial number of corrections for quantitative work involves injecting refractory particles in the plasma and following their movement (Seymour, 1964).

Gottschlich *et al.* (1966) and Baker (1964) have calculated peak jet velocities from spectroscopically measured temperature profiles and various conservation equations.

2. *Radial Profile Measurements*

The obvious refinement of the observations in the preceding section necessary for detailed quantitative calculations is to obtain, as functions of the radius, the temperature (Section II,C,2,*a*), heat flow and atomic composition (Section II,C,2,*b*), plasma density (Section II,C,2,*c*) and velocity (Section II,C,2,*d*).

a. Temperature Measurement. Because of the luminosity of the plasmas, the obvious approach to their diagnostics is through quantitative emission spectroscopy. The spectroscopic diagnostics of arcs is a well-known art for which we refer the reader to the standard works (Griem, 1964; Unsöld, 1955). The writer has, in fact, recently published a thoroughly pragmatic introduction to the topic and review of the more successful methods elsewhere (Freeman, 1969a). The fact is, however, that this process is very demanding, and it is only in the last few years that really credible work has been done on streaming plasmas. For a number of reasons (Freeman, 1968a) streaming plasmas are sufficiently more difficult to diagnose than stationary arcs that with the exception of the argon plasma, they have remained quantitatively intractable to the present time. Note that until recently it was not even known whether plasma jets of simple materials are in LTE. Measurements have now been refined to the point that we know atmospheric pressure argon is in LTE, but only in the center of the stream. When the plasma density falls below about 10^{15} cm^{-3}, apparent discrepancies show up between electron excitation temperatures and those inferred from the absolute density of emitters were they in equilibrium (Bott, 1966). For similar work on expanded supersonic jets the reader is referred to the work of McGregor and Brewer (1965, 1968).

Beside their general difficulty, spectroscopic diagnostic techniques have another problem in practice—namely, that jet luminosity falls off very rapidly as the wall is approached, so that actually the bulk of the flow cannot be seen unless great care is used. Another related approach that does cover the entire jet has been tried by Lamb and Schreiber (1966). It involves a measurement of the change of index of refraction over the cross section by means of a differential interferometer. By no means a routine measurement as yet, it nevertheless shows promise.

b. Heat Flow and Composition Profiles. Although numerous attempts have been made to develop calorimetric probes for STP diagnostics, the most significantly successful are those developed by Grey and co-workers. They are now being produced commercially by the Greyrad Corporation, Princeton, N. J. For an introduction to probe diagnostics and an excellent collection of references the reader is referred to a review by Grey (1965). Figure 21 shows one of Greyrad's latest model calorimetric probes capable of withstanding the maximum heat flux of a plasma jet. The $\frac{1}{16}$-in. o.d. permits sampling the flow as a function of position, while the double water jacket permits calorimetric measurements on a sample of plasma withdrawn through the tip from the undisturbed flow. The radial heat flux profile follows directly from the measurement. The cool gas emerging from the probe may be analyzed for atomic composition, although, except for the simplest mixtures, this will not reflect molecular composition in the plasma itself because of molecular rearrangement in the quenching process, albeit the quench is extremely rapid in this size probe (Erbe *et al.*, 1948). Further, by arresting the gas flow, it is possible to measure a stagnation pressure for the flow so that this type probe may also be used to measure a detailed velocity profile.

c. Plasma Density Measurements. The radial distribution of plasma density is another datum of some interest. At least for argon, this is the

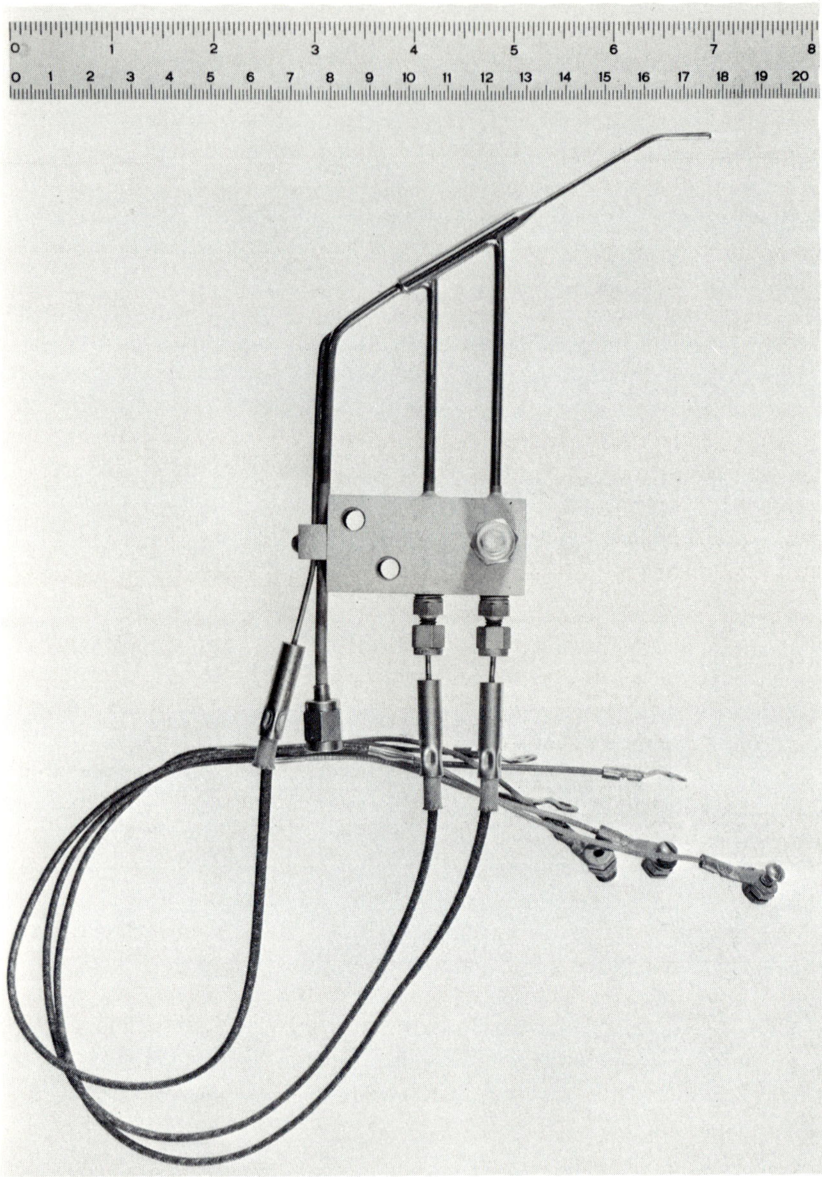

Fig. 21. The calorimetric probe. (Courtesy of Greyrad Corporation, Princeton, New Jersey.)

easiest quantity to measure spectroscopically, as it requires only measurement of continuum intensity over the cross section (Freeman, 1969a). A number of workers have carried through such measurements satisfactorily. However, for workers equipped to use probes it is probably easier to use the so-called electrostatic probe after the manner of Cabannes and Joubert (1965) and of Grey and Jacobs (1966).

d. Radial Velocity Profiles. In Section II,C,2,*b* a probe method of establishing radial velocity profiles was mentioned. Two methods of optically obtaining the same information are worth mentioning for the sheer experimental virtuosity involved. Chen (1966) has focused the beam of a giant-pulse ruby laser down to a spot in the plasma, causing a bright spark to form. The attendant shock waves dissipate immediately and the spark travels in the undisturbed flow. Its progress can be followed with a streak camera or with electrostatic probes. The other method ingeniously contrived for determining the velocity profile in open jets is that of Salzmann (1966). He rapidly swings a taut transverse wire through the plasma. By means of a synchronized electric pulse he explosively evaporates material from the surface of the wire just as it reaches dead center, thus releasing a fine line of easily seen metal ions. The deformation of the line as it travels with the jet gives the detailed information desired.

3. *Arc Diagnostics*

Diagnostic methods used on the arc processes giving rise to streaming plasmas involve, in addition to the optical methods used on the jet itself, the measurement of electric field determined from the voltage drop between consecutive segments in a segmented anode, and visual observation of the arc luminosity under special experimental arrangements. Of special interest from a diagnostic point of view is the excellent work performed by Pfender and co-workers [(1964–1966) see also Pfender and Cremers (1965), Pfender and Eckert (1965)] at the University of Minnesota.

4. *Chemical Diagnostics*

Whereas most workers have been content to accept whatever products emerge from their quenching processes, a very few have tried to gain some insight into the chemical species actually involved in the high temperature process. The methods used are time-of-flight mass spectroscopy (O'Halloren *et al.*, 1964) (a method of great promise, but thus far of unfortunately limited development), fine-probe quenching from the plasma [e.g., Bronfin and Hazlett (1966); Ammann and Timmins (1966)] and spectroscopic identification of species (Kana'an *et al.*, 1966). This last approach has interesting potential, but it is hard to make quantitative, and in the writer's opinion the approach cited here lacks the necessary sophistication. Indeed, one can argue that the really reactive species will not be seen at all, for they will be consumed as rapidly as they are formed, so that it may be that one can see only species diverted from the main stream of chemistry. The probe studies are of interest, but here we have for streaming plasmas the unknown stagnation effects at the entrance to the probe [e.g., Reis and Fenn (1963)] that complicate interpretation of the data.

In the writer's opinion, all of these pioneering attempts await only the proper development to become useful adjuncts to chemistry. The cited probe studies, intelligently interpreted as they were, have already contributed much to our understanding of plasma processes.

REFERENCES

Ahlborn, B. (1965). *Z. Naturforsch.* **20a**, 466.

Ammann, P. R., and Timmins, R. S. (1966). *A.I.Ch.E. J.* **12**, 956.

Ammann, P. R., Timmins, R. S., and Krukonis, V. (1968). *Am. Chem. Soc. Div. Fuel Chem. Preprints* **11**, (2), 364.

Anderson, J. E., and Case, L. K. (1962). *Ind. Eng. Chem. Process Design Develop.* **1**, 161.

Anon. (1968). *Eng. Mining J.* **169**, 13.

Aust, T., Ladner, W. R., and McConnell, G. I. T. " The Decomposition of Coal at High Temperatures." To be published.

AVCO (1968). OCR Contract No. 14-01-0001-493, R&D Rep. No. 34.

Baddour, R. F., and Bronfin, B. R. (1965). *Ind. Eng. Chem. Process Design Develop.* **4**, 162.

Baddour, R. F., and Timmins, R. S., eds. (1967). " The Application of Plasmas to Chemical Processing." M.I.T. Press, Cambridge, Massachusetts.

Baker, S. C. (1964). *J. Opt. Soc. Am.* **54**, 1204.

Bauder, U. (1968). Personal communication.

Beguin, C. P., Ezell, J. P., Salvemini, A., Thompson, J. C., Vickroy, D. G., and Margrave, J. L. (1967). *In* " The Application of Plasmas to Chemical Processing " (R. F. Baddour and R. S. Timmins, eds.), p. 35. M.I.T. Press, Cambridge, Massachusetts.

Bond, R. L., Galbraith, I. F., Ladner, W. R., and McConnell, G. I. T. (1963). *Nature* **200**, 1313.

Bond, R. L., Ladner, W. R., and McConnell, G. I. T. (1966). *Coal Sci.* **41**, 650.

Bott, J. F. (1966). *Phys. Fluids* **9**, 1540.

Bray, K. N. C. (1959). *J. Fluid Mech.* **6**, 1.

Bronfin, B. R. (1967). *In* " The Application of Plasmas to Chemical Processing " (R. F. Baddour and R. S. Timmins, ed.), p. 157. M.I.T. Press, Cambridge, Massachusetts.

Bronfin, B. R., and Cohen, L. S. (1969). *Proc. 3rd Inter. Symp. High Temp. Technol. Asilomar Conf. Grounds, Calif.*, 1967. Butterworths, London and Toronto.

Bronfin, B. R., and Hazlett, R. N. (1966). *Ind. Eng. Chem. Fundamentals* **5**, 472.

Bryson, R. J., and Fröhlich, J. P. (1966). AEDC-TR-65-268.

Buettner, G. (1965). *Schweisstech. (Vienna)* **15**, 171.
Burhorn, F., Maecker, H., and Peters, T. (1951). *Z. Physik* **131**, 28.
Burkhardt, H., and Schnell, G. (1966). *Z, Naturforsch.* **21a**, 485.
Cabannes, F., and Joubert, F. (1965). *J. Phys. (Paris)* **26**, 466.
Cabannes, F., Chapelle, J., Decroisette, M., and Sy, A. (1965). *Rev. Intern. Hautes Temp. Refract.* **2**, 297.
Chase, J. D. (1969). *J. Appl. Phys.* **40**, 318.
Chen, C. J. (1966). *J. Appl. Phys.* **37**, 3092.
Cohen, L. S. (1966). *AIAA Preprint* 66–617.
Dean, Jr., R. C., and Runstadler, P. W. (1966). *AIAA Preprint* 66–162.
DuPont (anon.) (1964). *Chem. Week* **94**, 64.
Eckert, H. U. (1968). *AIAA Preprint* 68–711.
Erbe, P. E., Grey, J. T., and Beal, J. L. (1948). SQUID Tech. Memo CAL-8, No. DD-420-4-9.
Eubank, P. T., and Johnson, J. R. (1967). *Symp. High Temp. Technol., 60th Ann. A.I.Ch.E. Meeting, New York, November* 1967.
Fedoseev, S. D., and Shteiner, E. V. (1965). *Tr. Mosk. Khim. Tekhnol. Inst.* **48**, 172.
Finkelnburg, W., and Maecker, H. (1956). *In* " Handbuch der Physik " (S. Flügge, ed.), Vol. XXII, p. 254. Springer, Berlin.
Freeman, M. P. (1962). *In* " Temperature—Its Measurement and Control in Science and Industry," Vol. 3, Pt. 2, p. 969. Reinhold, New York.
Freeman, M. P. (1965). Belgian Patent No. 651,280.
Freeman, M. P. (1968a). Chemical reactions in electrical discharges. *Advan. Chem. Ser.* **80**, 406.
Freeman, M. P. (1968b). *J. Quant. Spectry. Radiative Transfer* **8**, 435.
Freeman, M. P. (1969a). Introduction to the theory and practice of spectroscopic diagnostics of high density plasma sources of aerospace and chemical synthetic interest. *Progr. High Temp. Phys. Chem.* **3**, p. 255.
Freeman, M. P. (1969b). *Proc. 3rd Intern. Symp. High Temp. Technol. Asilomar Conf. Grounds, Calif.*, 1967. Butterworths, London and Toronto.
Freeman, M. P., and Chase, J. D. (1968). *J. Appl. Phys.* **39**, 180.
Freeman, M. P., and Fussman, F. (1966). U. S. Patent No. 3,275,411.
Freeman, M. P., and Longfield, J. E. (1968). *Ind. Eng. Chem. Fundamentals* **7**, 640.
Freeman, M. P., and Mentzer, C. C. (1969). Production of hydrogen cyanide from methane in a nitrogen plasma jet. *Ind. Eng. Chem. Process Design Develop.* In press.
Freeman, M. P., and Skrivan, J. F. (1962). *A.I.Ch.E. J.* **8**, 450.
Freeman, M. P., Li, S. U., and Von Jaskowsky, W. (1962). *J. Appl. Phys.* **33**, 2845.
Freeman, M. P., Li, S. U., and Von Jaskowsky, W. (1966). *J. Appl. Phys.* **37**, 448.
French Patent (1965). No. 1,425,828.
Galyaev, V. G., Melamed, V. G., Polak, L. S., and Khait, Y. L. (1965). *In* " Kinetika i Termodinamika Khimicheskikh Reaktsii v Nizkotemperaturnoi Plasme " (L. S. Polak, ed.), p. 60. Inst. Neftekhim. Sinteza Akad. Nauk SSR, Moscow.
German Patent (1966). No. 1,217,986, assigned to W. C. Heraeus G.m.b.H.
Gibson, J. O., and Weidman, R. (1963). *Chem. Eng. Progr.* **59**, 53.
Gladish, H. (1967). Personal communication.
Goldberger, W. M. (1964). *Ann. Meeting, 57th, Boston*, Massachusetts, December 1964. *A.I.Ch.E. Preprint.*
Goldberger, W. M. (1966). *Chem. Eng.* **73**, No. 6, 173; No. 7, 125.

Gottschlich, C. F., Enright, J. A., and Cudek, F. C. (1966). AIAA J. **4**, 1085.
Grey, J. (1965). *ISA* (*Instr. Soc. Am.*) Trans. **4**, 102.
Grey, J., and Jacobs, P. F. (1964). AIAA J. **2**, 433.
Grey, J., and Jacobs, P. F. (1966). AIAA Preprint 66–73.
Griem, H. R. (1964). " Plasma Spectroscopy." McGraw-Hill, New York.
Grosse, A. V., McGonigal, P. J., Murphy, W. J., and Stokes, C. S. (1965). *Mater. Res. Std*. **5**, 173.
Happel, J., and Kramer, L. (1967). *Ind. Eng. Chem*. **59**, 39.
Harnish, H., Heymer, G., and Schullus, E. (1963). *Chem. Ing. Tech*. **35**, 7.
Hirayama, C., and Maniero, D. A. (1967). Am. Chem. Soc. Div. Fuel Chem., Preprints 11, No. 2, 470.
Hulburt, H. M., and Freeman, M. P. (1963). *Trans. N. Y. Acad. Sci., Ser*. II 25, 770.
Incropera, F., and Lippert, G. (1966). AIAA J. **4**, 1087.
Jacobs, P. F. (1962). ARPA Order No. 206-61, AEL Rep. No. 625. Princeton Univ., Princeton, New Jersey.
Jones, M. (1967). Personal communication.
Kana'an, A. S., Beguin, C. P., and Margrave, J. L. (1966). *Appl. Spectry*. **20**, 18.
Kawa, W., Graves, R. D., and Hiteshue, R. W. (1966). *U.S. Bur. Mines Rep. Invest*. **6829**.
Kawana, Y., Makino, M., and Kimura, T. (1967). *Intern. Chem. Eng*. **7**, 359.
Kelly, A. J. (1965). AIAA J. **3**, 372.
Khait, Y. L. (1965). *In* " Kinetika i Termodinamika Khimicheskikh Reaktsii v Nizkotemperaturnoi Plazme " (L. S. Polak, ed.), p. 167. Inst. Neftekhim. Sinteza Akad. Nauk SSR, Moscow.
Kirk, R. E., and Othmer, D. F., eds. (1963). " Encyclopedia of Chemical Technology," 2nd ed., Vol. I, p. 180. Wiley (Interscience), New York.
Kirk, R. E., and Othmer, D. F., eds. (1967). " Encyclopedia of Chemical Technology," 2nd ed., Vol. 13, p. 271. Wiley (Interscience), New York.
Korman, S. (1964). *Intern. Sci. Technol*. p. 90.
Korman, S., and Sheer, C. (1966) Electrothermics and Met. Div. Abstr. 4, p. 62. Electrochem. Soc. Philadelphia.
Kruesi, P. (1968). Personal communication.
Krukonis, V. (1968). OCR Contract No. 14-01-0001-493, R&D Rept. No. 34. AVCO.
Lamb, J. D., and Schreiber, P. W. (1966). ARL-66-0222.
Leutner, H. W. (1962). *Ind. Eng. Chem. Process Design Develop*. **1**, 166.
Leutner, H. W. (1963). *Ind. Eng. Chem. Process Design Develop*. **2**, 315.
McGregor, W. K., and Brewer, L. E. (1965). AEDC-TR-65-131.
McGregor, W. K., and Brewer, L. E. (1968). AEDC-TR-68-187.
Maecker, H. (1959). *Z. Physik* **157**, 1.
Maniero, D. A., Kienst, P. F., and Hirayama, C. (1966). *Westinghouse Engr*. **26**, 66.
Melamed, V. G., Mukhtarova, T. A., Polak, L. S., and Khait, Y. L. (1965). *In* " Kinetika i Termodinamika Khimicheskikh Reaktsii v Nizkotemperaturnoi Plazme" (L. S. Polak ed.), p. 12, Inst. Neftekhim. Sinteza Akad. Nauk SSR.
Meyer, H. (1964). *Ber. Deut. Keram. Ges*. **41**, 112.
Moss, M., Smith, D. L., and Leferer, R. A. (1964). *Appl. Phys. Letters* **5**, 120.
Netherlands Patent Appl. (1966). No. 6,512,001 assigned to Gebr. Boehler & Co. A.-G.
O'Halloren, G. J., Fluegge, R. A., Betts, J. F., and Everett, W. L. (1964). AD 600,996.
Olsen, H. N. (1963a). *J. Quant. Spectry. Radiative Transfer* **3**, 59.
Olsen, H. N. (1963b). *J. Quant. Spectry. Radiative Transfer* **3**, 305.

Peters, T. (1964). Personal communication.
Pfender, E., and Cremers, C. J. (1965). AIAA J. **3**, 1345.
Pfender, E., and Eckert, E. R. G. (1965). *Proc. Heat Transfer Fluid Mech. Inst.*
Pfender, E., Wutzke, S. A., and Eckert, E. R. G. (1964). NASA Rept. No. CR 54080.
Pfender, E., Wutzke, S. A., and Eckert, E. R. G. (1965). NASA Rept. No. CR 54465.
Pfender, E., Eckert, E. R. G., and Raithby, G. D. (1966). *Proc. Intern. Conf. Ionization Phenomena Gases 7th*, 1966 (B. Perovic and D. Tosic, eds.), **I**, p. 691. Gradevinska Knjiga, Belgrade, Yugoslavia.
Pfender, E., Gruber, S., and Eckert, E. R. G. (1969). *Proc. 3rd Intern. Symp. High Temp. Technol., Asilomar Conf. Grounds, Calif.*, 1967. Butterworths, London and Toronto.
Polak, L. S., ed. (1965). "Kinetika i Termodinamika Khimicheskikh Reaktsii v Nizkotemperaturnoi Plazme." Inst. Neftekhim Sinteza Akad. Nauk SSR, Moscow.
Rao, V. V., Mackay, D., and Trass, O. (1965). *Can. J. Chem. Eng.* **43**, 183.
Rao, V. V., Mackay, D., and Trass, O. (1967). *Can. J. Chem. Eng.* **45**, 61.
Reed, T. B. (1961). *J. Appl. Phys.* **32**, 821.
Reed, T. B. (1966). Personal communication.
Reed, T. B. (1967). *Advan. High Temperature Chem.* **1**, 259.
Reed, T. B. (1969). *Proc.* 3rd *Intern. Symp. High Temp. Technol., Asilomar Conf. Grounds, Calif.*, 1967. Butterworths, London and Toronto.
Reis, V. H., and Fenn, J. B. (1963). *J. Chem. Phys.* **39**, 780.
Roseberry, T. J., Headman, M. L., and Parkinson, F. L. (1964). NASA Acc. No. N65-22107.
Salvemini, A. (1966). Personal communication.
Salzmann, H. (1966). AEC No. 20001, Rept. No. IPP 3136.
Seymour, E. V. (1964). NASA Acc. No. N 65-22776.
Sheer, C. (1965). U. S. Patent No. 3,214,623.
Sheer, C., and Korman, S. (1952). U. S. Patent No. 2,616,843.
Sheer, C., and Korman, S. (1965). Personal communication.
Sherman, M. P. (1964). OAR Contract AF 33(657)-9962.
AEL Rept. No. 707, Princeton Univ., Princeton, New Jersey.
Sherman, M. P., and Grey, J. (1964). AEL Rept. No. 707.
Guggenheim Lab., Princeton Univ., Princeton, New Jersey.
Skrivan, J. F., and Von Jaskowsky, W. (1965). *Ind. Eng. Chem. Process Design Develop.* **4**, 371.
Sprouse, J. A. (1965). AEDC-TR-65-146.
Stanford Res. Inst. (1966). "Chemical Economics Handbook."
Stokes, C. S. (1967). *Am. Chem. Soc. Div. Fuel Chem., Preprints* **11**, No. 2, 312.
Stokes, C. S., and Streng, L. A. (1965). *Ind. Eng. Chem. Prod. Res. Develop.* **4**, 36.
Strigl, A., and Bildstein, H. (1964). NASA Acc. No. N65-21916.
TAFA (1966). Appl. Data Bull. 10–A1.
Thorpe, M. L. (1968). Personal communication.
Tikhomirov, V. B., and Varlamov, I. V. (1964). *Inzh. Fiz. Zh. Akad. Nauk Belorussk. SSR* **7**, 16.
Timmins, R. S., and Ammann, P. R. (1967). *In* "The Application of Plasmas to Chemical Processing" (R. F. Baddour and R. S. Timmins, eds.), p. 99. M.I.T. Press, Cambridge, Massachusetts.
Unsöld, A. (1955). "Physik der Sternatmosphären," 2nd ed. Springer, Berlin.

U. S. Patent (1966a). No. 3,247,017, assigned to Agfa A-G.

U. S. Patent (1966b). No. 3,257,196, assigned to Centre Nationale de la Recherche.

U. S. Patent (1966c). No. 3,264,226, assigned to Phillips Petroleum Co.

Von Engel, A., and Steenbeck, M. (1931). *Wiss. Veröffentl. Siemens-Konzerns.* **10**, 155.

Vurzel, F. B., and Polak, L. S. (1965a). *In* " Kinetika i Termodinamika Khimicheskikh Reaktsii v Nizkotemperaturnoi Plazme " (L. S. Polak, ed.), p. 238. Inst. Neftekhim. Sinteza Akad. Nauk SSR, Moscow.

Vurzel, F. B., and Polak, L. S. (1965b). *In* " Kinetika i Termodinamika Khimicheskikh Reaktsii v Nizkotemperaturnoi Plazme " (L. S. Polak, ed.), p. 100. Inst. Neftekhim. Sinteza Akad. Nauk SSR, Moscow.

Warren, I. H., and Shimizu, H. (1965). *Can. Mining Met. Bull.* **58**, 551.

Whyman, D. (1967). *J. Sci. Instr.* **44**, 525.

Witt, A. (1967). Ph.D. Dissertation, Dept. of Chem. Eng., M. I. T., Cambridge, Massachusetts.

Studies of the Sublimation Mechanism of Solids

G. A. Somorjai

INORGANIC MATERIALS RESEARCH DIVISION, LAWRENCE RADIATION LABORATORY AND DEPARTMENT OF CHEMISTRY, UNIVERSITY OF CALIFORNIA, BERKELEY, CALIFORNIA

I. Introduction 204
II. Principles of Vaporization Studies 205
III. Experimental Considerations 207
IV. Studies of the Sublimation Mechanism of Sodium Chloride Single Crystals 209
A. Effect of Dislocations on the Evaporation Rate 210
B. Effect of Divalent Impurities (Ca^{2+}) on the Evaporation Rate .. 212
C. Discussion 214
V. Studies of the Sublimation Mechanism of Cadmium Sulfide Single Crystals 216
A. Effect of Doping of CdS on Its Evaporation Rate 216
B. The Effect of Light on the Evaporation Rate of CdS 219
C. Surface Concentration Dependence of the Evaporation Rate of CdS 219
VI. Studies of the Sublimation Mechanism of Gallium Arsenide Single Crystals 221
A. Results 222
B. Mass Spectrometric Studies 224
C. Effect of Liquid Gallium on the Vaporization of GaAs 226
D. Discussion 227
VII. Low Energy Electron Diffraction (LEED) Study of the Vaporizing Silver (100) Surface 228
VIII. Conclusions 231
A. Composition of the Surface 231
B. Charge Transfer 232
C. Impurities 232
D. Concurrent Vaporization Mechanisms 232
E. Invariance of Vapor Composition 232
References 233

I. Introduction

Investigations of the chemical reactions that control the vaporization of solids has become one of the most rapidly growing fields of thermochemistry. In these studies the vaporization rates and vapor compositions of solids are measured far from equilibrium, in order to obtain information about their kinetics of sublimation (Somorjai and Lester, 1967). Vaporization takes place under nonequilibrium conditions for most systems of technological importance. It is only reasonable to assume that an understanding of the kinetics and mechanism by which the vaporizing species break away from the surface of the condensed phase will permit one to control the rate of vaporization, reducing or increasing it, as desired.

The vaporization rates of many solids into vacuum have already been found to be much smaller than the maximum vaporization rates that can be calculated from equilibrium vapor pressure data (Somorjai and Lester, 1967). For many solids [As (Brewer and Kane, 1955), P (Kane, 1955), GaAs] the presence of liquid metals on the vaporizing surface may increase the sublimation rate by orders of magnituαe to that of the maximum rate. For other solids [NaCl (Lester and Somorjai, 1968), CdS (Somorjai and Lyon, 1965)] the presence of trace impurities (in part-per-million quantities) or small changes in defect concentration can decrease the sublimation rates by orders of magnitude. All of these striking changes in the vaporization rates are due to changes in the vaporization mechanism. By making minute changes in the composition of the vaporizing solid or in the conditions of sublimation we may change the rate-limiting reaction step in a series of reactions (Somorjai and Lester, 1967) that finally lead to the desorption of the vaporizing species.

Recent studies of vaporization have uncovered a great deal of information about the vaporization mechanism of different monatomic and diatomic solids (Somorjai, 1968a; Rosenblatt and Lee, 1968). It is hoped that as more materials are investigated we shall be able to identify the important structural or electronic parameters that control the sublimation process. We can then use this information to develop a unified theory of vaporization that can be used to predict vaporization mechanisms for different groups of materials.

We have been studying the vaporization characteristics of several solids to obtain information about the relationship between their chemical bonding and their sublimation mechanism. For these investigations we have selected binary compounds from groups IA–VIIA, IIB–VIA and IIIA–VA in the periodic table. In this chapter we shall discuss and compare the sublimation characteristics of sodium chloride, cadmium sulfide, and gallium arsenide. It is hoped that other compounds with similar bonding characteristics have similar vaporization mechanisms.

II. Principles of Vaporization Studies

Consider a perfectly flat face of a monatomic single crystal in which the binding energies of all surface atoms are equal to each other. Since all atoms in such a crystal face have equal probability of vaporizing, the overall *evaporation rate* is given by (Somorjai and Lester, 1967)

$$J^{\mathrm{v}}_{\max} \ (\text{moles/cm}^2 \text{ sec}) = k_{\mathrm{v}}(A)^{\max}_{\mathrm{s}} \tag{1}$$

where k_{v} is the rate constant of the vaporization reaction and $(A)^{\max}_{\mathrm{s}}$ is the total concentration of surface atoms. The maximum *rate of condensation* of the vapor flux on such a face is given by

$$J^{\mathrm{c}}_{\max} \ (\text{moles/cm}^2 \text{ sec}) = k_{\mathrm{c}}(A)_{\mathrm{g}} = (RT/2\pi M_A)^{1/2}(A)_{\mathrm{g}} \tag{2}$$

where $(A)_{\mathrm{g}}$ is the vapor density, R is the gas constant, T is the absolute temperature, and M_A is the atomic weight of $(A)_{\mathrm{g}}$. In equilibrium, the vaporization and condensation rates are equal $[k_{\mathrm{v}}(A)_{\mathrm{s}}=k_{\mathrm{c}}(A)_{\mathrm{g}}]$. Thus, the evaporation rate can be expressed in terms of the condensation rate and the well-known equation (Hertz, 1882; Knudsen, 1915; Langmuir, 1913)

$$J^{\mathrm{v}}_{\max} \ (\text{moles/cm}^2 \text{ sec}) = P_{\mathrm{eq}}(2\pi M_A RT)^{-1/2} \tag{3}$$

is obtained assuming the ideal gas law. If the condensation rate is smaller than the evaporation rate due to external " pumping," there is a net rate of removal R of vapor flux from the vaporizing surface, which is given by

$$R = J^{\mathrm{v}} - J^{\mathrm{c}} \tag{4}$$

In equilibrium the net rate is zero ($R=0$), and for vaporization into vacuum $R=J^{\mathrm{v}}$.

Now let us consider a " real " surface, in which there are many atomic positions. These are distinguishable by their different numbers of nearest neighbors. Surface atoms have different binding energies depending on which atomic positions they occupy. Under these conditions only a small fraction of the total number of surface atoms will occupy positions of lowest binding energy, from which vaporization can most readily proceed. We may rewrite Eq. (1) as

$$J^{\mathrm{v}} \ (\text{moles/cm}^2 \text{ sec}) = k_0(A)_{\mathrm{s}} \exp(-E^*/RT) \tag{5}$$

where $(A)_s$ is the concentration of atoms in surface sites from which vaporization takes place, and k_0 is a constant related to the frequency of attempted motion of vaporizing molecules over the energy barrier E^*. Thus the evaporation rate of a " real " crystal surface can be different from the evaporation rate of a " homogeneous " surface because of variations of the surface concentration $(A)_s$ or on account of a different activation energy of vaporization E^*. If the concentration of surface sites from which vaporization can occur is smaller than the total concentration of surface atoms $[(A)_s<(A)_s^{max}]$, the evaporation rate will be smaller than if all surface atoms could vaporize with equal probability.

A heterogeneous surface, however, has many atomic ledges and steps and exhibits " roughness " on an atomic scale. Therefore, surface area of a " real " crystal could be appreciably larger than the geometrical surface area. Although not every surface atom may have low enough binding energy to vaporize, the concentration of those that vaporize can be sufficiently greater on a rough surface than on a " homogeneous " surface. Melville (1936) and Winterbottom (1966) have shown that no matter how rough the surface becomes, the evaporation rate may reach but cannot exceed the maximum sublimation rate from a crystal with the smooth geometrical surface area (if the condensation rates are near to maximum). Thus the surface concentration of atoms of weakest binding energy in a " real " surface may increase and become equal to the surface density of atoms in a homogeneous surface. The vacuum evaporation rate would simultaneously increase until the maximum rate J^{v}_{max} is reached.

It is customary (Somorjai and Lester, 1967) to express the deviation of the vacuum evaporation rate J^{v} from the maximum equilibrium rate J^{v}_{max} in terms of the evaporation coefficient α, which is given by,

$$\alpha(T) = J^{v}(T)/J^{v}_{max}(T) \tag{6}$$

If the vacuum evaporation rate is equal to the maximum rate $\alpha(T)=1$ or if $J^{v} \ll J^{v}_{max}$, then $\alpha(T) \ll 1$.

The activation energy of vaporization E^* can be larger or smaller than the equilibrium heat of vaporization ΔH_V or can be equal to it. Its magnitude reflects the overall energy necessary for the surface atoms to participate in the rate-limiting reaction step (or steps) (Somorjai and Lester, 1967). Vaporization is a multistep process in which atoms that break away from their lattice sites may diffuse on the surface until they are ready to vaporize. For many solids the surface atoms must undergo rearrangements such as association, dissociation, or charge transfer prior to desorption from the vaporizing surface (Somorjai and Lester, 1967). The rate of any of these reaction steps may control the overall sublimation process. Certain solids may vaporize by two different reaction paths, which can operate simultaneously at different surface sites (Lester and Somorjai, 1968).

The activation energy of vaporization may change as the experimental vaporization conditions are varied. In order to be able to compare activation energies of different sublimation reactions, values are determined from measurements using the pure solids and carried out in vacuum ($J^c = 0$; $R = J^v$). Such values are often listed (Somorjai and Lester, 1967).

III. Experimental Considerations

The experimental techniques that are most frequently used in sublimation studies are (1) total weight loss measurements using microbalance or torsion balance and (2) mass spectrometric studies of the composition of the vapor flux. The two techniques are complementary; weight loss measurements give absolute evaporation rates. From these studies, as a function of temperature, the activation energy of vaporization (per mole of solid) is determined. Mass spectrometric investigation of the vapor flux allows one to determine and monitor the vapor composition. The activation energies of vaporization (per mole of vapor) of each vaporizing species can be calculated from the variation of the composition with temperature. Typical microbalance and mass spectrometer systems used in our vaporization studies (Lester and Somorjai, 1968) are shown in Figs. 1 and 2.

For purposes of definitive studies of sublimation mechanisms one face of a single crystal should be used. Evaporation rates obtained using polycrystalline samples are very difficult to interpret due to their unknown free surface area (Brewer and Kane, 1955; Rosenblatt, 1963). These samples contain a large inner surface from which the vaporizing atoms will undergo many collisions before leaving the solid. Thus a certain fraction of the powder surface acts as an equilibrium source.

The crystal face that is to be used in the sublimation studies should be well characterized. It should be a " stable " surface that retains its crystallographic orientation throughout the experiments (Somorjai and Stemple, 1953). It is important to determine its dislocation density using etch-pit count techniques (Lester and Somorjai, 1968) and to investigate its surface topology using optical or electron microscopy (Searcy, 1968). It has been shown that the surface structure can markedly influence the evaporation rates of many solids (Lester and Somorjai, 1968; Rosenblatt and Lee, 1968; Searcy, 1968). The bulk properties of the vaporizing crystal samples should also be well characterized. There is, in general, communication between the

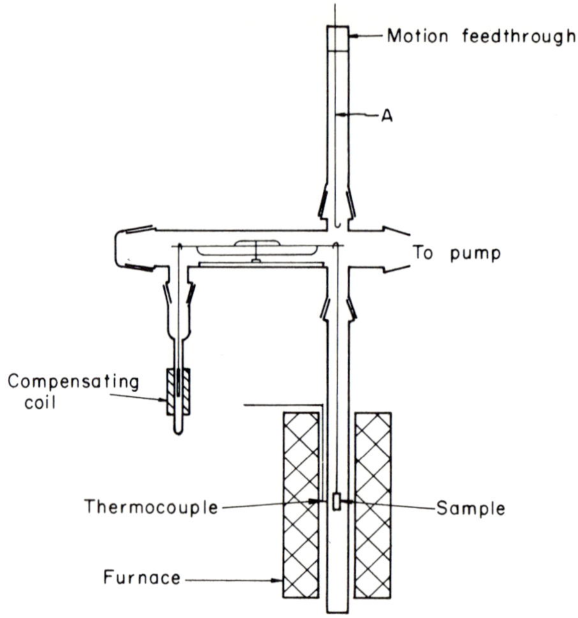

Fig. 1. Scheme of the experimental arrangement used in the microbalance studies of the total vaporization rate.

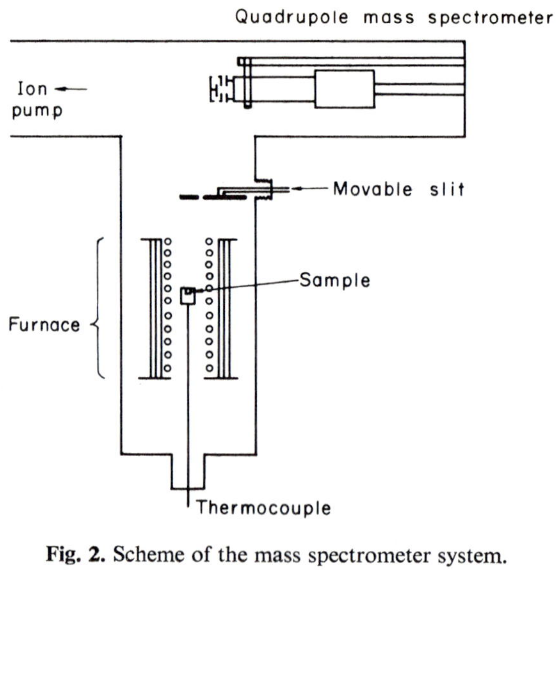

Fig. 2. Scheme of the mass spectrometer system.

bulk and the vaporizing surface via atom or defect diffusion or via charge transfer (Somorjai and Lester, 1967; Somorjai and Lyon, 1965). Thus changes in bulk properties may greatly influence the evaporation rate. The defect, impurity, and charge concentrations in the bulk solid should be determined.

Since the structure, composition, and other physical chemical properties of both the surface and the bulk depend on the thermal history of the solid, it is necessary to devise a suitable heat treatment to standardize these parameters at the onset of the vaporization studies (Somorjai and Lester, 1967; Somorjai and Lester, 1965a).

Several additional techniques that can be applied to studies of vaporization kinetics have become available in recent years. These include low energy electron diffraction (Morabito and Somorjai, 1968), Auger spectroscopy (Palmberg and Rhodin, 1968), ellipsometry (Morabito *et al.*, 1969), and scanning electron microscopy (Searcy, 1968).

IV. Studies of the Sublimation Mechanism of Sodium Chloride Single Crystals

Sodium chloride vaporizes according to the dominant net reaction (Lester and Somorjai, 1968; Berkowitz and Chupka, 1958).

$$\mathrm{NaCl(solid)} \rightarrow (1-x)\ \mathrm{NaCl(vapor)} + x/2\ \mathrm{Na_2Cl_2(vapor)}$$

In the temperature range of our investigation, 450 to 650°C, and under the different vaporization conditions, the dimer is roughly 5–30 mole % of the vapor.

The majority of the crystals used in these vaporization studies were obtained from the Harshaw Chemical Company. These crystals are 99.95% pure. The major impurities are anions containing bromine and oxygen (probably Br^-, OH^-, and O_2^-). Polyvalent cation impurities were present at levels not exceeding 5 ppm.

Dislocation densities were determined by counting chemically formed etch pits on a (100) cleavage face, which was used in all of the vaporization studies. Most of the pure NaCl crystals had etch pit densities of $0.8–4\times10^6$ cm^{-2}. We have assumed that a constant proportionality exists between observed etch pits and dislocations that intersect the surface under observation, and that the proportionality factor is close to one. The dislocation density of a crystal could be increased by plastically deforming it. Crystals with dislocation densities (etch pit counts) as high as 1.5×10^7 cm^{-2} were produced in this way. Etch pits in the unstrained crystals generally were uniformly distributed; however, in the strained crystals they tended to appear in " slip bands." Therefore, the etch pit density was counted in several areas on the surface and the results averaged to obtain the reported figure for these crystals.

A. Effect of Dislocations on the Evaporation Rate

The (100) face of sodium chloride was used in these studies. This face remained stable throughout the vaporization run, i.e., it did not exhibit faceting to other low-index crystal planes. Thus steady state evaporation rates could readily be obtained at a given temperature. Initial experiments showed large (factor of 3) variations in the total evaporation rates from crystal to crystal. In every case, however, steady state vaporization rates for a given sample could readily be achieved and maintained. It was discovered that the total evaporation rate of sodium chloride crystals depends on the dislocation density in the sample. Figure 3 shows a plot of the logarithm of the evaporation rates of sodium chloride crystals with two different dislocation densities ($\sim 1 \times 10^6$ cm^{-2}" and $\sim 1 \times 10^7$ cm^{-2}) as a function of reciprocal temperature. Here we can also plot the maximum evaporation rate as calculated from equilibrium vapor-pressure measurements. The steady state evaporation rate was approximately doubled by increasing the dislocation

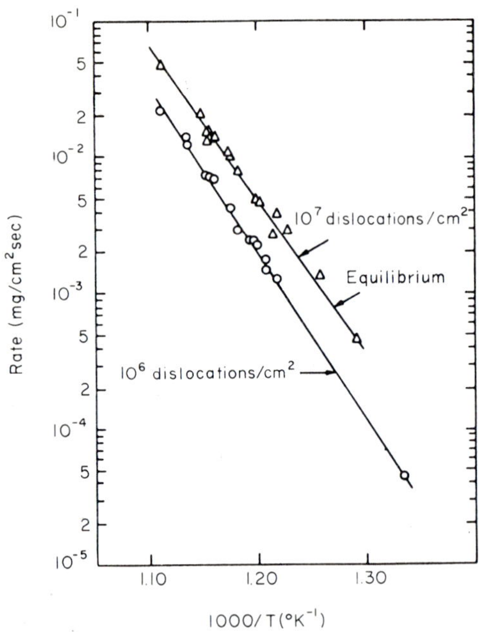

Fig. 3. Evaporation rate of low and high dislocation density sodium chloride single crystals as a function of temperature. Etch pit densities: ○, 1×10^6 cm^{-2}; △, 1×10^7 cm^{-2}. Equilibrium data from Zimm and Mayer) (1944).

density by an order of magnitude. Sodium chloride crystals with dislocation densities of $\sim 10^6$ cm^{-2} gave evaporation rates that were lower than the equilibrium rate by about a factor of 2. The crystals with larger etchpit counts ($\sim 10^7$ cm^{-2}) had evaporation rates equal to the maximum equilibrium rate. Except for dislocation density these crystals should be identical (impurities and preparation technique). These curves give an average activation energy of vaporization of approximately 55 kcal/mole. It should be noted that this value is a weighted average of the activation energies of vaporization of the monomer and dimer molecules for the different dislocation density crystals.

The composition of the vapor flux emanating from the subliming (100) crystal face of sodium chloride was monitored by a quadrupole mass spectrometer as a function of temperature. The ion peaks detected in the mass spectrum of sodium chloride vapor were $m/e = 23$ (Na^+), 58,60 ($NaCl^+$) and 81,83 (Na_2Cl^+). A representative curve showing the logarithm of the ion intensities (Na^+, $NaCl^+$, Na_2Cl^+) as a function of the reciprocal temperature is given in Fig. 4. These slopes were used to calculate the activation energies of vaporization for both NaCl(vapor) and Na_2Cl_2(vapor). We also plot representative values of the computed monomer to dimer ratio on a semilogarithmic scale as a function of temperature.

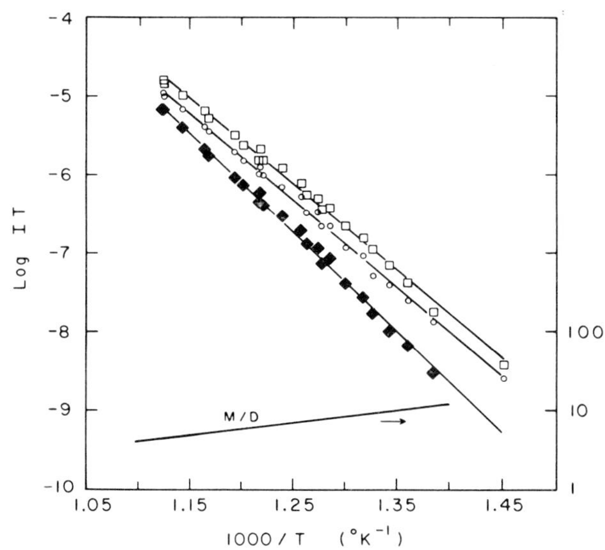

Fig. 4. Representative curves of the ion intensities Na^+ (□), $NaCl^+$ (○), and Na_2Cl^+ (■) and the monomer to dimer ratio M/D as a function of temperature.

We have also monitored the vapor composition from the (111) crystal face of sodium chloride as a function of temperature. Within the accuracy of our measurements both the monomer to dimer ratio and the activation energies of vaporization $E_D{}^*$ and $E_M{}^*$ were identical to those for the (100) face of sodium chloride crystals of similar dislocation density. Inspection of the (111) face after the vaporization studies revealed the development of facets of (100) orientation. It is likely that after a short transient the unstable (111) surface has virtually disappeared and the largest fraction of the vapor flux came from the freshly formed (100) facets. In Table I we list the activation energies of vaporization for the monomer and the dimer from crystals of different dislocation density. It is apparent that the activation energies vary somewhat for crystals with different etch pit densities. There seems to be a small increase in the activation energies with increasing dislocation density, which is not significantly outside our experimental error. The monomer to dimer ratio appears not to change appreciably with increasing dislocation density in the crystals. There may be a small increase in the monomer concentration relative to the dimer for the highly strained samples 1×10^7 cm^{-2}).

TABLE I. Activation Energies of Vaporization for NaCl Crystals with Different Dislocation Densities

Dislocation density	$E_M{}^*$ (kcal/mole)	$E_D{}^*$ (kcal/mole)
0.8×10^6 cm^{-2}	50.7 ± 1	58.2 ± 1
1×10^8 cm^{-2}	52.4 ± 1.5	61.5 ± 1.6
2×10^6 cm^{-2}	54.0 ± 1	62.0 ± 4
$\sim4\times10^6$ cm^{-2}	55.6 ± 0.6	65.6 ± 0.6
$\sim6\times10^{-6}$	53.8 ± 1.1	63.3 ± 0.8
$>15\times10^6$ cm^{-2}	56.6 ± 1.1	62.8 ± 0.9

B. Effect of Divalent Impurities (Ca^{2+}) on the Evaporation Rate

Sodium chloride single crystals containing 330 ± 50 ppm Ca^{2+} ions used in these studies show the effect of divalent impurities on the vacuum evaporation rate of NaCl.† Figure 5 shows the evaporation rates of these crystals as a function of the reciprocal temperature on a semilogarithmic plot. There is a large decrease of the total evaporation rate (by a factor of 5) for crystals

†In addition to Ca^{2+}, sodium ion vacancies in excess of their steady state concentration have also been introduced in the sodium chloride crystal lattice.

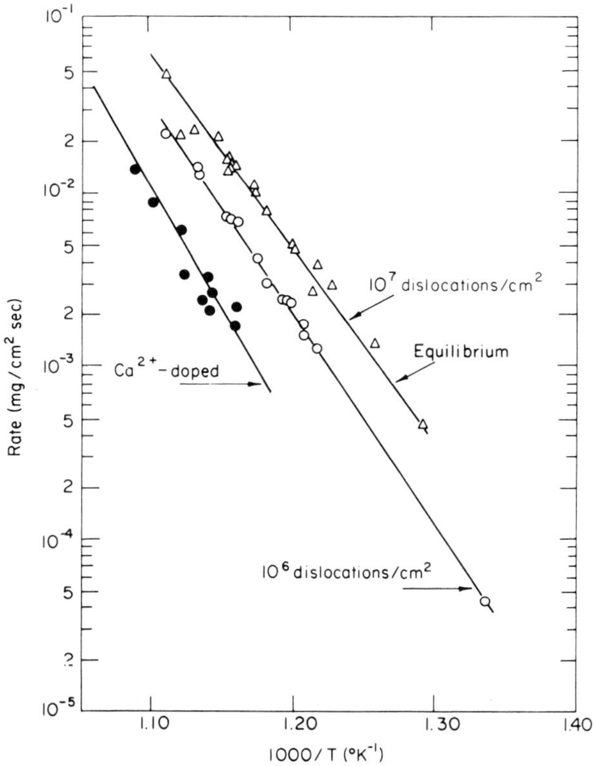

Fig. 5. Evaporation rates of calcium-doped sodium chloride crystals (●) and of pure crystals with different dislocation densities.

doped with calcium. These Ca^{2+} doped crystals not only show drastically reduced evaporation rates but also a marked increase of the average activation energy of vaporization. While the average vaporization coefficient has dropped to $\alpha \approx 0.1$ from $\bar{\alpha} \approx 0.5$ obtained for the pure low dislocation density sodium chloride crystals, the average activation energy of vaporization has increased to $\bar{E}^* \simeq 63$ kcal/mole with respect to that obtained for the pure crystals ($\bar{E}^* \simeq 55$ kcal/mole).

There is experimental evidence indicating that the concentration of calcium ions slowly increases at the vaporizing sodium chloride surface as a function of time during vaporization. The evaporation rate was found to decrease slowly for a given sample during the experiment. Simultaneously, the apparent activation energy of vaporization also increases. The vaporization data given in Fig. 5 are representative of the results obtained early in the measurements using several calcium-doped samples.

Sodium chloride samples doped with Br^- ion showed vaporization characteristics that were identical to undoped high purity single crystals of the same dislocation density. The presence of oxygen or OH^- ions in the crystals also had no observable effect on the evaporation rates in the temperature range of this study.

C. Discussion

We find that the *steady state* vacuum vaporization rate of the solid is determined by the dislocation density. These results indicate that the dislocation density in the sodium chloride crystals remains largely unchanged throughout the vaporization and that dislocations control the concentration of surface sites from which vaporization is to occur. An order of magnitude increase in the dislocation density increased the evaporation rate by a factor of 2 or the average evaporation coefficient from $\bar{\alpha} \approx 0.5$ to $\bar{\alpha} \approx 1.0$. The activation energies of vaporization exhibit a small upward trend with increasing dislocation densities and corresponding increased rates of vaporization. This change in activation energy is less than 3 kcal. It should be noted that an increase of 2 kcal in E^* would decrease the evaporation rate of NaCl by a factor of 3 at 800°K if there were no accompanying changes in the preexponential factor. We observed an increase in the rate by a factor of 2 ($2\times$). Thus the preexponential factor would have to increase by a factor of 6 to accommodate the increased evaporation rate in this case. It is apparent from our data that an etch pit density of 10^7 cm^{-2} corresponds to a dislocation density sufficient to establish the maximum equilibrium vaporization rate. We cannot say what is the lower limit on the vaporization rate of high purity crystals as we were not able to prepare samples with dislocation densities lower than $\sim 0.8 \times 10^6$ cm^{-2}.

The monomer to dimer ratios were essentially unchanged for all of the sodium chloride crystals within the accuracy of the measurement (with the exception of the crystals of the highest dislocation density, which showed larger relative monomer concentrations). If the dimers formed by an equilibrium association of two monomer molecules adsorbed on the surface then the monomer to dimer ratio M/D should decrease as a function of the increasing monomer concentration; i.e., $M/D \propto 1/M$, at a given temperature. This does not happen. For crystals that have high vaporization fluxes (and high dislocation densities) there are more monomer molecules on the surface, but, if anything, the monomer to dimer ratio increases with respect to the lower vaporization rate crystals (low dislocation density). Thus there is no equilibrium between the monomer and dimer molecules on the vaporizing sodium chloride surface. It seems that the monomer to dimer ratio is fixed by the relative energies necessary to remove a monomer or dimer, respectively, from a surface site. The absence of any appreciable equilibrium interaction between the absorbed monomer and/or dimer molecules could be due to the short residence time of these molecules on the vaporizing surface prior to desorption into vacuum. The mean free path ($\bar{X}$) of molecules away from the ledge depends on the activation energies of surface diffusion E^*_{diff} and of desorption E_{des} from the free surface and can be estimated by the equation

$$\bar{X} = a \exp[(E_{\text{des}} - E^*_{\text{diff}})/2kT]$$

where a is the interatomic distance (Burton *et al.*, 1951). For KCl these energies have been calculated to be $E_{\text{des}} = 0.35$ eV and $E^*_{\text{diff}} = 0.23$ eV (Hove, 1955). These energies should be similar for NaCl. Using these values, at $T = 800°$K, we have $\bar{X} = 0.4a$. Thus, at the vaporization temperature, the mean free path of the sodium chloride molecule could not be more than a few lattice spacings. This implies that most of the monomer and dimer molecules vaporize near ledges and that surface diffusion cannot play an important role in the vaporization reaction. The lack of any appreciable surface diffusion also inhibits the establishment of a steady state ledge concentration at the vaporizing surface that could be approximated by the equilibrium value. It is apparent that the removal of a sodium chloride molecule from a surface site became more difficult in the presence of neighboring divalent ions or sodium ion vacancies that are introduced in excess of the steady state concentration. Such a marked effect on the evaporation rate by doubly charged impurities indicates that the breaking away of sodium chloride molecules from lattice sites (kinks) on the vaporizing surface is a rate-controlling step.

These vaporization characteristics of the sodium chloride single crystals can be explained in terms of the stepwise model of vaporization of the heterogeneous solid surface (Stranski, 1928). We postulate a possible mechanism of vaporization of sodium chloride that can explain most of the data.

(1) Ledges (of approximately monatomic height) are formed in large concentrations at positions where dislocations intersect the surface (and at the edges of the crystal).

(2) Both monomer and dimer molecules are formed at the same sites (kinks) on these ledges.

(3) The desorption rate of both monomer and dimer molecules from the proximity of the ledges is rapid.

The first step explains the observed increase of the evaporation rate of sodium chloride surfaces with increased density of dislocations. Dislocations have been shown to be nucleation sites for crystal growth, and thermal etch pits have been observed on alkali halide crystals at the point of emergence of dislocations at the surface.

The second and third steps, the breaking away of both monomer and dimer molecules independently from the kinks and the apparent absence of any surface diffusion, explain the insensitivity of the monomer to dimer ratio to increased dislocation density. The marked decrease of the evaporation rate (factor of 5) of calcium-doped crystals and the simultaneous large increase in the activation energy of vaporization indicate that the breaking away of sodium chloride molecules at ledges is a rate determining step in the vaporization process.

V. Studies of the Sublimation Mechanism of Cadmium Sulfide Single Crystals

Cadmium sulfide vaporizes according to the dominant reaction (Jeunehomme, 1962)

$$\mathrm{CdS(solid)} \rightarrow \mathrm{Cd(vapor)} + \tfrac{1}{2}\,\mathrm{S_2\ (vapor)}$$

Crystals are available in high purity, crystallize in the wurtzite structure, and show electronic conductivity. They are n-type (electrons are the majority charge carriers) under all conditions of stoichiometry. This is thought to be due to ionized sulfur vacancies V_S^+, which predominate over ionized cadmium vacancies V_{Cd}^- (Somorjai and Jepsen, 1964b; Woodbury, 1964). It has been found (Somorjai and Stemple, 1953) that the c–[(0001)]–face is stable under conditions of vacuum evaporation, while the other low index face, the $(10\bar{1}0)$ or a-face, is not. The c-face remains invariant during the experiments, therefore, this face was used in most of the studies.

The temperature dependence (Somorjai, 1964a,b; Somorjai and Jepsen, 1964a) of the vacuum evaporation rate of the (0001)-face of CdS has been measured in the temperature range 650–800°C (Fig. 6). The rates are more than an order of magnitude lower than the maximum evaporation rates calculated from equilibrium vapor pressure measurements (Somorjai, 1964a,b; Neuhaus and Retting, 1958). The activation enthalpy and the heat of evaporation are $\Delta H^* = 50.3$ kcal and $\Delta H_V = 75.2$ kcal/mole of vaporizing solid, respectively.

A. Effect of Doping of CdS on Its Evaporation Rate

1. *Excess Cadmium and Sulfur in* CdS

The free carrier concentration (electrons and holes) can be changed by more than 10 orders of magnitude by doping the CdS crystals with cadmium ($\sigma \approx 0.1$ mho cm^{-1}) or sulfur ($\sigma \approx 10^{-10}$ mho cm^{-1}) at elevated temperatures

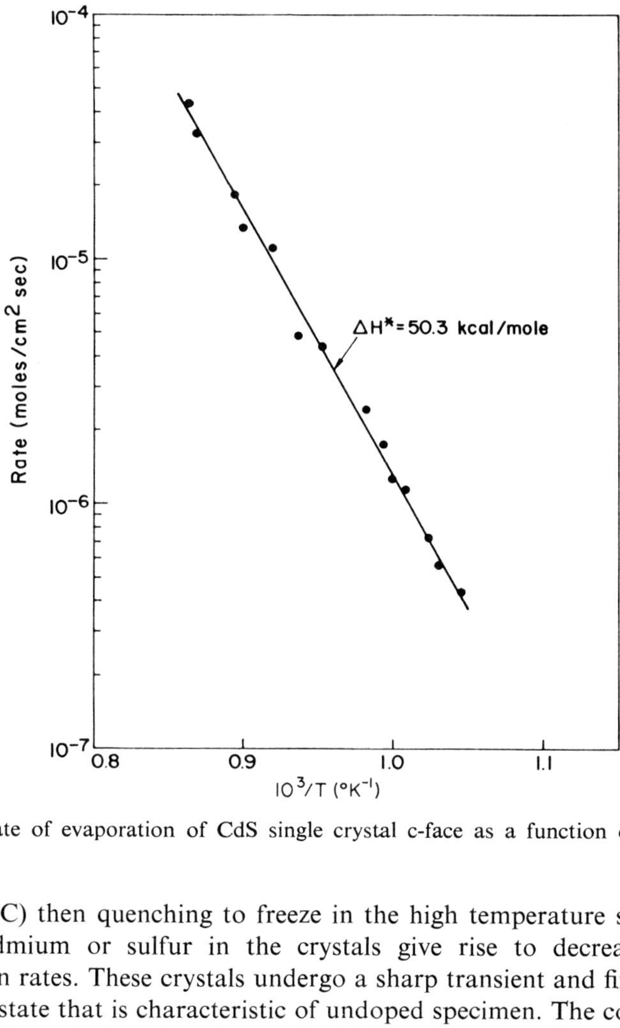

Fig. 6. Rate of evaporation of CdS single crystal c-face as a function of reciprocal temperature.

(900–1100°C) then quenching to freeze in the high temperature solubilities. Excess cadmium or sulfur in the crystals give rise to decreased initial evaporation rates. These crystals undergo a sharp transient and finally reach the steady state that is characteristic of undoped specimen. The conductivity change of the cadmium-doped crystals indicates the out-diffusion of excess cadmium with a corresponding three to four orders of magnitude decrease in carrier concentration. The loss of excess cadmium is also indicated by the color change in the crystals. Cadmium-doped samples are black to gray due to precipitated cadmium in the crystal lattice (Woodbury, 1964). Crystals showing steady state evaporation rate are yellow.

For cadmium- or sulfur-doped CdS crystals the evaporation is controlled by the out-diffusion of excess cadmium or sulfur from the crystal lattice.

As the excess surface cadmium evaporates it is replenished from the bulk by diffusion. As long as the diffusion rate of excess cadmium or sulfur is of the same order of magnitude or greater than the evaporation rate at a given temperature, there is a constant " communication " between the bulk and the surface of the vaporizing crystal. Thus bulk diffusion, which controls the surface concentration of excess cadmium or sulfur, is the rate-limiting step in this case.

If the diffusion rate of electrically active impurities is smaller than the evaporation rate, the impurity could not diffuse out of the crystal. Thus there would be no transient evaporation, the constant impurity surface concentration would establish the evaporation rate. This is the case for copper impurity in CdS.

2. *Effect of Copper-Doping of* CdS

CdS single crystals can be uniformly doped by diffusing the surface evaporated copper into the crystal. When the CdS crystal is uniformly saturated with copper at the temperature of evaporation, the evaporation rate of Cu-doped CdS is decreased by more than 50%, as shown in Fig. 7,

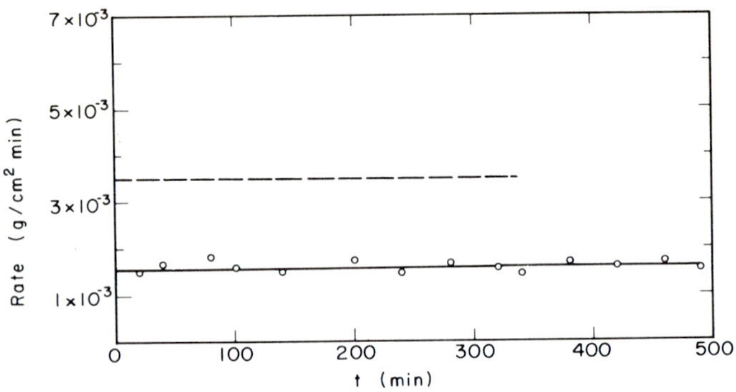

Fig. 7. Evaporation rate of the (0001)-face of CdS at 700°C. The crystal was copper doped at 700°C. Dotted line shows the evaporation rate of undoped crystal.

over that of the evaporation rate of the undoped crystal (Somorjai and Lyon, 1965).

In order to investigate the dependence of CdS evaporation rate on the surface concentration of diffused copper, copper is vacuum deposited on only the (000$\bar{1}$)-face of the undoped crystals. Then the (0001)-face is allowed to vaporize while copper diffuses into the crystal from the back surface (Somorjai and Lyon, 1965). This way the vaporizing surface retreats while the copper diffusion front moves toward the vaporizing face. As the copper profile reaches the vaporizing surface a rapid change in the copper surface concentration occurs, giving rise to an exponential decrease of the vaporization rate. When the copper surface concentration reaches its maximum, which corresponds to the solubility of copper in CdS at the evaporation temperature, a new lower steady state evaporation rate is established that is characteristic of the constant surface concentrations of copper ions.

B. The Effect of Light on the Evaporation Rate of CdS†

The room temperature band gap of CdS is 2.41 eV and the temperature dependence of the gap is -5×10^{-4} eV/deg. Photons of energy greater than or equal to the band gap energy create electron–hole pairs at the vaporizing CdS surface, in excess of the equilibrium carrier concentration. If the vaporization is controlled by charge transfer, an increase of the evaporation rate is expected in light with respect to the steady state rate in the dark, at a given temperature. For light to have an observable effect on the evaporation rate it is necessary that the electron–hole pair concentration produced by illumination should be of the same order of magnitude or larger than the free carrier concentration in the dark at the evaporation temperature (680–740°C). This criterion was met by using light intensities in the range 5.0×10^{3} to 2.0×10^{5} μW/cm^2. In addition, the trapping lifetime of photoelectrons and holes must be long enough, relative to the residence time of the neutral cadmium and/or sulfur species on the surface, to avoid direct electron–hole recombination prior to the desorption of the neutral species.

These conditions have been met for sulfur-doped CdS ($\sigma\approx10^{-10}$ mho cm^{-1}). The effect of light on the evaporation rate of these crystals is shown in Fig. 8. The evaporation rate increases linearly with the light intensity in the studied range, as shown in Fig. 9.

C. Surface Concentration Dependence of the Evaporation Rate of CdS‡

The surface concentration of the final species of the reaction were varied using a molecular beam of sulfur and/or an atomic beam of cadmium (Somorjai and Jepsen, 1964a). Care was taken in all of the molecular beam experiments to maintain a sufficiently large mean free path of all vapor species in order to exclude possible gas phase interactions between the beam and species being evaporated from the CdS.

†Somorjai (1964a, b), Somorjai and Lester (1965a, b).
‡Somorjai and Jepsen (1964b).

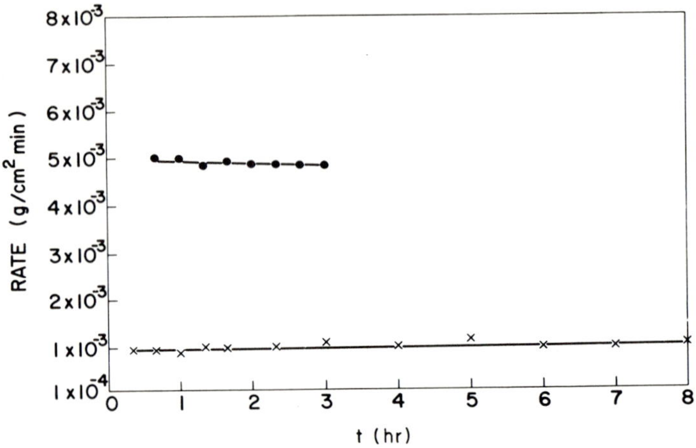

Fig. 8. Evaporation rate of sulfur-doped (1200°C, 20 atm) CdS single crystal c-face in dark (X) and in light (●) at 696°C.

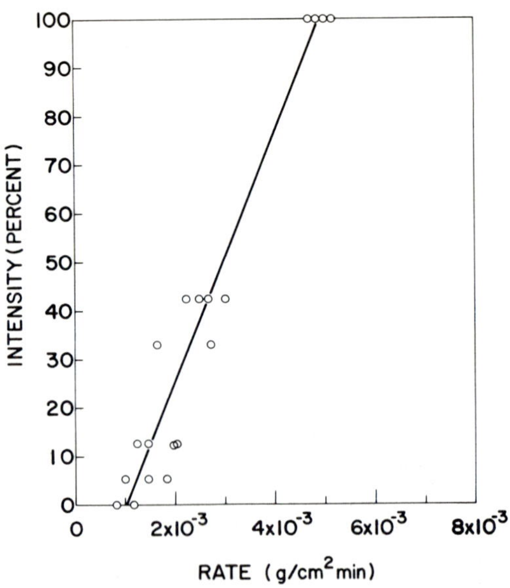

Fig. 9. Light intensity dependence of the evaporation rate of sulfur-doped (1200°C, 20 atm) CdS single crystal c-face at 696°C. 100% = 2.0×10^5 μW/cm^2.

The molecular beams of S_2(vapor) or Cd(vapor) were prepared using a double-oven arrangement that allowed preheating the impinging beam to that of the surface temperature of the vaporizing CdS crystal. The weight loss was monitored with a microbalance, while the vapor flux was determined in separate measurements (Somorjai and Jepsen, 1964a). At low impingement rates, there is no observable effect. As the sulfur flux increases and becomes the same order of magnitude as that of the evaporation rate of CdS, the evaporation rate becomes proportional to the $-\frac{1}{2}$ power of the sulfur flux ($J_V \approx J_{S_2}^{-1/2}$).

Cadmium atomic beams had no appreciable effect on the evaporation rate of CdS in the range of impingement rates studies (Somorjai and Jepsen, 1964a).

From the results of the experiments we can now deduce the mechanism of CdS evaporation. The presence of excesses of cadmium, sulfur, and copper in CdS decreased the evaporation rates markedly. These impurities have low solubilities in cadmium sulfide, but they can change the free carrier concentration in the vaporizing crystals by orders of magnitude. It is apparent that the concentration of charge carriers plays an important role in the vaporization process. The effects of electrically active impurities and of light on the evaporation rate indicate that the *vaporization is controlled by a charge transfer process.* The vaporization reaction steps that can be deduced from the available experimental information are (a) diffusion of excess cadmium or sulfur to the surface, (b) electron transfer to neutralize the cadmium ions at the surface, (c) hole transfer to neutralize the sulfur ions at the surface, (d) association of sulfur atoms, (e) desorption of sulfur molecules, and (f) desorption of cadmium atoms. Under vacuum evaporation conditions for undoped CdS crystals in dark or in light, steps (b) and (c) are indistinguishable and rate limiting. For sulfur- and cadmium-doped crystals, reaction (a) controls the evaporation rate. For the vaporization of copper-doped CdS, step (b) is the possible slow step.

VI. Studies of the Sublimation Mechanism of Gallium Arsenide Single Crystals

Gallium arsenide has zinc blende structure and a melting point of 1238°C. It is a semiconductor† and is available in the highest purity.‡ The compound

†The crystals used in the experiments were n-type, had room temperature resistivities of 0.1 to 0.2 Ω cm and mobility in the range 5×10^3 cm^2/V sec.

‡Purchased from Cominco American Inc., Spokane, Washington.

vaporizes according to the dominant net reaction (Gutbier, 1961; Arthur, 1967)

$$GaAs(solid) \rightarrow Ga(vapor) + \tfrac{1}{2}x\ As_2(vapor) + \tfrac{1}{4}(1-x)\ As_4(vapor)$$

in the temperature range 750–900°C. The tetramer arsenic concentration is about 25 to 65 mole % of the total arsenic flux, and it increases with increasing temperature.

The vaporization characteristics of both, the (111) or gallium face and the $(\bar{1}\bar{1}\bar{1})$ or arsenic face were studied in this temperature range. These two low index faces should be composed dominantly of only one type of atom and have been shown to have different oxidation (Miller *et al.*, 1961) and etching rates (Russell *et al.*, 1966; Gatos and Levine, 1960) under a variety of experimental conditions.

A. Results

1. *Microbalance Studies*

The weight loss of the sample was measured as a function of time at a given temperature. From these data and the geometric surface area, the evaporation rate (mg/cm^2 sec) was calculated. The surface area was not flat, of course, during the vaporization. The presence of small droplets of liquid gallium are readily discernible on both crystal faces (Lou and Somorjai, 1969). The area under the liquid is always smooth and deeper lying, which indicates that a larger vaporization flux is emanating from the gallium coated area than from the uncoated portion of the surface.

A reproducible steady state vaporization rate could eventually be obtained† at any temperature in the range of our study, regardless of the thermal history of the vaporizing crystal. There is, however, always an induction period of transient evaporation at any given vaporization temperature before the steady state evaporation rate is obtained, and this transient behavior very much depends on the thermal history. Figure 10 shows the vacuum evaporation rate of the gallium (111) and arsenic $(\bar{1}\bar{1}\bar{1})$ faces as a function of time at 800°C. The sample was (a) heated or (b) cooled to 800°C prior to the onset of the vaporization run. After *heating* to 800°C there is an initial induction period of slow vaporization, after which the evaporation rate increases to attain the higher steady state rate. After *cooling* to 800°C there is a somewhat shorter initial induction period of rapid vaporization, after which the evaporation rate decreases to attain the lower steady state rate. Steady state vacuum evaporation rate can be established for the (111) and $(\bar{1}\bar{1}\bar{1})$ crystal faces even though that both of these faces of gallium arsenide

† for the $(\bar{1}\bar{1}\bar{1})$ face.

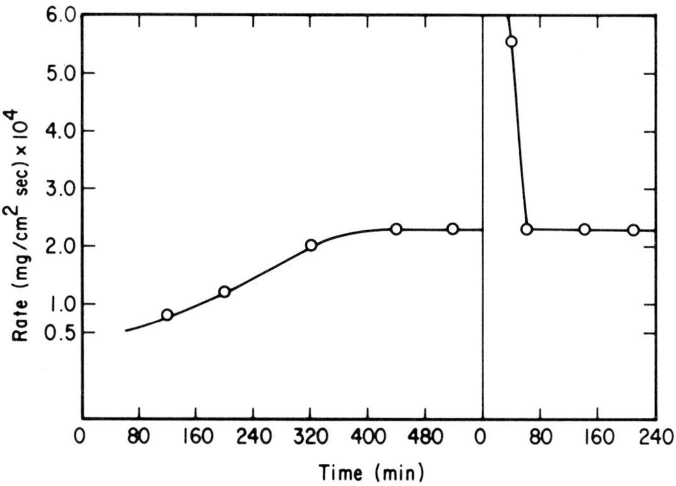

Fig. 10. Evaporation rates of the (111) and ($\bar{1}\bar{1}\bar{1}$) faces of gallium arsenide as a function of time at $T = 750 \pm 2°C$. The left-hand side is heating and the right-hand side is cooling.

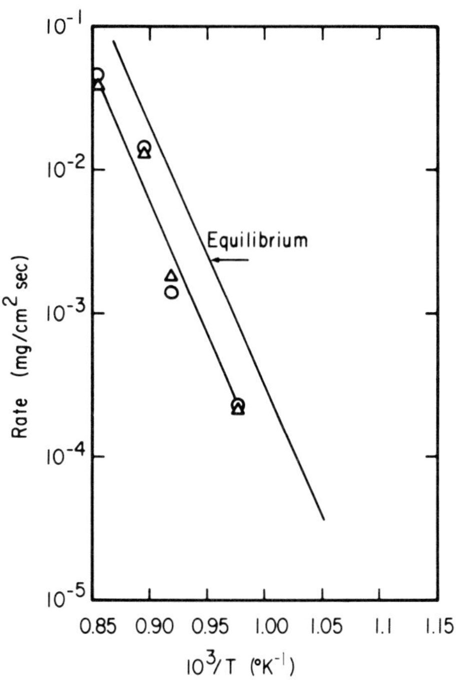

Fig. 11. The evaporation rates of the gallium (111) (○) and arsenic ($\bar{1}\bar{1}\bar{1}$) (△) faces of gallium arsenide as a function of temperature. Equilibrium data from Arthur (1967).

vaporize incongruently to give liquid gallium and arsenic molecules. The initial transient period during which the vaporization rate is changing is longer for the arsenic $(\bar{1}\bar{1}\bar{1})$ face than for the gallium (111) face.

The steady state vacuum evaporation rates were measured for both crystal faces of gallium arsenide as a function of temperature and are plotted in Fig. 11. The log R (mg/cm^2 sec) vs $1/T$ (°K) curves give straight lines above 750°C for both crystal faces, which seem to overlap within the accuracy of the measurements. Here, we have also plotted the maximum evaporation rates obtained from equilibrium studies (Arthur, 1967) as a function of temperature. These curves are parallel, and their slopes give $E^* \simeq \Delta H_v{}^\circ = 90 \pm 4$ kcal/mole of solid. The vacuum evaporation rates however are less than one third of the maximum rates in the temperature range 750–900°C. Thus the average evaporation coefficient $\bar{\alpha}(T)$ is about $\bar{\alpha}(T) \simeq 0.27$.

B. Mass Spectrometric Studies

During the vaporization of gallium arsenide the ions readily detectable in the mass spectra were As_4^+, As_2^+, As^+, and As_3^+. The dimer and tetramer ion peaks had the largest intensities, and they were of comparable magnitude in the temperature range of our study. The experimental geometry (Fig. 2) has the advantage of allowing us to sample directly the composition of the vapor that emanates from the vaporizing gallium arsenide surface. Thus any association reaction that may take place on the heater or chamber walls (Arthur, 1967) would only change the background intensity distribution, which can be measured independently and has been subtracted from the signal. Both the As^+ and As_2^+ can be identified (Lou and Somorjai, 1969) as fragmentation products of the electron impact ionization of As_2 and As_4 molecules.

The intensities of the As_2^+ and As_4^+ peaks that were detected during the vaporization of the gallium face (111) of gallium arsenide are plotted as a function of temperature in Fig. 12. The activation energies of vaporization are $E_D^* = 94.5$ kcal/mole and $E_T^* = 95.9$ kcal/mole of vapor respectively. The intensity ratios As_2^+/As_4^+ that were obtained at different temperatures once steady state vaporization was reached are given in Table II. The log I^+T vs $1/T$ (°K) plots for the arsenic $(\bar{1}\bar{1}\bar{1})$ face are shown in Fig. 13. The activation energies of vaporization are $E_D^* = 99.6$ kcal/mole and $E_T^* = 142$ kcal/mole of vapor, respectively. The intensity ratios As_2^+/As_4^+ that were obtained at different temperatures in steady state are also given in Table II. The results indicate that the activation energies of vaporization of tetramer arsenic molecules are markedly different for the two crystal faces.

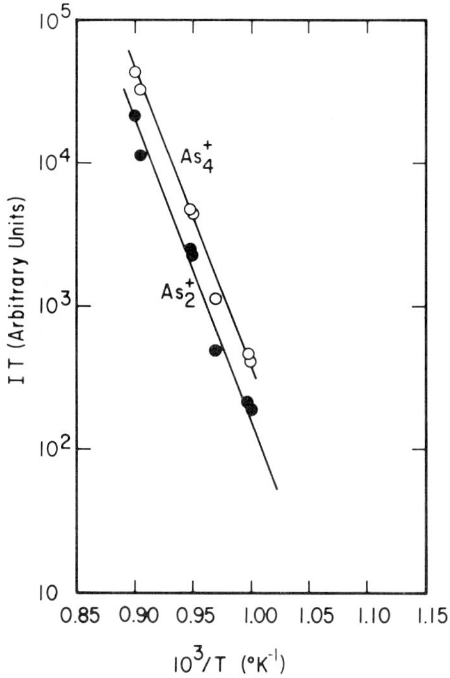

Fig. 12. Representative curves of the ion intensities As_2^+ and As_4^+ as a function of temperature for the (111) face of GaAs.

TABLE II. INTENSITY RATIOS As_2^+/As_4^+ FOR THE GALLIUM (111) AND ARSENIC ($\bar{1}\bar{1}\bar{1}$) FACES AT DIFFERENT TEMPERATURES

Ga(111) face		As($\bar{1}\bar{1}\bar{1}$) face	
T (°C)	As_2^+/As_4^+	T (°C)	As_2^+/As_4^+
728	0.47	705	3.57[a]
730	0.48	746	0.81
758	0.44	798	0.78
780	0.49	809	0.73
782	0.51	842	0.71
832	0.44	—	—
838	0.48	—	—

[a] May not have reached steady state.

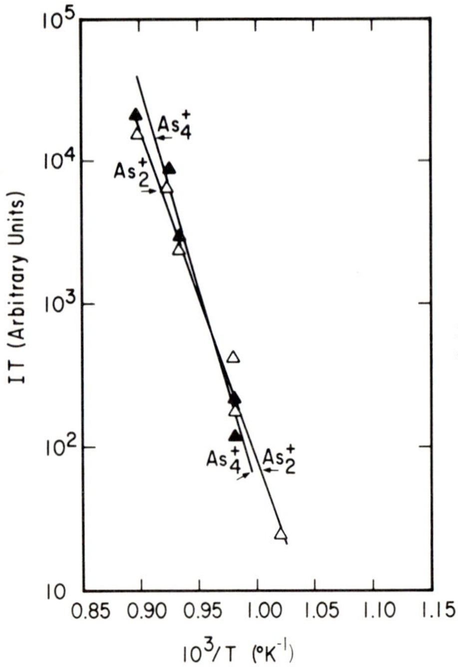

Fig. 13. Representative curves of the ion intensities As_2^+ and As_4^+ as a function of temperature for the $(\bar{1}\bar{1}\bar{1})$ face of GaAs.

The activation energy for As_4 vaporization is larger for the arsenic $(\bar{1}\bar{1}\bar{1})$ face than for the gallium (111) face $[E_T^*(\bar{1}\bar{1}\bar{1}) > E_T^*\ (111)]$, while for As_2 it is similar for both faces. This is clearly reflected in the tabulated intensity ratios.

C. Effect of Liquid Gallium on the Vaporization of GaAs

When the (111) crystal faces were covered with excess liquid gallium and were then vaporized in vacuum, the total evaporation rates (microbalance runs) increased more than three fold and reached the maximum equilibrium value throughout the studied temperature range. Mass spectrometer studies also showed that the intensities of the As_2^+ and As_4^+ peaks increased markedly for the (111) crystal faces with respect to the intensities detectable in the absence of liquid gallium. However, the intensity ratios As_2^+/As_4^+ and the activation energies of vaporization for all of the arsenic species remained the *same* as that for the pure samples. Thus the presence of liquid gallium on the vaporizing surface merely increases the evaporation rates of all of the vaporizing molecules. It does not change the vapor composition or the energetics of the vaporization surface reactions.

D. Discussion

The vacuum evaporation rates from both faces are roughly one-third the maximum evaporation rate ($J^{\mathrm{v}}(T)/J^{\mathrm{v}}_{\max}(T)=\bar{\alpha}=0.28$), while the average activation energy of vaporization (kcal/mole of solid) is the same as the equilibrium heat of vaporization (kcal/mole of solid), $E^* \simeq \Delta H_{\mathrm{V}}=90 \pm 4$ kcal/mole of solid. The vaporization rate seems to be controlled by the magnitude of the preexponential factor of the overall rate equation J (mole/sec) $= K_0(S) \exp(-E^*/RT)$, where K_0 is a constant related to the frequency of attempted motion of a molecule over the energy barrier E^*, and (S) is the concentration (moles/cm²) of the molecules in the surface sites from which vaporization proceeds.

The vapor compositions (As_2/As_4) over the vaporizing (111) and ($\bar{1}\bar{1}\bar{1}$) faces of gallium arsenide were found to be different. This appears to be due to the markedly different activation energies of vaporization of the tetramer (As_4) molecules from the gallium ($E^*=95.9$ kcal/mole of vapor) and the arsenic ($E^*=142$ kcal/mole of vapor) crystal faces, respectively. This result indicates that the formation of the tetramer molecules is dependent on some physical-chemical or structural property of the particular crystal face. Differences in dislocation densities (Lester and Somorjai, 1968), charge density distributions (Somorjai and Lester, 1967), or impurity concentrations (Somorjai and Lester, 1967) in the (111) and ($\bar{1}\bar{1}\bar{1}$) faces may play an important role in determining the surface concentration of the vaporizing species.

In the presence of liquid gallium the total vaporization rate of the (111) face increases to obtain the maximum equilibrium rate. At the same time, however, the vapor composition remain unaffected by the presence of the liquid metal. It appears that liquid gallium merely increases the number of surface sites from which vaporization can proceed. It does not change the mechanism of vaporization, which would be reflected in the lowering of the activation energy of vaporization or in the changing of the vapor composition. Thus, Ga(l) does not shift the equilibrium in any way between gallium and arsenic species on the gallium arsenide (111) surface. There is evidence from other vaporization studies that different liquid metals of low vapor pressure [In and Ga on GaN (Schoonmaker *et al.*, 1965), Tl on As (Brewer and Kane, 1955)] can also accelerate the evaporation of IIIA–VA or VA compounds. Therefore, liquid gallium may not be the only liquid metal that can increase the evaporation rate of gallium arsenide; other metals probably could also speed up the evaporation process.

Several observations indicate that there is no equilibrium between the dimer (As_2) and tetramer (As_4) arsenic molecules on the vaporizing surface. If the tetramers were formed via the equilibrium association of dimers,

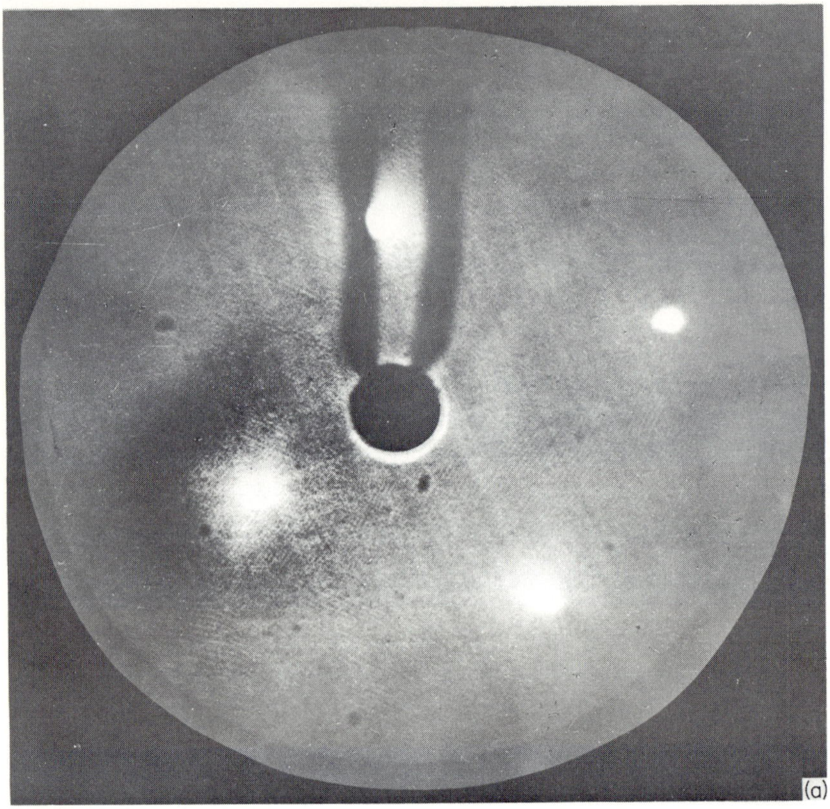

Fig. 14. (a) Diffraction pattern of an annealed face centered cubic metal (100) face at $E = 58$ eV.

$2D \rightleftharpoons T$ (an exothermic reaction [Arthur, 1967] $\Delta H = -62$ kcal/mole), then the dimer to tetramer ratio As_2/As_4 would have decreased while the evaporation rate increases in the presence of liquid gallium: $D/T \approx 1/D$. This does not happen. The presence of the liquid metal does not change the vapor compositions on the two studied surfaces. In addition, the experimental fact that the dimer to tetramer ratios are different on the two crystal faces indicates that equilibrium between these molecules cannot be established at least on one of the crystal faces.

VII. Low Energy Electron Diffraction (LEED) Study of the Vaporizing Silver (100) Surface

The diffraction of low energy electrons (5–500 eV) from single crystal surfaces can be used to obtain information about the structure of surfaces

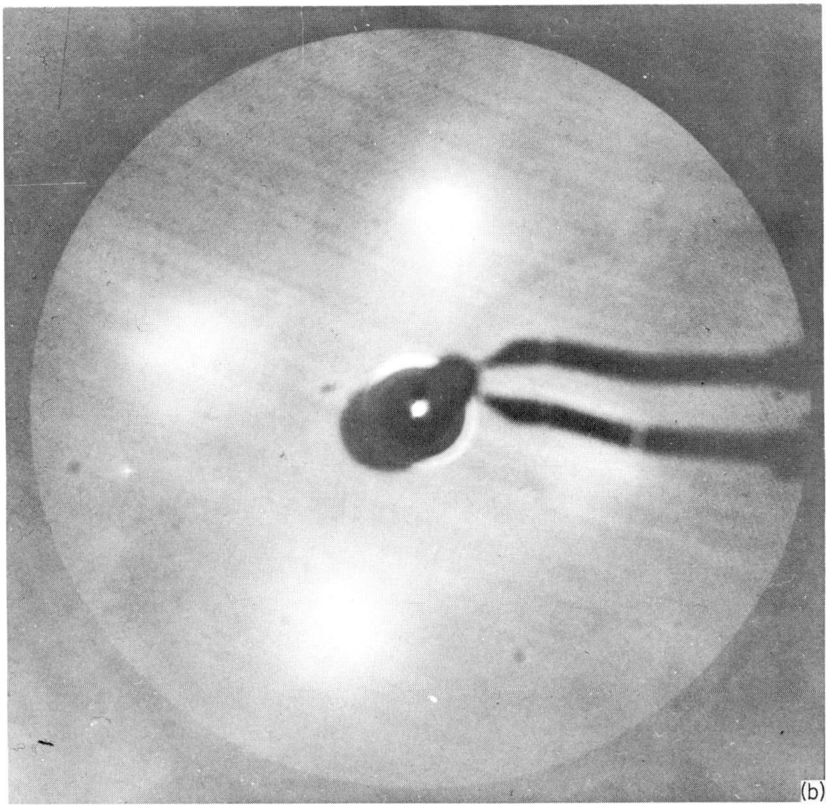

Fig. 14. (b) Pattern after ion bombardment.

on an atomic scale (Somorjai, 1968b). The diffraction pattern is representative of the periodic arrangements of atoms in the surface. In these experiments a monochromatic electron beam is back-scattered from one face of a single crystal and the elastically scattered fraction of electrons post-accelerated onto a fluorescent screen where the diffraction pattern is displayed. The studies have to be carried out in ultrahigh vacuum ($<10^{-8}$ Torr) to avoid possible contamination of the sample surface by the adsorption of gases from the ambient (Morabito *et al.*, 1969; Somorjai, 1968b).

One important discovery of low energy electron diffraction studies in recent years is that surface atoms may reside in different kinds of surface structures (Somorjai, 1968b). The presence of these structures is indicated by the appearance of extra diffraction features in the LEED pattern, which are superimposed on the diffraction pattern of the surface unit mesh that is predicted by the bulk unit cell. The surface structures can be characterized by lattice parameters that are integral multiples of the unit cell dimensions that characterize the bulk unit cell. Most semiconductors (Si, Ge, GaAs, CdS) and insulators (Al_2O_3) that have been studied so far exhibit surface rearrangements (Somorjai, 1968b). Among the metals only Au, Pt, Bi, and Sb were found to rearrange, while Al and Ni have surface unit meshes that are expected from their bulk unit cell (Somorjai, 1968b).

The purpose of these continuing studies is to obtain information about the atomic structure of the vaporizing surface. We would like to determine whether vaporization occurs from (a) an ordered surface where most atoms occupy equilibrium lattice sites or (b) a liquid-like surface layer in which long range order, which is characteristic of the bulk crystal, has been destroyed. LEED should be sensitive to detect and monitor changes in the surface structure.

It is well-known (Somorjai and Lester, 1967; Lester and Somorjai, 1968) that the surface that vaporizes into vacuum becomes " rough," i.e., exhibits high density of steps, pits, and ledges. These macroscopic changes in the surface structure cannot, in general, be detected by low energy electron diffraction. The lateral coherence length of low energy electrons is of the order of $\sim 10^2$ Å. As long as the distance between steps is larger than the coherence length, the conditions for constructive interference between the scattered electrons is not affected. Thus the presence of steps in the surface would not alter the diffracted electron beam intensities.

If the step or defect densities in the surface become so high that the ordered domains are reduced in size below $(\sim 10^2)^2$ Å^2, the intensity of the diffraction features decreases markedly; i.e., LEED becomes sensitive to surface roughness. Figure 14 shows diffraction patterns of the (100) face of platinum (a) before ion bombardment (by 300 eV argon ions) and (b) after ion bombardment. As a result of ion bombardment damage, the spots become less intense and large. It should be noted that the spot width is proportional to $1/N$, where N is the number of atoms in the ordered domains.

Ordered arrangements of surface defects or steps could, of course, also be detectable by low energy electron diffraction. Recently the presence of ordered rows of steps in the (553) face of uranium dioxide has been reported by Ellis (1968).

Silver is a face centered cubic metal; it has a melting point of 961°C. The metal has a high vapor pressure at the melting point ($>10^{-3}$ Torr), and its vapor is composed primarily of monatomic gaseous species. Vacuum sublimation studies (Winterbottom and Hirth, 1964) indicate that after a short induction period, the maximum sublimation rate is obtained within the accuracy of the experiments. The activation energy of sublimation is equal to the heat of sublimation ($E^* \simeq \Delta H_V$). It appears that equilibrium can be established in all the surface reaction steps that lead to vaporization, and that the desorption of silver atoms from the vaporizing surface requires no activation energy.

The (100) crystal face of silver was vaporized in a low energy electron diffraction chamber under conditions of free vaporization in the temperature range of 600 to 930°C. The diffraction pattern was detected and monitored during vaporization. The sample was then quenched to room temperature and the intensities of the diffraction spots were recorded. It should be noted that the intensity of the diffracted beams decreases exponentially with increasing temperature of the solid. This well-known Debye–Waller effect is due to the increased mean displacement of surface atoms with increasing temperature (Somorjai, 1968b).

The results of this low energy electron diffraction study indicate that the vaporizing Ag(100) surface retains its structure and long range order. No significant changes in either the arrangement or the intensity of the diffraction spots were found to occur during vaporization. Thus it appears that the vaporizing silver surface is ordered and certainly not liquid-like, as far as it can be determined by the LEED technique.

VIII. Conclusions

The sublimation mechanisms of sodium chloride, cadmium sulfide, and gallium arsenide simple crystals have been studied in considerable detail. Although the experimental information is insufficient to develop a more general theory of vaporization, we can identify several important parameters that seem to control the sublimation rate of different groups of compounds.

A. Composition of the Surface

Minute deviations from the solid state composition (excess vacancies) that produced a steady state sublimation rate have markedly changed the evapoıation rate. Changes in the surface concentration of gallium in GaAs and of cadmium or sulfur in CdS give rise to transient evaporation rates that approach the steady state value only slowly. It is likely that the marked lowering of the sublimation rates of calcium-doped NaCl may also be due to changes in the composition of the surface. The presence of divalent ions in the alkali halide crystal lattice creates a large concentration of sodium ion vacancies in excess of their steady state concentration in the pure crystals. In addition to point defects, line defects such as dislocations could also markedly influence the sublimation rate (NaCl). This effect should be detectable in solids in which there is a low mobility of dislocations so that they cannot anneal out at the sublimation temperature.

B. Charge Transfer

It is apparent that the capture of mobile charge carriers (electrons and holes) by the cadmium and sulfur ions to form neutral species that may vaporize is a rate-controlling step in the vaporization of CdS. Charge transfer may also play an important role in the sublimation of gallium arsenide, although conclusive experimental evidence is lacking at present. The effect of liquid metals that accelerate the sublimation process of IIIA–VA compounds may be through their providing a source of free electrons that facilitates both the dissociation of the crystal and the association of the vaporizing species (As_2, As_4). In alkali halides in which the electrons are tightly held and sublimation occurs by ion pairs, charge transfer may not be expected to be an important reaction step.

C. Impurities

Impurities that are electrically active (copper in CdS) or change the solid composition (calcium in NaCl) can markedly change the sublimation rates. It is likely that similar effects will be found in GaAs as well.

Several observations seem to hold for all three groups of compounds we have investigated.

D. Concurrent Vaporization Mechanisms

For NaCl and GaAs the vapor fluxes are composed of more than one type of vapor molecule. There seems to be *no equilibrium* on the vaporizing surface between NaCl and Na_2Cl_2 or As_2 and As_4 molecules. Sublimation occurs by two independent reaction paths under all conditions we have investigated. Surface diffusion, which would allow reaction between the molecules, does not seem to take place in these surfaces.

E. Invariance of Vapor Composition

We have studied all three compounds under a variety of experimental conditions that have changed their evaporation rates by one to two orders of magnitude. The steady state vapor compositions, however, remained virtually unchanged within the accuracy of the experiments under all conditions of sublimation.

ACKNOWLEDGMENT

This work was performed under the auspices of the United States Atomic Energy Commission.

REFERENCES

Arthur, J. R. (1967). *J. Phys. Chem. Solids* **28**, 2257.
Berkowitz, J., and Chupka, W. A. (1958). *J. Chem. Phys.* **29**, 653.
Brewer, L., and Kane, J. S. (1955). *J. Phys. Chem.* **59**, 105.
Burton, W. K., Cabrera, N., and Frank, F. C. (1951). *Phil. Trans. Roy. Soc. London Ser. A* **243**, 299.
Ellis, W. P. (1968). *Surface Sci.* **11**, 82.
Gatos, H. C., and Levine, M. C. (1960). *J. Electrochem. Soc.* **107**, 427.
Gutbier, H. B. (1961). *Z. Naturforsch.* **16a**, 268.
Hertz, H. (1882). *Ann. Physik* **17**, 177.
Hove, J. E. (1955). *Phys. Rev.* **99**, 480.
Jeunehomme, M. (1962). Ph.D. Dissertation, Univ. Libre de Bruxelles, Belgium.
Kane, J. S. (1955). Ph.D. Dissertation, Univ. of California, Berkeley, California.
Knudsen, M. (1915). *Ann. Physik* **47**, 697.
Langmuir, I. (1913). *J. Am. Chem. Soc.* **35**, 931.
Lester, J. E., and Somorjai, G. A. (1968). *J. Chem. Phys.* **49**, 2940.
Lou, K., and Somorjai, G. A. (1969). *Conf. Effusion Vaporization, Los Alamos, April 1969.*
Melville, H. (1936). *Trans. Faraday Soc.* **32**, 1017.
Miller, D. P., Harper, J. G., and Perry, T. R. (1961). *J. Electrochem. Soc.* **108**, 1123.
Morabito, J. M., and Somorjai, G. A. (1968). *J. Metals* **20**, 17.
Morabito, J. M., Steiger, R. F., Muller, R. H., and Somorjai, G. A. (1969). *Proc. Intern. Materials Symp. 4th Univ. of California, Berkeley, California, June 1969.* Wiley, New York. To be published.
Neuhaus, A., and Retting, W. (1958). *Z. Elektrochem.* **62**, 33.
Palmberg, P. W., and Rhodin, T. N. (1968). *J. Appl. Phys.* **39**, 2425.
Rosenblatt, G. (1963). *J. Electrochem. Soc.* **110**, 563.
Rosenblatt, G. M., and Lee, P.-K. (1968). *J. Chem. Phys.* **49**, 2995.
Russell, G. J., Lip, H. K., and Haneman, D. (1966). *J. Appl. Phys.* **37**, 3328.
Schoonmaker, R. C., Buhl, A., and Lemley, J. (1965). *J. Phys. Chem.* **69**, 3455.
Searcy, A. W. (1968). *Pacific Conf. Chem. Spectry., San Francisco, California, November 1968.*
Somorjai, G. A. (1964a). *In* "Condensation and Evaporation of Solids" (E. Rutner, P. Goldfinger, and J. P. Hirth, eds.). Gordon and Breach, New York.
Somorjai, G. A. (1964b). *Surface Sci.* **2**, 298.
Somorjai, G. A. (1968a). *Science* **162**, 755.
Somorjai, G. A. (1968b). *Ann. Rev. Phys. Chem.* **19**, 251.
Somorjai, G. A., and Jepsen, D. W. (1964a). *J. Chem. Phys.* **41**, 1389.

Somorjai, G. A., and Jepsen, D. W. (1964b). *J. Chem. Phys.* **41**, 1394.
Somorjai, G. A., and Lester, J. E. (1965a). *J. Chem. Phys.* **42**, 4140.
Somorjai, G. A., and Lester, J. E. (1965b). *J. Chem. Phys.* **43**, 1450.
Somorjai, G. A., and Lester, J. E. (1967). *Progr. Solid State Chem.* **4**, 1.
Somorjai, G. A., and Lyon, H. B. (1965). *J. Chem. Phys.* **43**, 1456.
Somorjai, G. A., and Stemple, N. R. (1953). *J. Appl. Phys.* **35**, 3398.
Stranski, I. N. (1928). *Z. Phys. Chem.* (*Leipzig*) **136**, 259.
Winterbottom, W. L. (1966). Ph.D. Dissertation, Carnegie Inst. of Technol., Pittsburgh, Pennsylvania.
Winterbottom, W. L., and Hirth, J. P. (1964). *In* " Condensation and Evaporation of Solids " (E. Rutner, P. Goldfinger, and J. P. Hirth, eds.). Gordon and Breach, New York.
Woodbury, H. H. (1964). *Phys. Rev.* **134**, 492.
Zimm, B., and Mayer, J. (1944). *J. Chem. Phys.* **12**, 362.

High Temperature Phase Studies With a Solar Furnace

Tetsuo Noguchi

SOLAR RESEARCH LABORATORY, GOVERNMENT INDUSTRIAL RESEARCH INSTITUTE, HIRATE-MACHI, KITA-KU, NAGOYA, JAPAN

I. Introduction 235
II. The Nature of a Solar Furnace 236
A. The Optical System of a Solar Furnace 236
B. Structure and Types of Solar Furnaces 238
C. Attainable Temperatures 239
D. Irradiation Techniques 240
III. Temperature and Emissivity Measurement 243
A. Specular Reflection Method 243
B. Centrifugal Blackbody Furnace 244
IV. Freezing Point Determination of High Melting Refractory Oxides .. 244
V. Liquidus Curve Measurement 248
VI. Phase Studies at High Temperatures 250
A. High Temperature Reactions 250
B. Vaporization and Rate Studies 251
C. High Temperature Phase Diagrams 252
VII. Summary and Conclusions 259
References 260

I. Introduction

A great deal of research effort has been concentrated on establishing the phase diagrams of high melting refractory materials and on phase studies and measurement of properties at elevated temperatures. In recent developments of high temperature chemistry, various kinds of heating devices have been utilized in the temperature range beyond 2000°C. Induction and arc furnaces, the plasma torch, electron beam bombardment apparatus, and solar and arc imaging furnaces (called thermal imaging devices) have been applied to the heat treatment of specimens in varied atmospheres. Magnetic or electric fields may also be introduced in most of these furnaces, in vacuum, neutral, or reducing atmospheres. However, contamination from the crucible and wall materials or the heating elements is inevitable in all these heating devices except those that employ thermal imaging, and heating in an oxidizing atmosphere such as air at higher temperatures can be achieved only with thermal imaging techniques, with which there is easy atmosphere control.

In earlier work with solar furnaces some refractory systems have been studied by melting specimens and quenching from the melt without determination of a heating temperature. In the thermal imaging techniques temperature measurement has been widely investigated to find a means of separating the emitted and reflected radiation impinging on the pyrometer from the specimen. It was not until 1962 that methods for accurate temperature measurement were established.

This chapter will describe temperature and emissivity measurements in the thermal imaging technique, freezing and melting point measurements of metal oxides, and phase studies on binary oxide systems at higher temperatures with a solar furnace.

II. The Nature of a Solar Furnace

A. The Optical System of a Solar Furnace

The solar furnace is not actually a " furnace " but rather an optical system designed to concentrate solar radiation at the focal plane of a paraboloidal reflector or lens. A specimen at the focus is heated by absorption of the concentrated radiation energy reflected from the concave mirror; a part of energy may be reflected or transmitted through the target specimen. Solar radiation may be replaced by the radiation from a plasma torch or high intensity arc as in the case of an arc imaging furnace. A xenon lamp may also be used as the radiation source.

Radiation impinging on a paraboloidal mirror from the sun is not strictly parallel to its optical axis but has a solid angle of about 32′. Accordingly the image at the focal plane of an ideal paraboloid has an optional size depending on the focal length of the paraboloid. The size of the image is given by $d = 2 \tan 16' = 0.0093 f$, where d is the diameter of the sun's image and f is the focal length of the paraboloid. The ratio of the diameter of the paraboloid D to the focal length is called the aperture ratio n of the paraboloidal reflector and is expressed from the geometry as

$$n = D/f = 4[(1 - \cos \theta)/(1 + \cos \theta)]^{1/2}$$

where θ is the angle between the optical axis and the reflected light as shown in Fig. 1.

The highest heat flux density E_d obtained in the area of diameter d at the focal plane may be expressed as

$$E_d = [4RC_aS/(0.0093)^2] \sin^2 \theta$$

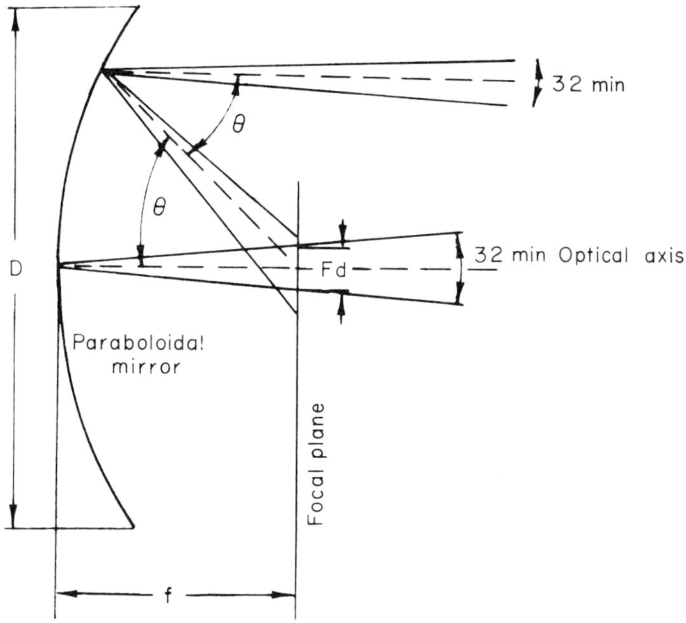

Fig. 1. Cross section of a paraboloidal reflector.

where R is the reflectivity of the paraboloid (in percent), C_a is the absorption coefficient of the air mass perpendicular to the surface of the earth, and S is the solar constant (0.135 W/cm^2).

From these equations the relations among θ, n, and E_d are calculated. Examples based on the assumption that $R=85\%$ and $C_aS=0.1$ W/cm^2 are shown in Figs. 2 and 3, respectively. In Fig. 2 curve (1) indicates the calculated data for a blackbody, and curve (2) represents a body in which one-half of the radiation energy is absorbed. Curve (3) is a plot of $E_d/[4RC_aS/(0.0093)^2]$.

Figure 3 shows the energy distribution at the focal plane with varied aperture ratios. The longer the focal length of the paraboloid mirror, the larger the sun's image becomes; but the highest attainable temperature remains fixed for any given aperture ratio. To obtain a higher temperature for the target specimen the value of the aperture ratio must be increased, but not to a value greater than 4 where the rim angle reaches to 90°, because of the difficulty in sample positioning. The aperture ratio for most solar furnaces is approximately 2.3–2.4.

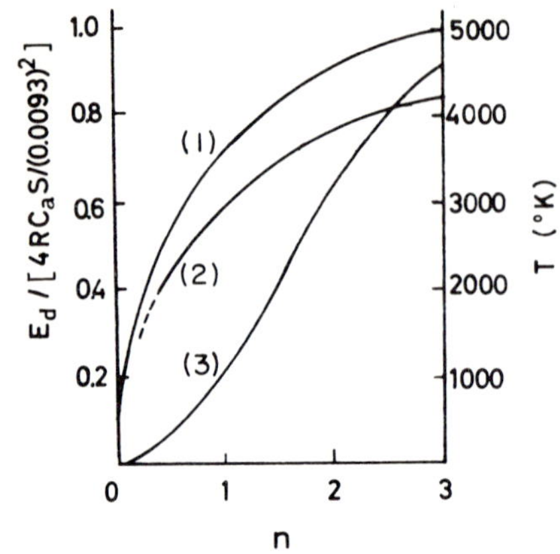

Fig. 2. The relations between aperture ratio n and the maximum heat flux density or temperature.

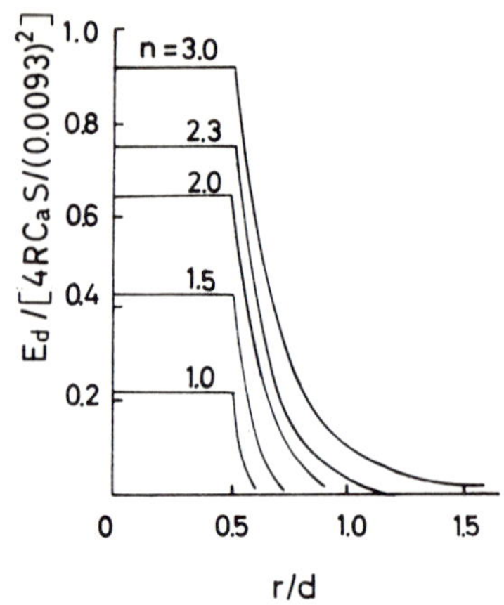

Fig. 3. Energy distribution at the focal plane with varied aperture ratios. n is the aperture ratio. r is the distance from the center of the sun's image. d is the diameter of the sun's image.

B. Structure and Types of Solar Furnaces

The reflecting mirror of a solar furnace is mounted in different ways depending on the purpose of the experiment. Several mirror arrangements are shown in Fig. 4.

The highest heat flux density is attainable in the direct type solar furnace, while a lower energy input might be expected in the heliostat type solar furnace because of the multiple reflection. However, the heliostat type shown in (b), where the optical axis is horizontal, has been adopted for the convenience of the experimental techniques in most solar furnaces. The heliostat type shown in (c), where the optical axis is vertical, is recommended in order to obtain a uniform temperature distribution on the target specimen, especially in single crystal growth studies.

In the smaller solar furnaces, surplus searchlight mirrors have often been used. The paraboloidal mirror of such a searchlight consists of a back-silvered glass reflector or rhodium coated copper or stellite mirror. Mirror surfaces coated with aluminum are also used because of their high reflectivity. In large installations the mirror is composed of segments that are aligned to generate the paraboloid.

A lens type solar furnace was built at California Institute of Technology in 1936, but the difficulty in obtaining a lens of large aperture has made the construction of such solar furnaces impractical. The movement of the sun is

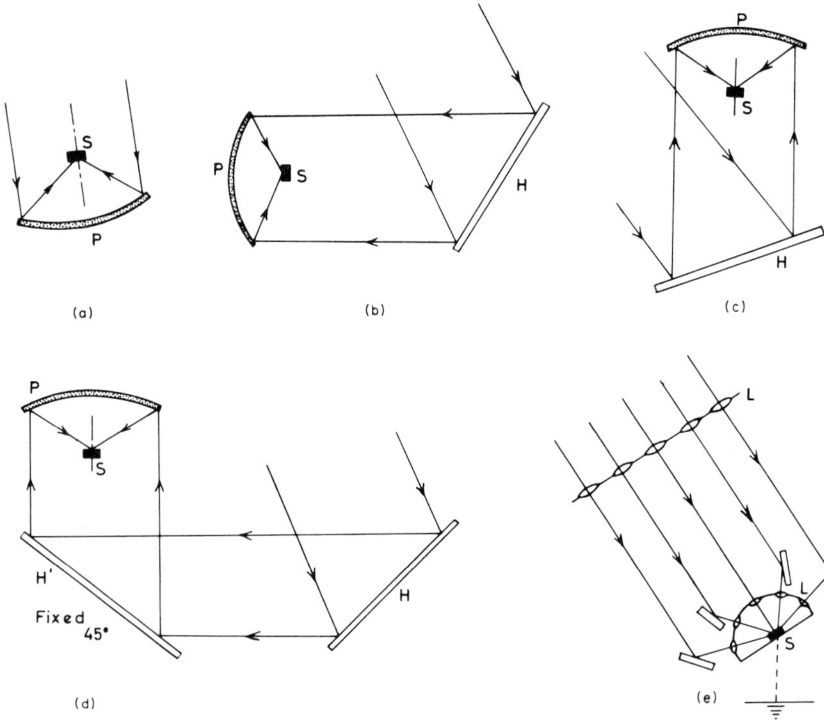

Fig. 4. Schematics of the solar furnace. *P*, paraboloidal mirror; *S*, sample; *H*, auxiliary plane mirror; *L*, lens. (a) Direct type. (b) Heliostat type (optical axis horizontal). (c) Heliostat type (optical axis vertical). (d) Heliostat type (optical axis vertical). (e) Lens type.

tracked by a combination of phototubes, photocells, and servomechanisms, or by synchronous motors. A summary of image furnace techniques used throughout the world has been published by Laszlo (1965).

C. Attainable Temperatures

The sun is considered as a blackbody radiator the temperature of which is calculated from Wien's displacement law as about 5800°K. The highest attainable temperature for a practical solar furnace will be lower than this value because of attenuation by the air mass—i.e., absorption by water vapor, carbon dioxide gas, dust and air pollution—as well as geometrical imperfection and nonuniformity of the reflecting surface of the paraboloid. A calculation showed that the theoretical attainable temperature for the heliostat type solar furnace at the Solar Research Laboratory, Nagoya, was estimated to be 4220°K for a blackbody target; however, it is reasonable to estimate that the highest temperatures being reached empirically by most solar furnaces might be about 3500°C.

The temperature profile of a molten ZrO_2 specimen heated at the focal plane is shown in Fig. 5 with varied heat flux density as reported by Noguchi

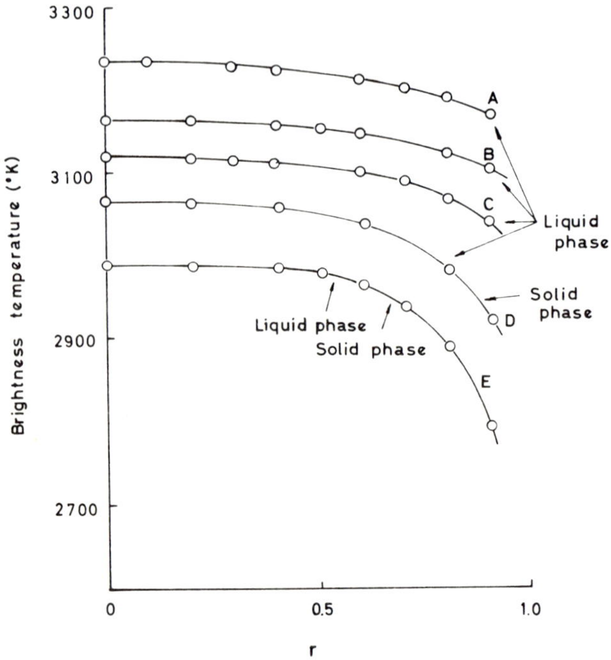

Fig. 5. Temperature profile of a melted ZrO_2 specimen. Shutter openings: A, 90° (100%); B, 88° (98%); C, 85° (93%); D, 70° (67%); E, 66° (61%).

and Kozuka (1966a). From the temperature distribution across the sun's image, the temperature difference between the center and periphery of the image is estimated to be more than 50°C for the ZrO_2 specimen.

D. Irradiation Techniques

A specimen may be heated in any atmosphere at the focal plane of a solar furnace—oxidizing, reducing, neutral, or in vacuo or in air. Trombe *et al.* (1961) reported that zone melting was possible when the specimen was heated along the optical axis as shown in Fig. 6, by moving the target specimen within the focal plane. They also proposed a centrifugal blackbody furnace (Fig. 7) to obtain a relatively uniform temperature inside a cavity

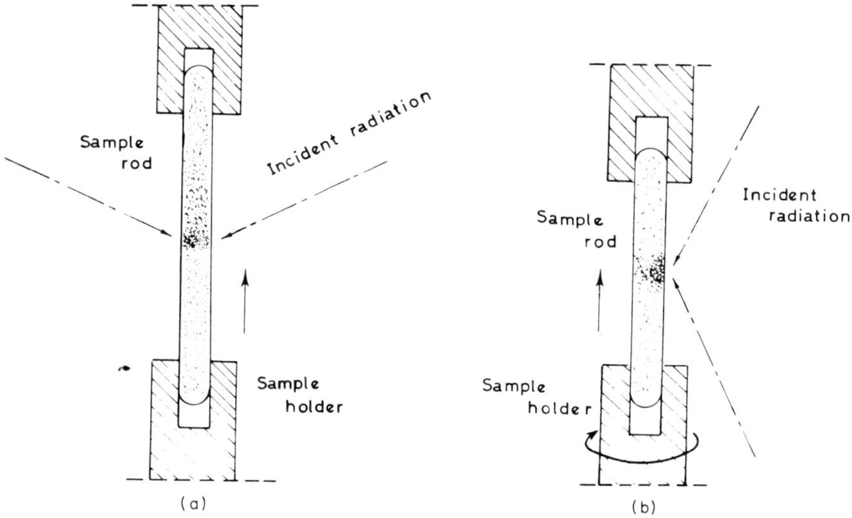

Fig. 6. Schematics of zone melting with solar furnace.

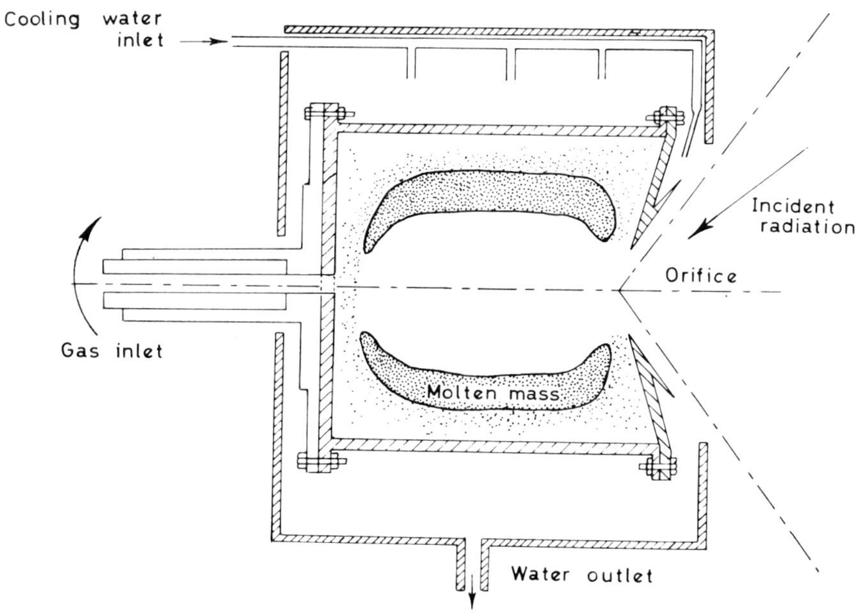

Fig. 7. Centrifugal blackbody cavity.

by realizing a blackbody condition. A high frequency plasma torch attached to a Verneuil type sample feeder was combined with an imaging furnace by Foex (1964). He suggested the use of plasma torch as an auxiliary heating source through the rear side of a centrifugal cavity.

In order to control the heating atmosphere, a Pyrex or fused silica container such as a tube, flask, or hemispherical cover is often used. Noguchi *et al.* (1961) reported a continuous vacuum fusion apparatus. Within the fused silica tube of 50 mm dia, an oxide specimen was fused at 10^{-6} Torr. Laszlo (1961) and Glaser (1961) reported the hemispherical devices. Figure 8 shows one of these containers.

Any type of sample holder should be equipped with a three dimensional positioner controlled either by a servomechanism or a manual adjustment so that a specimen may be located exactly at the focal plane. The heating

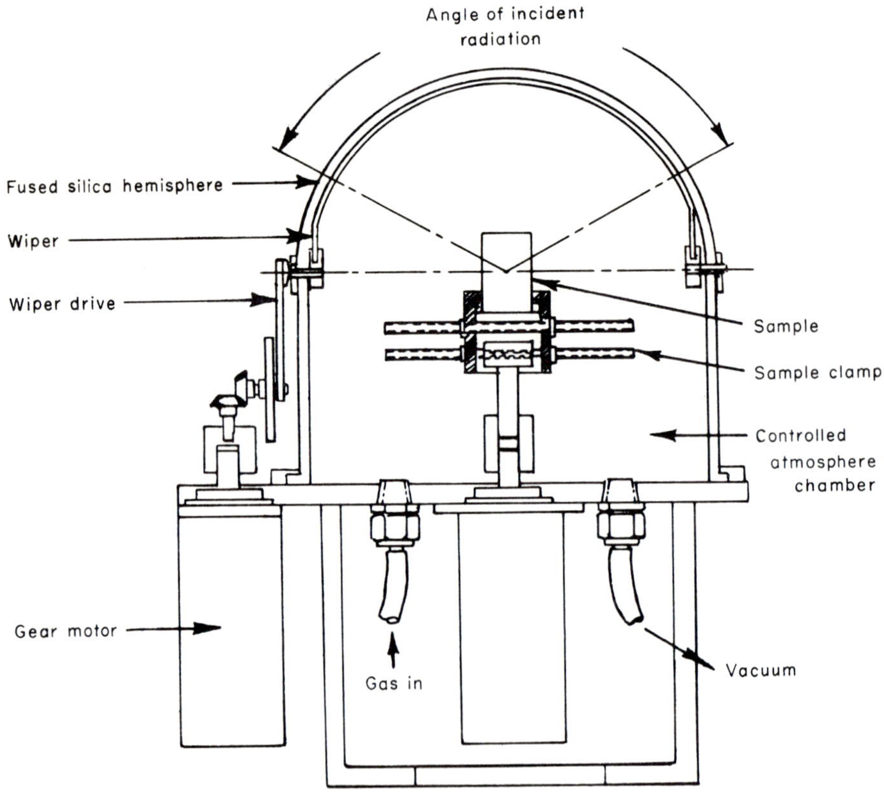

Fig. 8. Controlled atmosphere sample holder with wiper (Laszlo, 1961).

temperature is easily controlled by the operation of a Venetian-blind attenuator, shutter, door, or curtain in front of the paraboloid. Moving a specimen away from the focal plane to achieve temperature control is not recommended. A satisfactory window material for irradiation under high pressure has not yet been developed, although Cook (1964) designed a pressurized arc-imaging furnace that was installed inside a pressure vessel at 150 psi.

III. Temperature and Emissivity Measurement

The emission from a specimen heated by a solar furnace, impinging on a pyrometer, consists of the temperature radiation and reflected solar radiation from the specimen surface. The temperature radiation must be separated from the reflected in order to obtain the brightness temperature of the specimen at the focal plane. Methods of accurate temperature measurement in thermal imaging techniques have been sought by many investigators by the separation of the reflection from the emission of the heated specimen. Conn and Braught (1954) proposed using a rotating sector to separate the incident and reflected radiations from the specimen surface in his solar furnace work. Since then Glaser (1958), Blau (1960), Comstock and Camus (1964), and Laszlo *et al.* (1964) have reported the use of a specific filter, and use of a rotating sector, a chopper, and a synchronous shutter to give a periodic heat flow to a target specimen. In temperature measurement Kamada (1964) and Mann (1965) have reported the selection of a wavelength in water absorption bands on the earth at 1.38 μm, and Noguchi *et al.* (1967a) have made measurements at 2.72 μm. The specular reflection surface of a molten specimen has been used to separate the two radiations by Diamond and Schneider (1960) and Noguchi and Kozuka (1966a). Foex (1965) has demonstrated the use of a cavity or centrifugal blackbody furnace. The latter two methods have been applied to measure the temperature in transient phenomena such as heating or cooling processes, while the other methods were used either in temperature measurement in steady state condition or to obtain the brightness temperature. The methods of temperature and emissivity measurement in recent high temperature phase studies with a solar furnace will be described in the following paragraphs.

A. Specular Reflection Method

On the specularly reflecting surface of an opaque body fused with a heliostat type solar furnace one can observe the distorted image of the heliostat plane mirror. If a shadowing plate is inserted between the heliostat and paraboloid, the image of the plate is also visible within the image of the heliostat. Thus the reflected solar radiation is rejected, and the brightness temperature of a heated specimen is determined when the pyrometer is focused within the area of the image of the shadowing plate.

When the shadowing plate is transferred within the effective area of the paraboloid, the reflectance joins with the emitted radiation and the difference between the reflected plus emitted radiation and emitted radiation alone gives the spectral reflectivity as compared with energy impinging on the paraboloid. The spectral emissivity is thus calculated from Kirchhoff's equation assuming that the spectral transmittance is negligible. This assumption was verified by Noguchi and Kozuka (1966a) using a specimen of 6 mm thickness and also by Diamond and Dragoo (1966) for a molten alumina drop.

The true temperature of a specimen heated with a solar furnace is then calculated from the data on the brightness temperature and spectral emissivity. The measurement is made with a brightness pyrometer having a built-in reference lamp calibrated with a tungsten filament standard lamp.

B. Centrifugal Blackbody Furnace

By the use of a water-cooled centrifugal blackbody furnace in which the refractory oxide powder is placed inside a cylindrical metal container coated with copper, the blackbody condition is approximated with the cavity, the orifice of which is about 11 mm in diameter. Any type of optical pyrometer may be used for temperature measurement in this device; however, the signal from the pyrometer must be calibrated by a known temperature standard such as the melting point of alumina. The effect of the reflected solar radiation is thus cancelled, and the true temperature of the heated specimen is determined by this method.

IV. Freezing Point Determination of High Melting Refractory Oxides

The heating temperatures and spectral emissivity of ceramic oxides at temperatures near but above their melting points can be measured by the methods already described. When the solar radiation is intercepted by an attenuator, a cooling curve of a specimen is observed and the brightness temperature of the freezing point can be measured from these curves. The use of an oscilloscope or a rapid-response instrument is recommended in the observation of such transient phenomena. On the other hand, if the melted specimen is quenched at a very high speed from high temperatures, the freezing might be considered to proceed in a nonequilibrium state. In such a dynamic method special attention must be paid to the thermal history, cooling rate, and structure of the specimen as well as to the establishment of equilibrium before quenching. This method will be unsuitable for poorly or sluggishly crystallizing material such as silicates.

Examples of the cooling curves for ZrO_2 and Al_2O_3 derived by a dual beam cathode ray oscilloscope equipped with a Polaroid camera are shown in Fig. 9. In the rapid cooling of a pure liquid compound, supercooling is

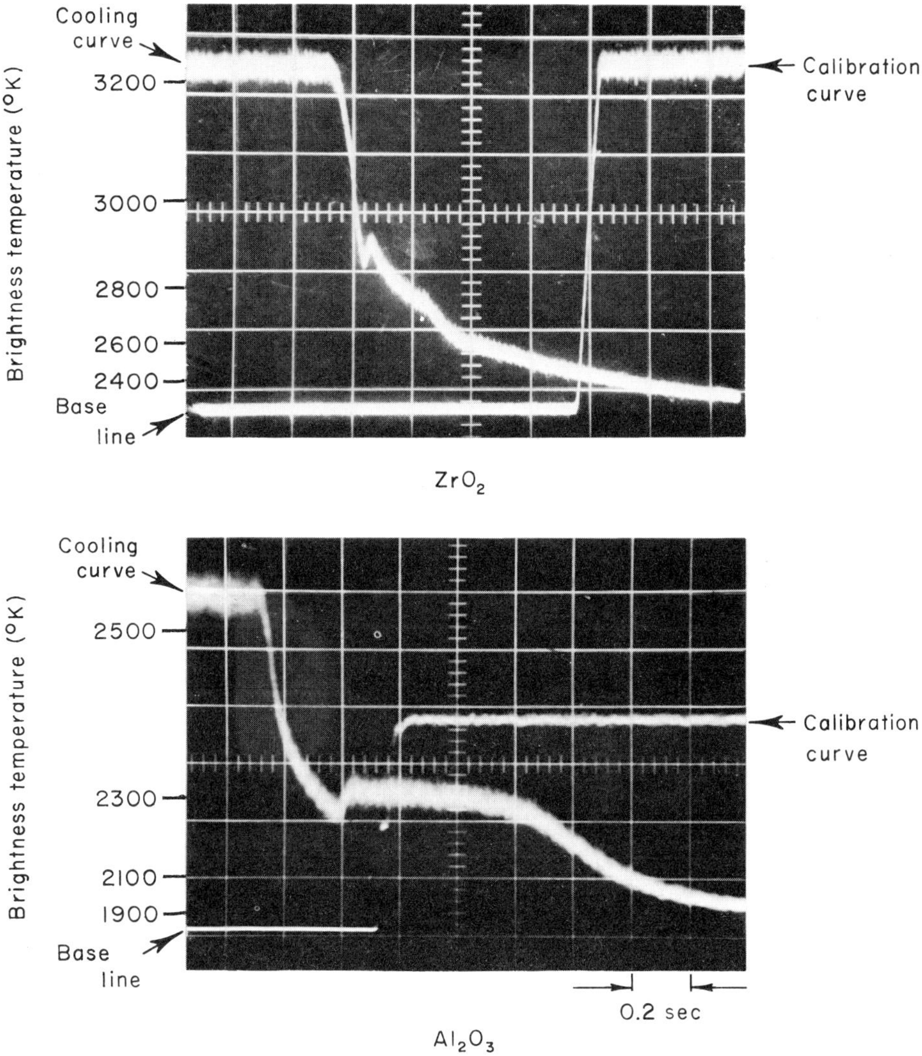

Fig. 9. Oscillograms of the cooling curves for ZrO_2 and Al_2O_3.

almost always observed. α-Al_2O_3 gives a typical cooling curve as shown in the oscillogram. ZrO_2 shows the sharp exothermic peak ascribed to the heat liberated on crystallization, followed by two weak exothermic peaks, which might possibly be due to transformations in the solid state. HfO_2 shows the same type of cooling curve. The cooling rate of the ZrO_2 specimen is about 2300 deg/sec in the range from 3200° to less than 1000°K as in the figure. The accuracy of the brightness pyrometer used in the measurement was found by Takada *et al.* (1967) to be within an error of $\pm 7°$ at 3237°K, and that of the derived temperature is estimated to be within $\pm 20°$ at 3000°K including the reading errors of oscillograms. The spectral emissivities of ZrO_2 and Al_2O_3 were found to be 0.81 and 0.97 at their respective freezing points within an error of ± 0.005.

Effects of the variation of heating temperature and the position of the measurement on the specimen surface during freezing point determination were found to be negligible, and the results agreed within the error of the measurement. The results for several metal oxides are presented with those of fifteen lanthanide sesquioxides in Table I. Noguchi and Kozuka (1966a)

TABLE I. FREEZING POINT DATA OF CERAMIC OXIDES

Oxide	Freezing point (± 20°C)	Spectral emissivity (0.65 μm)	Purity (%)
ZrO_2	2706	0.81	99.8
HfO_2	2753	0.91	99.4
Al_2O_3	2065	0.97	99.99
CaO	2585	0.92	99
SrO	2599	0.95	99
Sc_2O_3	2403	0.96	99.9
Y_2O_3	2376	0.96	99.9
La_2O_3	2256	0.95	99.9
Pr_6O_{11}	2183	0.98	99.9
Nd_2O_3	2233	0.97	99.9
Sm_2O_3	2269	0.98	99.9
Eu_2O_3	2291	0.95	99.9
Gd_2O_3	2330	0.98	99.9
Tb_4O_7	2303	0.97	99.9
Dy_2O_3	2228	0.98	99.9
Ho_2O_3	2330	0.98	99.9
Er_2O_3	2344	0.94	99.9
Tm_2O_3	2341	0.98	99.9
Yb_2O_3	2355	0.98	99.9
Lu_2O_3	2427	0.93	99.9

gave the freezing point of ZrO_2 as 2979°±20°K, HfO_2 as 3026°±20°K, and α-Al_2O_3 as 2343°±15°K; those of Ln_2O_3 by Noguchi and Mizuno (1967a) are shown in the table. The brightness temperatures of the two weak exothermic peaks for ZrO_2 were 2767°K and 2598°K, and those of HfO_2 were 2875°K and 2697°K, respectively. The latter peaks for both oxides might be related to the cubic to tetragonal transformation as claimed by Smith and Cline (1962). Both ZrO_2 and HfO_2 specimens quenched from the melt in air showed the monoclinic structure, while alumina included less than 2–3% of the gamma alumina of cubic type in the quenched specimens. The effects of the existence of this compound and its behavior near the melting point of alumina on its freezing point should be examined.

By the use of a heliostat type solar furnace, Foex (1965) determined the freezing points of several oxides, using alumina as a basis for comparison at 2042°C: he reported 2710°±10°C for ZrO_2, 2910°C for CaO, 2660°C for SrO, and 2010°C for BaO. The freezing points of lanthanide oxides and lanthanide chromites are also given by the same author (1966) as shown in Fig. 10.

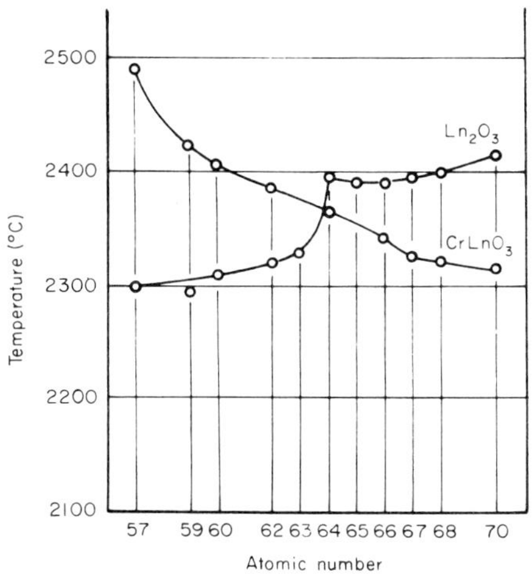

Fig. 10. Freezing points of lanthanide oxides and lanthanide chromites.

In his compilation of the melting points of metal oxides, Schneider (1963) observed wide variation in reported values. Since 1966 research efforts have been concentrated on establishing temperature standards by the Task Force on Secondary Temperature Standards under the High Temperature Commission of IUPAC. The first reaction to be measured is the melting point of alumina. The solar furnace is to be used for this purpose, among other heating equipment.

Heating curve measurement was also investigated by Noguchi and Kozuka (1966b) with a heliostat type solar furnace. By the specular reflection method the melting points of ZrO_2 and Y_2O_3 are given as $2995° \pm 20°K$ and $2686° \pm 20°K$ with spectral emissivities of 0.84 and 0.96, respectively. However, the separation of emitted and reflected radiations was rather difficult as compared with the cooling curve measurement, and alumina failed to give melting point data.

V. Liquidus Curve Measurement

The freezing point of a single oxide is readily determined from the sharp exothermic peak in the cooling curve. For an intermediate composition of a polycomponent system, particular care must be taken with the interpretation of a cooling curve in such a dynamic method because the cooling process might proceed in a nonequilibrium manner. The formation of a compound with congruent or incongruent melting point, eutectic mixture, and solid solution have often been observed when new phases appear. The thermal history, cooling rate, and impurities contained will have an effect on the freezing behavior of a fused composition even if equilibrium is established before quenching. The peritectic reaction, segregation, and stability of a compound might also affect the accuracy in the liquidus curve measurement.

In the cooling process of a typical composition, the relations between the phase diagrams and cooling curves for two compositions are shown schematically in Fig. 11. The heat effects of phase changes are indicated by the shaded area. It is important to note that the liquidus line will be obtained with the solidus line when correct cooling curve data are plotted.

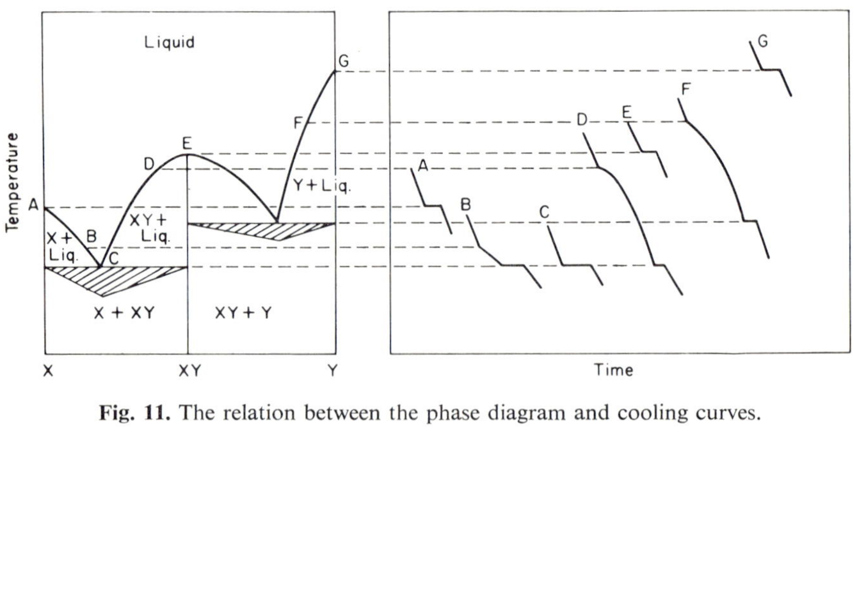

Fig. 11. The relation between the phase diagram and cooling curves.

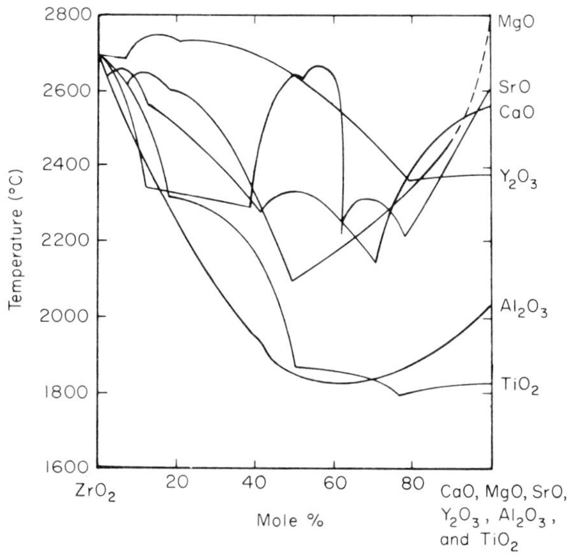

Fig. 12. Liquidus curves of binary oxide systems containing ZrO_2.

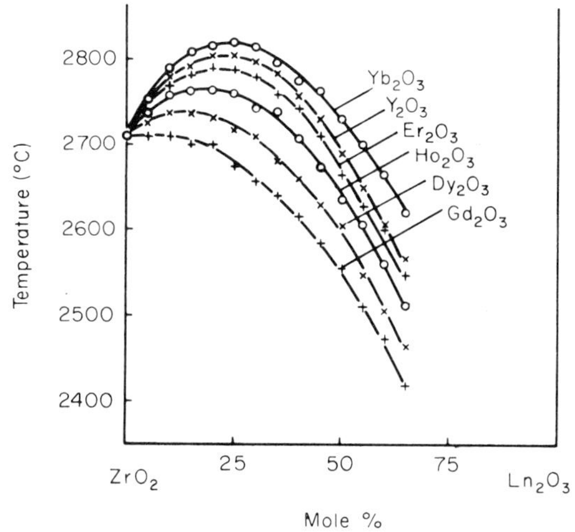

Fig. 13. Liquidus curves of the ZrO_2–Ln_2O_3 system.

The solar furnace has been used in the liquidus curve measurement of high melting refractory oxide systems. In earlier work by Conn and Braught (1954) the system Al_2O_3–SiO_2 was investigated. Noguchi (1969) reported the liquidus curves of binary systems containing ZrO_2 as one of the end members—ZrO_2–CaO, ZrO_2–MgO, ZrO_2–Y_2O_3, ZrO_2–TiO_2, ZrO_2–SrO, and ZrO_2–Al_2O_3—as well as Al_2O_3–Y_2O_3. Anomalies not reported in the previous papers are observed in each system as shown in Fig. 12. The La_2O_3–Cr_2O_3 system was examined by Foex (1962), showing the liquidus curve together with the solidus line in the system. Foex and Rouanet (1967) also published the liquidus curves in the ZrO_2–Ln_2O_3 systems; these results are presented in Fig. 13. A maximum point appears in ZrO_2 solid solution region at higher temperatures than the freezing point of ZrO_2. Thus the use of an image technique has recently shown the high temperature behavior of refractory oxide svstems as a dynamic method.

VI. Phase Studies at High Temperatures

A. High Temperature Reactions

By the use of a solar furnace all solid materials can be fused within a few seconds; a few more seconds are necessary to reach thermal equilibrium. The specimen can be kept for prolonged heating at desired temperatures up to about 3500°C. A heating experiment in clear sky showed that a specimen could be heat treated at 3000°K for about 1 hr within an accuracy of $\pm 10°$ (Noguchi *et al.*, 1965). On the other hand, the vapor pressure of a component increases with temperature. Vaporization and sometimes decomposition may result from prolonged heating. Therefore, repeated fusion for long periods may not be desirable in the preparation of a material at very high temperatures.

Solid–solid and liquid–solid reactions are of importance in the field of high temperature synthesis. Chemical reaction and synthesis of a compound in polycomponent systems are expected to proceed very rapidly at high temperatures, especially by fusion, unless decomposition occurs or metastable structures are formed. A heliostat type solar furnace has been used in the synthesis of boron carbide from graphite and boron in the small vacuum chamber by Vuillard (1961). The reduction of Eu_2O_3 in high vacuum, resulting in the formation of the monoxide was reported by Achard (1961). The reaction between sesquioxide and carbon or boron was also studied by the same author, and europium protoxide was found to be an intermediate product. Hexaboride was reported as the stable compound in the reduction of the sesquioxide with boron.

Many composite minerals have been produced at the Solar Energy Laboratory, Mont-Louis (C.N.R.S.) for example, production of ZrO_2 by fusion of zircon, and spinel ($Al_2O_3 \cdot MgO$), forsterite ($2MgO \cdot SiO_2$), chromite ($Cr_2O_3 . MgO$), etc. were studied with the large solar furnace for high temperature refractory materials. Muster (1962) reported that sixteen artificial minerals such as gehlenite ($2CaO \cdot Al_2O_3 \cdot SiO_2$), akermanite ($2CaO \cdot MgO \cdot 2SiO_2$), diopside ($CaO \cdot MgO \cdot 2SiO_2$), etc. have been successfully synthesized by fusion with a solar furnace from high purity MgO, CaO, Al_2O_3, and SiO_2 without contamination.

B. Vaporization and Rate Studies

Kinetic studies of decomposition reactions, oxidation, and nitridation of materials are very important in high temperature technology. Vaporization studies at elevated temperatures have been widely made using balances of various types in vacuo and in controlled atmospheres. These techniques are applicable in solar furnace work for rate determination in varied atmospheres. The weight changes of several refractory oxides in air at their respective melting temperatures were presented by Noguchi and Mizuno (1967b), and Noguchi *et al.*, (1967b) as shown in Table II. These data were

TABLE II. Vaporization Rate of Ceramic Oxides at the Melting Temperatures

Oxide	Weight loss		Melting point
	(g/cm² min)	(mole/cm² min)	(±20°C)
ZrO_2	5.3×10^{-4}	4.3×10^{-6}	2706
SrO	1.7×10^{-2}	1.6×10^{-4}	2599
Al_2O_3	1.8×10^{-4}	1.8×10^{-6}	2065
Y_2O_3	4.2×10^{-4}	1.9×10^{-6}	2376
MgO	9.0×10^{-3}	2.2×10^{-4}	(2800)
$SrZrO_3$	3.4×10^{-3}	1.5×10^{-5}	2646
$MgO\cdot ZrO_2$	2.3×10^{-3}	1.4×10^{-5}	2121

obtained to examine the stoichiometry of specimens in the Al_2O_3–Y_2O_3 and ZrO_2–SrO systems after fusion for 10 min. A thermobalance is under design by Noguchi at the Solar Research Laboratory in combination with a heliostat type solar furnace.

The vaporization rates of metal oxides such as TiO_2, Nb_2O_3, Ta_2O_3, and La_2O_3 as well as of mixtures of the binary systems La_2O_3–Nb_2O_3, La_2O_3–TiO_2, and La_2O_3–Ta_2O_3 were studied by Segui and Trombe (1968). The weight decreases of four compounds were given as $TiO_2=4$, $Nb_2O_3>2.5$, $Ta_2O_3=1$, and $La_2O_3=3$ mg/cm² min. Vaporization studies on the Ln_2O_3–Cr_2O_3 system are also being made at the Ultra-Refractory Laboratory, Mont-Louis.

1. ZrO_2–TiO_2 *System*

In the liquidus curve measurements a new peritectic point was announced by Noguchi and Mizuno (1968a) for the 17.5 mole % TiO_2 composition at 2343°±20°C. The compound $ZrTiO_4$ showed an incongruent melting point at 1878°±15°C, and a quenched specimen from the melt included from 60 to 70% $ZrTiO_4$ together with ZrO_2 and TiO_2. The solid solution boundaries for ZrO_2, $ZrTiO_4$, and TiO_2 were determined with x-ray techniques as shown in Fig. 14. The diagrams proposed by Brown and Duwez (1954) and Cocco and Torriano (1965) on the ZrO_2–TiO_2 system are revised, and the existence of a new zirconia solid solution stable at above 2500°C as well as ZrO_2 cubic solid solution is suggested with the aid of cooling curve data.

2. ZrO_2–SrO *System*

The compound $SrZrO_3$ is investigated for use as a duct wall refractory material in magnetohydrodynamic power generation; however, few papers have been published on phase studies in the ZrO_2–SrO system at higher temperatures. The phase diagram of the system in lower temperature regions was studied by Tilloca and Jorba (1964) and also by Cocco and Chiacigh (1965). The tentative diagram proposed by Noguchi *et al.* (1967b) is shown in Fig. 15. A peritectic point for 12 mole % SrO at 2343°±20°C, and four eutectic points for 38, 51, 61, and 77 mole % SrO at 2295°, 2622°, 2202°, and

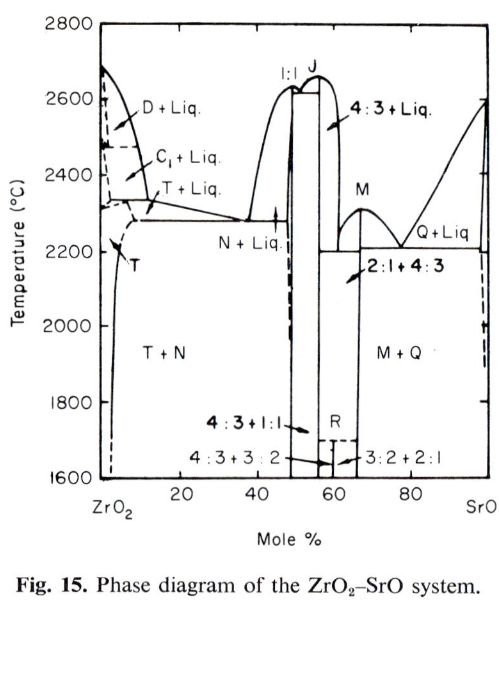

Fig. 15. Phase diagram of the ZrO_2–SrO system.

$2224° \pm 20°C$, respectively, were found. The compounds $Sr_nZr_{n-1}O_{3n-2}$ as $n=4$, 3, and 2 were also confirmed with $SrZrO_3$. The freezing points of $SrZrO_3$, $Sr_4Zr_3O_{10}$, and Sr_2ZrO_4 were found as 2646°, 2674°, and $2325° \pm 20°C$, respectively, and $Sr_3Zr_2O_7$ was stable below 1700°C.

3. ZrO_2–CaO *System*

The ZrO_2–CaO system is a most important zirconia refractory, and many papers have been published up to the present, including some inconsistent results. The phase diagram of the system by Duwez *et al.* (1952) has been generally accepted and partially revised by Barbariol (1965).

The liquidus curve of the system was presented by Noguchi *et al.* (1967c), and anomalies not reported previously were found for 3.5, 7.0, and 10.0 mole % CaO at 2603°, 2622°, and $2578° \pm 20°C$, respectively. A maximum point was observed at 7.0 mole % CaO as shown in Fig. 16. The composition quenched from the melt was found to be a single phasic of apparent rhombic form with an optic angle of $2V = -76°$ and birefringence $\gamma - \alpha = 0.013$, rather than a distorted cubic lattice. Further, the quenched composition from the melt was readily decomposed into the monoclinic and tetragonal forms when reheated at 1300°C. The solid solution boundaries were determined by x-ray diffraction measurements. However, the interpretation of

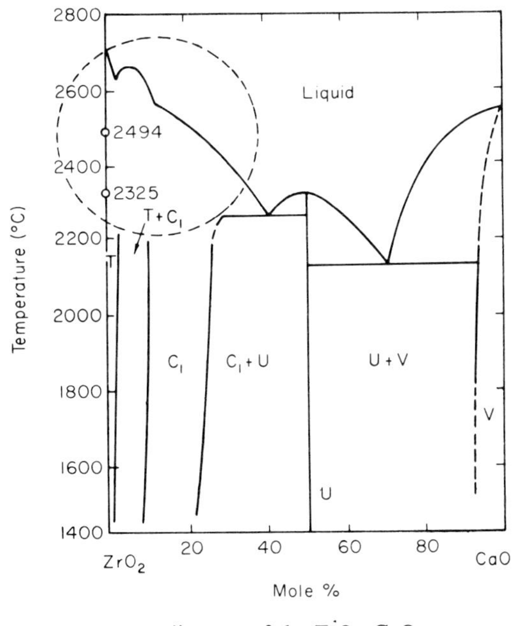

Fig. 16. Phase diagram of the ZrO_2–CaO system.

data in the region enclosed with the dashed line was difficult because the relations between cubic ZrO_2 and cubic zirconia solid solution as well as a stable region for $Ca_{0.07}Zr_{0.93}O_{1.93}$ are not clear at present. These regions should be investigated with a dynamic method such as high temperature x-ray diffractometry in the thermal imaging technique. Analogies in the liquidus curves among the ZrO_2–CaO, ZrO_2–MgO (Noguchi and Mizuno, 1968b), and ZrO_2–Y_2O_3 systems in the ZrO_2 rich region are also indicated in Fig. 12, showing the respective maximum points.

4. ZrO_2–La_2O_3 *System*

The ZrO_2–La_2O_3 system was first investigated by Brown and Duwez (1955). Later the existence of $La_2Zr_2O_7$, a face centered cubic structure of pyrochlore type, was established by Roth (1956) and Jorba and Collongues (1959). Rouanet (1968) reported on the high temperature phase diagram of the system from the results of thermal analysis with a solar furnace and x-ray diffractometry as shown in Fig. 17.

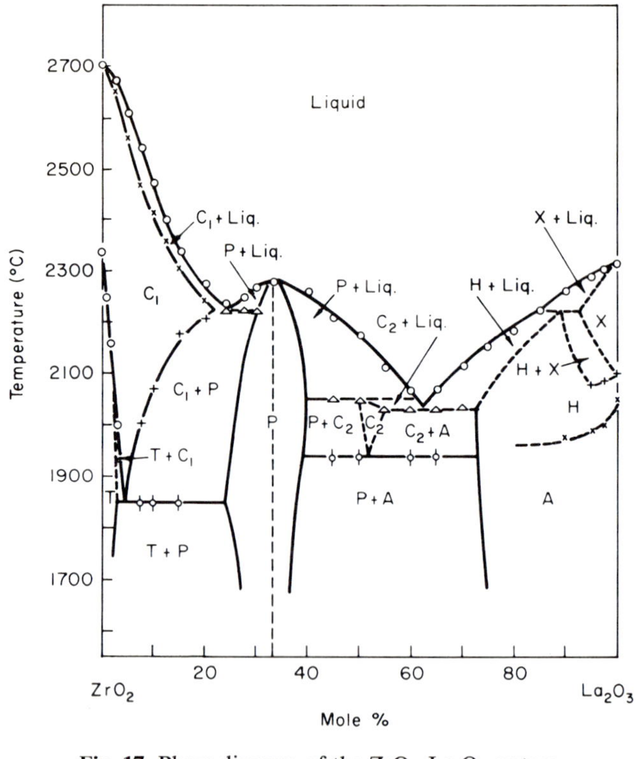

Fig. 17. Phase diagram of the ZrO_2–La_2O_3 system.

In the liquidus curve of the system, two eutectic points were found for 25 mole % La_2O_3 at 2220°C and 62.5 mole % at 2030°C. The freezing points of ZrO_2, La_2O_3, and $La_2Zr_2O_7$ are announced as 2710°, 2310°, and 2280°C, respectively. Rouanet claimed an invariant point as eutectoid for 5 mole % La_2O_3 at 1850°C, where a cubic phase of CaF_2 type and a two phase region of tetragonal ZrO_2 solid solution plus pyrochlore exist contiguously. A cubic structure of Tl_2O_3 type was suggested for the equimolecular mixture $ZrO_2{\cdot}La_2O_3$ at temperatures above 1900°C, and a new cubic phase *X* claimed by Foex and Traverse (1966) was also included for the La_2O_3 side at high temperatures.

5. ZrO_2–Sm_2O_3 *and* ZrO_2–Gd_2O_3 *Systems*

Only a few papers on the ZrO_2–Sm_2O_3 and ZrO_2–Gd_2O_3 systems have been published. Jorba (1962) examined both systems, and later Rouanet and Foex (1968) reported on the high temperature phase diagram by the use of a solar furnace. Their results are shown in Figs. 18 and 19, respectively.

In the ZrO_2–Sm_2O_3 system, a eutectic point was found for 75 mole % Sm_2O_3 at 2190°C. The existence of a compound $Sm_2Zr_2O_7$ of cubic structure

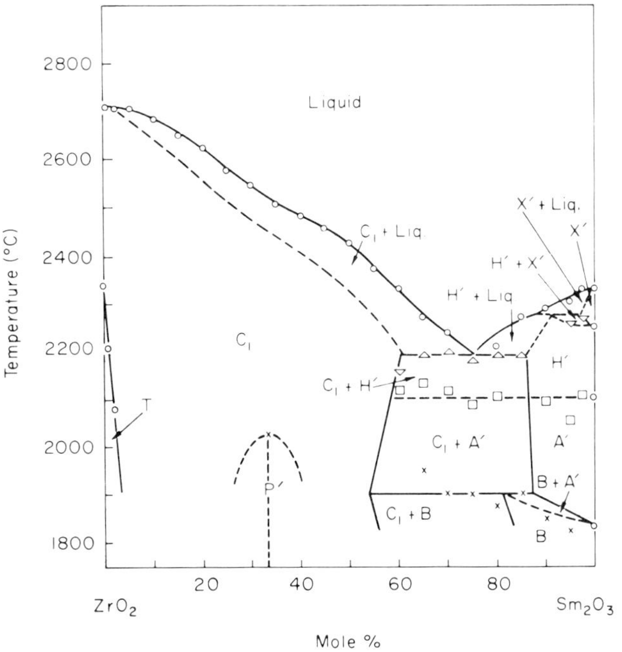

Fig. 18. Phase diagram of the ZrO_2–Sm_2O_3 system.

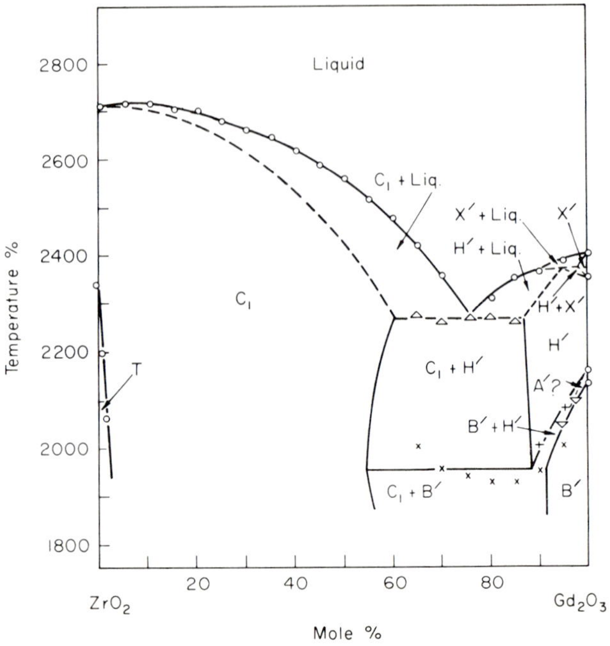

Fig. 19. Phase diagram of the ZrO_2–Gd_2O_3 system.

of pyrochlore type was assumed below 2025°C, while the liquidus curve shows the weak inflection at 33.3 mole % Sm_2O_3. The immiscible range of the cubic ZrO_2 solid solution and the hexagonal Sm_2O_3 solid solution as well as the Gd_2O_3 solid solution existed from about 60 to 87 mole % Ln_2O_3. The solubility limit of ZrO_2 in both Sm_2O_3 and Gd_2O_3 was about 13 mole % at 2200°C. In the ZrO_2–Gd_2O_3 system, a eutectic point was reported for 76 mole % Gd_2O_3 at 2260°C. The new cubic phase *X* for both sesquioxides are also indicated at higher temperatures.

6. Al_2O_3–Y_2O_3 *System*

The phase diagram of the Al_2O_3–Y_2O_3 system has been widely studied by Schneider *et al.* (1961), Olds and Otto (1964), and Toropov *et al.* (1964) giving different results. The three compounds in the system, $3Y_2O_3 \cdot 5Al_2O_3$, $Y_2O_3 \cdot Al_2O_3$, and $2Y_2O_3 \cdot Al_2O_3$, are confirmed by Yoder and Keith (1951), Roth (1957), and Reed and Chase (1962), respectively. The equimolecular compound of perovskite structure $Y_2O_3 \cdot Al_2O_3$ was reported as a metastable form by Warshaw and Roy (1959); however, conclusive data on the liquidus curve of the system have not been reported yet. The results obtained by solar heating are presented by Noguchi and Mizuno (1967b) in Fig. 20. The liquidus curve between 40 and 60 mole % Y_2O_3 showed a gentle peak that

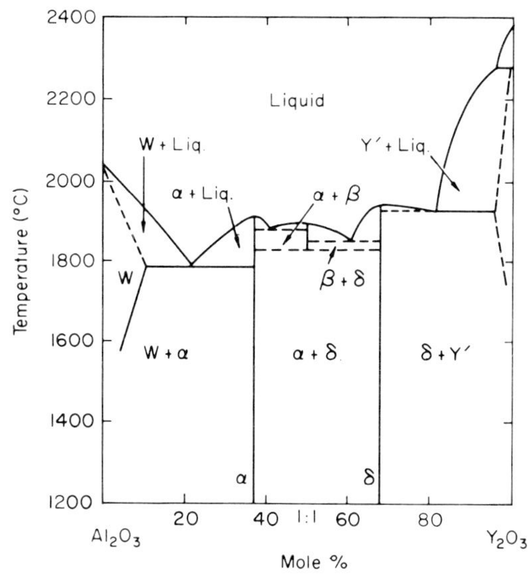

Fig. 20. Phase diagram of the Al_2O_3–Y_2O_3 system.

might signify the metastable compound $Y_2O_3 \cdot Al_2O_3$, which was readily decomposed into a mixture of garnet, perovskite, and monoclinic $2Y_2O_3 \cdot Al_2O_3$ at 1600°C. The solid solution boundaries in the Y_2O_3-rich side were estimated from the cooling curve data.

VII. Summary and Conclusions

This chapter presents briefly the state of the art regarding high temperature studies on solid–solid reactions and solid–liquid or condensed phase–gas equilibria with a solar furnace. Freezing point measurement of single oxides, liquidus curve measurements, and phase diagram studies on high temperature refractory oxides are summarized. One of the other promising fields is the crystal growth of high melting ceramics such as ThO_2 single crystals from the vapor phase as reported by Laszlo *et al.* (1967). Many papers have already been published on crystal growth by an arc imaging furnace in combination with the commercial Verneuil system (Glaser and Walker, 1964).

The solar furnace may also be used for heating specimens in high temperature x-ray diffractometry with a θ–θ goniometer that rotates in a horizontal plane about the vertical axis in which the sample is stationary. The equipment has been designed by Stecura (1968) at National Bureau of Mines, College Parks, Md., and also by Noguchi at the Solar Research Laboratory, Nagoya. Diffraction data of oxide systems above 2000°C in air will be obtained with these devices.

It should be pointed out that one of the most important fields in high temperature chemistry is property measurements of materials at high temperatures. An effort to obtain a quantitative heat of fusion for ZrO_2 has failed because of the lack of knowledge of thermal properties such as thermal conductivity, specific heat, and emissivity at very high temperatures in the heating curve measurement by Noguchi and Kozuka (1966b). The development of instrumentation and heating devices for measuring thermal and electrical conductivities, thermal diffusivity, heat content, specific heat, thermal expansion, etc. to high accuracy at elevated temperatures is an urgent necessity, and the solar furnace can accelerate such developments with pioneering endeavours in the future.

REFERENCES

Achard, J. (1961). *U. N. Conf. New Sources of Energy, Rome*, 1961. III-F. S 108, New York, United Nations, 1964. E/3577/rev. 1. Doc. ST/ECA/72.

Barbariol, I. (1965). *Ann. Chim. (Rome)* **55**, 321.

Bauer, G., and Littke, W. (1960). *J. Inorg. Nucl. Chem.* **16**, 67.

Blau, Jr., H. (1960). *Proc. Intern. Symp. High Temp. Technol., Asilomar Calif.*, 1959, pp. 45–53. McGraw-Hill, New York.

Brown, F. H., and Duwez, P. (1954). *J. Am. Ceram. Soc.* **37**, 129.

Brown, F. H., and Duwez, P. (1955). *J. Am. Ceram. Soc.* **38**, 95

Cocco, A., and Chiacigh, I. (1965). *Ann. Chim. (Rome)* **55**, 1341.

Cocco, A., and Torriano, G. (1965). *Ann. Chim. (Rome)* **55**, 153.

Comstock, D. F., and Camus, A. L. (1964). *In* " Thermal Imaging Techniques " (P. E. Glaser and R. F. Walker, eds.), pp. 271–277. Plenum Press, New York.

Conn, W. M., and Braught, G. (1954). *J. Opt. Soc. Am.* **44**, 45.

Cook, J. C. (1964). *In* " Thermal Imaging Techniques " (P. E. Glaser and R. F. Walker, eds.), pp. 55–76. Plenum Press, New York.

Diamond, J. J., and Schneider, S. J. (1960). *J. Am. Ceram. Soc.* **43**, 1.

Diamond, J. J., and Dragoo, A. L. (1966). *Rev. Intern. Hautes Temp. Refract.* **3**, 273.

Duwez, P., and Loh, E. (1957). *J. Am. Ceram. Soc.* **40**, 321.

Duwez, P., Odell, F., and Brown, F. H. (1952). *J. Am. Ceram. Soc.* **35**, 109.

Evans. P. E. (1960). *J. Am. Ceram. Soc.* **43**, 443.

Foex, M. (1964). *In* " Thermal Imaging Techniques " (P. E. Glaser and R. F. Walker, eds.), pp. 141–162. Plenum Press, New York.

Foex, M. (1965). *Solar Energy* **9**, 61.

Foex, M. (1966). *Rev. Intern. Hautes Temp. Refract.* **3**, 309.

Foex, M., and Rouanet, A. (1967). *Compt. Rend.* **264**, Sér. C, 947.

Foex, M., and Traverse, J. P. (1966). *Rev. Intern. Hautes Temp. Refract.* **3**, 429.

Glaser, P. E. (1958). *Solar Energy* **2**, 7.

Glaser, P. E. (1961). *U. N. Conf. New Sources of Energy, Rome*, 1961, III-F. S 16. New York, United Nations, 1964. E/3577/rev. 1. Doc ST/ECA/72.

Glaser, P. E., and Walker, R. F., eds. (1964). " Thermal Imaging Techniques." Plenum Press, New York.

Jorba, P. Y. (1962). *Ann. Chim. (Paris)* **7**, 479.

Jorba, P. Y., and Collongues, R. (1959). *Bull. Soc. Chim. France* p. 1967.

Kamada, O. (1964). *Appl. Opt.* **3**, 1397.

Laszlo, T. S. (1961). *U. N. Conf. New Sources of Energy, Rome*, 1961, III-F. S 5. New York, United Nations, 1964. E/3577/rev. 1. Doc ST/ECA/72.

Laszlo, T. S. (1965). " Imaging Furnace Techniques." Wiley (Interscience), New York.

Laszlo, T. S., Gannon, R. E., and Seehan, P. J. (1964). *Solar Energy* **8**, 105.

Laszlo, T. S., Seehan, P. J., and Gannon, R. E. (1967). *J. Phys. Chem. Solids* **28**, 313.

Mann, K. (1965). *Solar Energy* **9**, 136.

Muster, J. (1962). *Solar Energy* **6**, 129.

Noguchi, T., and Kozuka, T. (1966a). *Solar Energy* **10**, 125.

Noguchi, T., and Kozuka, T. (1966b). *Solar Energy* **10**, 203.

Noguchi, T., and Mizuno, M. (1967a). *Solar Energy* **11**, 90.

Noguchi, T., and Mizuno, M. (1967b). *Kogyo Kagaku Zasshi* **70**, 834.

Noguchi, T., and Mizuno, M. (1968a). *Bull. Chem. Soc. Japan* **41**, 2895.

Noguchi, T., and Mizuno, M. (1968b). *Bull. Chem. Soc. Japan* **41**, 1583.

Noguchi, T. (1969). *Solar Energy Conf. Phoenix, Arizona, March* 1965 and 3rd Intern. Conf. High Temp. Technol., Asilomar, Calif., 1967. To be published in *Proc. High Temp. Technol.* Butterworths, London (1969).

Noguchi, T., Mizuno, M., Nakayama, N., and Hayashi, H. (1961). *U. N. Conf. New Sources of Energy, Rome*, 1961. III-F. S 57. New York, United Nations, 1964. E/3577/rev. 1. Doc ST/ECA/72.

Noguchi, T., Mizuno, M., Kozuka, T., Yoshida, M., and Yamamoto, T. (1965). *Nagoya Kogyo Gijutsu Shikensho Hokoku* **14**, 28.

Noguchi, T., Mizuno, M., and Ito, H. (1967a). *Calorimetric Conf. of Japan, 3rd*, Osaka, November 1967, A1 13.

Noguchi, T., Okubo, T., and Yonemochi, O. (1967b). *Solar Energy Conf.* VII–3 *Tempe, Arizona*, May 1967 (*J. Am. Ceram. Soc.* **52**, 178).

Noguchi, T., Mizuno, M., and Conn, W. M. (1967c). *Solar Energy* **11**, 145.

Olds, L. E., and Otto, H. E. (1964). *In* " Phase Diagrams for Ceramists " (E. M. Levin, C. R. Robbins, and H. F. McMurdie, eds.). Am. Ceram. Soc., Columbus, Ohio.

Reed, J. W., and Chase, A. B. (1962). *Acta Cryst.* **15**, 812.

Roth, R. S. (1956). *J. Res. Natl. Bur. Std.* **56**, 17.

Roth, R. S. (1957). *J. Res. Natl. Bur. Std.* **58**, 75.

Rouanet, A. (1968). *Compt. Rend.* **267**, Sér. C, 395.

Rouanet, A., and Foex, M. (1968). *Compt. Rend.* **267**, Sér. C, 873.

Royère, C., and Trombe, F. (1968). *Compt. Rend.* **267**, Sér. C, 1275.

Schneider, S. J. (1963). *Natl. Bur. Std. (U.S.) Monograph* **68**.

Schneider, S. J., Roth, R. S., and Waring, J. L. (1961). *J. Res. Natl. Bur. Std. A* **65**, 345.

Segui, M., and Trombe, F. (1968). *Compt. Rend.* **267**, Sér. C, 1039.

Smith, D. K., and Cline, C. F. (1962). *J. Am. Ceram. Soc.* **45**, 249.

Stecura, S. (1968). *Rev. Sci. Instr.* **39**, 760.

Takada, S., Ito, H., Noguchi, T., and Kozuka, T. (1967). *Keiryo Kenkyu-sho Hokoku* **16**, 206.

Tilloca, G., and Jorba, P. Y. (1964). *Rev. Intern. Hautes Temp. Refract.* **1**, 331.

Toropov, N. A., Bondari, I. A., Galokov, F. Y., Nikogosyan, K. S., and Ninigradova, N. V. (1964). *Izv. Akad. Nauk SSSR Ser. Khim.* **7**, 1158.

Trombe, F., Foex, M., and Blanchetais, C. H. L. (1961). *U. N. Conf. New Sources of Energy, Rome*, 1961, III-F. S 35. New York, United Nations, 1964. E/3577/rev. 1. Doc ST/ECA/72.

Vuillard, G. (1961). *U. N. Conf. New Sources of Energy, Rome*, 1961. III-F. S 36. New York, United Nations, 1964. E/3577/rev. 1. Doc ST/ECA/72.

Wang, K., Dreger, L. H., Dadape, V. V., and Margrave, J. L. (1960). *J. Am. Ceram. Soc.* **43**, 509.

Warshaw, I., and Roy, R. (1959). *J. Am. Ceram. Soc.* **42**, 434.

Yoder, H. S., and Keith, M. L. (1951). *Am. Mineralogist* **36**, 519.

Author Index

Numbers in italics show the page on which the complete reference is listed.

A

Abadie, D., 62, *74*
Abuaf, N. A., 110, 112, 132, *140*
Achard, J., 250, *260*
Ackermann, R. J., 43, 60, 72, *74*
Ahlborn, B., 154, 165, 194, *198*
Akerlind, L., 10, 11, *74, 83,* 100, *103*
Akishin, P. A., *140*
Alcock, C. B., 47, 48, 49, 58, 59, 67, 72, *74, 83*
Alexander, C. A., 18, 54, 70, *74*
Alice, M. B., 130, *143*
Alkemade, C. Th. J., 40, *79*
Altman, R. L., 54, *74*
Amano, T., 10, *74*
Amend, W. E., 111, 119, *140*
Ames, L. L., 9, 43, 45, 46, 47, 48, 50, 51, 52, 53, 54, 57, 59, 62, 63, 64, 66, 69, 72, 73, *74, 83*
Ammann, P. R., 170, 179, 186, 191, 197, *198, 201*
Ancey-Moret, M. F., 63, *74*
Anderson, C. T., 44, *74*
Anderson, J. B., 109, 110, 111, 112, 114, 119, 121, 131, 132, 134, *140, 141, 143*
Anderson, J. E., 176, 184, 185, *198*
Anderson, P. D., 17, 18, 51, 53, 60, 64, *78*
Andres, R. P., 109, 110, 111, 112, 114, 121, 131, 132, 134, *140, 141, 146*
Andresen, S. G., 117, *141*
Angenmaier, P., 114, *141*
Anthrop, D. F., 73, *74*
Arkell, A., 9, 46, *74*
Arshadi, M., 128, *145*
Arthur, J. R., 222, 223, 224, 228, *233*
Ashkenas, H., 110, 112, 134, *141*
Aslund, N., 11, *76*
Atkins, P. W., 98, *103*
Audit, P., 114, 132, 140, *141*
Aust, T., 162, *198*

B

Baddour, R. F., 177, 183, *198*
Bailey, A. O., 128, *147*
Bailey, S. M., 17, 18, 22, 23, 24, 25, 37, 38, 42, 46, 48, 49, 55, 56, 59, 62, 63, *83*
Baker, F. B., 22, *74*
Baker, S. C., 194, *198*
Ballard, R., 114, *150*
Balwanz, W. W., 125, *141*
Barbariol, I., 255, *260*
Barber, M., 125, *141*
Barchewitz, P., 10, *80*
Barrett, A. H., 140, *141*
Barrow, R. F., 5, 9, 10, 11, 12, 15, 42, 48, 49, 63, 64, 65, 66, *74, 75, 76, 80, 82,* 85, 87, 89, 92, 100, *104*
Barton, Jr., G. W., 39, 40, *80*
Bartz, J. A., 119, *150*
Bass, A. M., 85, *103*
Bauchert, J., 132, *141*
Bauder, U., 162, *198*
Bauer, G., 252, *260*
Bauer, P. H., 113, 121, *149, 150*
Bauer, S. H., 96, *103,* 126, *146*
Bawn, C. E. H., 56, *74*
Baylis, A. B., 124, *149*
Beachey, J., 119, 121, 124, 125, 130, 132, 133, 134, *144, 147*
Beal, J. L., 195, *199*
Becart, M., 37, *74*
Beck, D., 130, *141*
Becker, E. W., 121, 132, *141*
Becker, G., 114, *141*
Bederson, B., 137, *141*
Beguin, C. P., 183, 197, *198, 200*
Belford, R. L., 126, *144*

Bell, G. D., 140, *141*
Bell, W. E., 22, 24, 51, 59, 60, 61, *74, 80, 81,* 124, *148*
Belton, G. R., 48, *74*
Bennett, J. E., 87, *103*
Bennewitz, H. G., 115, 116, *141*
Bentley, P. G., 121, 125, 130, *141*
Berg, R. A., 4, *74,* 96, *105,* 139, *141*
Berkowitz, J., 40, 41, 52, 53, 65, 66, 67, 68, 70, 73, *74, 75, 79, 82,* 108, 124, 125, 139, *141, 142, 149,* 209, *233*
Berkowitz-Mattuck, J. B., 126, *142*
Bernstein, R. B., 107, 114, 118, *142, 145*
Berry, R. S., 46, *74*
Betts, J. F., 128, *148,* 197, *200*
Biddlestone, A. J., 113, *148*
Bier, K., 114, 121, 131, *141, 142*
Bildstein, H., 175, *201*
Bills, J. L., 22, 48, *74*
Bishop, J., 125, *141*
Bivans, R. L., 109, *150*
Blackburn, P. E., 18, 38, *75*
Blanchetais, C. H. L., 240, *262*
Blau, Jr., H., 243, *260*
Blauer, J. A., 126, *142*
Blewett, J. P., 39, *75*
Blocher, Jr., J. M., 18, *82*
Blue, G. D., 52, 53, *74,* 139, *142*
Boers, A. L., 125, *150*
Bohme, D. K., 128, *142*
Bond, R. L., 177, *198*
Bondari, I. A., 257, *261*
Bonne, V., 129, *142*
Bossel, U., 111, 131, *142*
Bott, J. F., 195, *198*
Bowers, V. A., 87, *104*
Boyd, G. E., 24, *75*
Boyer, M. H., 116, 119, *142*
Boyle, B. J., 22, 23, 44, 57, *75*
Brackett, E., 2, *75*
Brackmann, R. T., 116, *143*
Braught, G., 243, 249, *260*
Bray, K. N. C., 169, *198*
Brewer, J., 121, *144*
Brewer, L., 1, 2, 3, 5, 9, 10, 11, 17, 22, 23, 24, 39, 47, 49, 53, 55, 57, 67, 68, 70, 71, 73, *75, 80,* 108, 113, 140, *142,* 204, 207, 227, *233*
Brewer, L. E., 195, *200*
Brink, G. O., 115, 128, *142*
Brix, P., 18, 56, 58, *75*
Broida, H. P., 85, *103*
Bronfin, B. R., 170, 183, 191, 197, *198*
Brooks, R., 6, 40, *75*
Brower, K., 119, *149*
Brown, F. H., 254, 255, 256, *260*
Brown, R. F., 114, 130, *142, 144*
Brundin, C. L., 109, *142*
Bryson, R. J., 194, *198*
Buchheit, K., 130, *142*
Büchler, A., 38, 52, *75,* 94, 96, *103,* 139, *141, 142, 143*
Buchler, J. B., 126
Buettner, G., 175, *199*
Buhl, A., 227, *233*
Bundel, A. A., 22, 24, 39, *76*
Burghoff, H., 121, 132, *141, 142*
Burhorn, F., 154, *199*
Burke, R. R., 128, *142*
Burkhardt, H., 194, *199*
Burns, R. P., 11, 18, 37, 42, 43, 44, 45, 50, 54, 55, 57, 58, 65, 70, 71, *75, 76, 77*
Burton, W. K., 215, *233*
Busol, F. I., 112, *142*
Byrne, M. A., 126, *145*

C

Cabannes, F., 194, 197, *199*
Cabrera, N., 114, *150,* 215, *233*
Calder, G. V., 92, 95, 96, *104*
Caldwell, W. C., 18, 57, *79*
Calia, V. S., 126, *148*
Callahan, W. R., 13, *75*
Callear, A. B., 56, *75*
Callinan, J. P., 128, *146*
Campargue, R., 112, 114, 134, *142*
Campbell, I. E., 18, *82*
Camus, A. L., 243, *260*
Cannon, M. C., 67, *74*
Carlson, K. D., 3, 69, *75,* 89, 96, 97, 100, *104, 105*
Carpenter, J. H., 65, 66, *79*
Carrera, N. J., 18, 58, 62, *75*

Carrigan, T., 128, *142*
Carrington, A., 9, *75*
Carter, A. J., 126, *148*
Case, L. K., 176, 184, 185, *198*
Cauley, J. M., 121, *141*
Chandler, G. G., 11, 15, 66, *75*
Chandrasekharaiah, M. S., 2, 9, 10, 11, *75*
Chanin, L. M., 128, *144*
Chapelle, J., 194, *199*
Charkin, O. P., 97, *104*
Charles, G. W., 15, *75*
Chase, A. B., 257, *261*
Chase, J. D., 160, 161, *199*
Cheetham, C. J., 5, 11, 66, *75*, 85, 89, *104*
Chekhovski, V. Ya., 71, *79*
Chen, C. J., 197, *199*
Chiacigh, I., 254, *260*
Christensen, A. V., 48, *79*
Christensen, Jr., A. V., 44, *79*
Chupka, W. A., 39, 40, 41, 52, 53, 54, 65, 66, 67, 68, 70, 73, *74, 75, 79, 81*, 125, *141*, 209, *233*
Claasen, A., 39, *75*
Claydon, C. R., 89, *104*
Clifton, J. R., 97, *104*
Cline, C. F., 247, *261*
Cobb, G. C., 51, 72, *75*
Cobble, J. S., 24, *75*
Cocco, A., 254, *260*
Cochran, E. L., 87, *104*
Code, R. F., 139, *142*
Cohen, L. S., 169, 191, *198, 199*
Coleman, E. H., 9, 50, *75*
Colin, R., 40, 42, 48, 59, 64, 65, *75, 76*
Collins, R. L., 125, *143*
Collomon, J. H., 9, 12, 38, *75*
Collongues, R., 256, *261*
Comstock, D. F., 243, *260*
Conn, W. M., 243, 249, 255, *260, 261*
Conway, D. C., 128, *145*
Conway, K. C., 22, 23, 44, 57, *75*
Cook, J. C., 243, *260*
Cooper, C. D., 38, 51, 72, *75*
Coppens, P., 38, 43, 52, 63, 68, 71, *75, 76*
Coquart, B., 58, *76*
Cordes, H., 58, *75*
Cosgrove, L. A., 22, 40, *75*
Cotton, F. A., 22, 48, *74*
Couet, C., 58, *76*
Coughlin, J. P., 23, 24, 57, 64, *76*
Craddock, M. K., 114, *150*
Crawford, C. K., 140, *142, 143*
Cremers, C. J., 197, *201*
Cubicciotti, D., 24, *76*
Cudek, F. C., 194, *200*
Curl, R. F., 140, *141*
Cuthbert, J., 130, *143*

D

Daane, A. H., 17, 18, 45, 57, *77, 82*
Dadape, V. V., 252, *262*
Daniels, J. M., 10, *76*
Darken, L. S., 47, *76*
Darnell, A. J., 67, *76*
Das Sarma, J. M., 10, 55, *76*
Datz, S., 107, 114, *143*
Dauermann, L., 127, *148*
Daum, F. L., 131, *143*
Dean, Jr., R. C., 154, *199*
Deckers, J. M., 112, 120, *144, 146, 150*
Declerck, F., 37, *74*
Decroisette, M., 194, *199*
De Galen, L., 39, *76*
Degreve, F., 48, *76*
De Kock, C. W., 97, *104*
De Leeuw, J. H., 109, *143*
Del Greco, F. P., 130, *144*
Della Luca, L., 128, *147*
De Maria, G., 11, 18, 37, 42, 50, 54, 55, 65, 70, 71, *75, 76*
Deutsch, E. W., 10, 15, 63, *74*
Deutsch, J. L., 10, *74*
Dever, D. F., 10, 17, 43, 52, 53, 57, 59, *83*, 94, *105*
Dhumwad, R. K., 9, *76*
Diamond, J. J., 18, 71, *82*, 243, 244, *260*
Dibeler, V. H., 46, *76*
Diesin, R. N., 126, *143*
Dimiduk, P. W., 124, *143*
Dodonov, A. F., 125, *149*
Dole, M., 130, *143*
Dorain, P. B., 10, *76*

D'Orazio, L. A., 22, 23, *76*
Doris, M. H., 140, *141*
Douglas, A. E., 17, 42, 55, *76*
Dove, J. E., 126, *143*
Dowling, J. M., 10, *77*
Dragoo, A. L., 244, *260*
Draper, J. S., 129, *145*
Dreger, L. H., 18, 60, 61, *76,* 252, *262*
Dressler, K., 58, *76*
Drowart, J., 11, 18, 37, 38, 39, 42, 43, 48, 50, 52, 54, 55, 59, 63, 64, 65, 68, 70, 71, *75, 76, 78, 82,* 108, *143, 145*
Dunkin, D. B., 128, *143*
Dunn, G. H., 140, *146*
Durie, R. A., 9, 10, 12, 13, 42, 43, 50, *76*
Duwez, P., 252, 254, 255, 256, *260*
Dwyer, R. J., 49, *76*
Dyatkina, M. E., 97, *104*
Dyer, P. N., 9, *75*

E

Eckert, E. R. G., 155, 165, 197, *201*
Eckert, H. U., 160, *199*
Edelson, D., 128, *146*
Edvinsson, E., 11, *76*
Edwards, J. G., 109, *143*
Edwards, R. H., 125, *143*
Edwards, R. K., 47, *80*
Efimova, A. G., 44, 51, *77*
Eick, H., 23, *82*
Eisler, B., 64, *76*
Ellis, W. P., 230, *233*
Engelke, J. L., 126, *142*
Engh, R. O., 128, *149*
Enloe, J. D., 127, *147*
Enright, J. A., 194, *200*
Erbe, P. E., 195, *199*
Esdaile, R. J., 37, *76*
Estermann, I., 107, *143*
Eubank, P. T., 167, *199*
Evans, A. G., 56, *74*
Evans, P. E., 252, *260*
Evans, W. E., 22, 23, *81*
Evans, W. H., 17, 18, 22, 23, 24, 25, 37, 38, 42, 46, 48, 49, 55, 56, 59, 62, 63, *83*
Everett, W. L., 128, *148,* 197, *200*
Exsteen, G., 42, 54, 59, 65, *76*
Eyring, L., 23, *82*
Ezell, J. P., 183, *198*

F

Faircloth, R. F., 60, *74*
Falconer, W. E., 139, *142, 143*
Farmer, J. B., 87, *104*
Farren, J., 125, *141*
Favin, S., 121, *150*
Feder, H. M., 63, *83*
Fedoseev, S. D., 178, *199*
Fehsenfeld, F. C., 128, *143*
Feiser, J., 59, *76*
Fenn, J. B., 109, 110, 111, 112, 114, 116, 119, 121, 131, 132, 134, *140, 141, 143, 150,* 198, *201*
Ferguson, E. E., 128, *143*
Ferguson, L. D., 130, *143*
Feugier, A., 125, *143*
Ficalora, P., 108, *150*
Fine, J., 56, *82*
Finkelnburg, W., 153, 162, *199*
Firsova, L. P., 59, *80*
Fisher, S. S., 121, 135, *143, 146*
Fite, W. L., 107, 116, 119, *143, 145*
Fitzgibbon, G. C., 22, 23, 24, 46, 52, 57, 62, *76, 78*
Flidlider, G. V., 22, 24, 39, *76*
Fluegge, R. A., 128, *142, 148,* 197, *200*
Foex, M., 240, 242, 243, 247, 250, 252, 257, *260, 261, 262*
Fogel, Ya. M., 132, *150*
Foner, S. N., 87, *104,* 116, 117, 124, 125, 126, *143*
Fontijn, A., 131, *143*
Fox, P. G., 37, *76*
Frank, F. C., 215, *233*
Frankenthal, S., 128, *143*
Franklin, J. L., 46, *76,* 128, *149*
Frantseva, K. E., 56, *82*
Fraser, R. G. J., 107, 114, 137, *143*
Freed, K. F., 5, *78*
Freeman, M. P., 152, 154, 160, 162, 163, 164, 165, 167, 169, 171, 177, 180,

184, 187, 190, 193, 194, 197, *199, 200*
Freeman, R. D., 109, *143*
French, J. B., 109, 112, 114, 119, *143, 146*
Freund, R. S., 119, *143*
Friedman, A. S., 9, 13, *77*
Frisch, M. A., 55, *76*
Fristrom, R. M., 124, 125, *150*
Fröhlich, J. P., 194, *198*
Fussman, F., 180, *199*

G

Galbraith, I. F., 177, *198*
Galokov, F. Y., 257, *261*
Galyaev, V. G., 167, *199*
Gannon, R. E., 243, 259, *261*
Gatos, H. C., 222, *233*
Gatterer, A., 9, 11, 50, *76, 77*
Gaur, J. P., 128, *144*
Gay, I. D., 126, *144*
Gaydon, A. G., 9, 47, 50, 55, 56, 63, 67, *75, 77*
Gazzola, C., 112, *144*
Gel'd, P. V., 23, 56, *77, 79*
Gerhardt, W., 24, *82*
Gerry, M. C. L., 87, *104*
Gersing, E., 114, *144*
Gerstein, B. C., 17, 46, *77*
Ghosh, C., 9, *77*
Ghosh, S. N., 128, *144*
Gibson, J. O., 155, 175, *199*
Giddings, J. C., 130, *144*
Giese, C. F., 40, 41, *75*
Gilbert, I. G. F., 43, *77*
Gilles, P. W., 22, 23, 67, 68, *74, 77, 83*
Gilmore, F. R., 58, *77*
Giordmaine, J. A., 114, *144*
Gissane, W. J. M., 5, 9, 12, 42, *74, 75*
Gladish, H., 171, *199*
Glaser, P. E., 242, 243, 259, *260*
Glemser, O., 43, *77*
Glockler, G., 46, *77*
Goldberg, R. N., 23, *77*
Goldberger, W. M., 162, 177, 182, *199*
Goldfinger, P., 40, 42, 65, *75,* 108, *143*
Goldstein, A. W., 43, 46, 47, 48, 51, 52, 54, 57, 59, 64, 66, 72, 73, *83*
Goldstein, H. W., 17, 57, *77, 83*
Goldstein, S. N., 126, *142*
Golomb, D., 130, *144*
Good, R. E., 130, *144*
Goodings, J. M., 128, *142*
Goodman, L. S., 139, *141*
Gordon, R. J., 139, *144*
Gordy, W., 137, 140, *144, 148*
Gorokhov, L. N., 44, 51, *77*
Gottschlich, C. F., 194, *200*
Govers, T. R., 112, *144*
Gowers, J. R., 112, *146*
Graves, R. D., 178, *200*
Grebenshchikov, R. G., 40, *77*
Green, D. W., 3, *75,* 140, *142*
Greene, F. T., 109, 113, 116, 117, 119, 121, 122, 123, 124, 125, 127, 129, 130, 131, 132, 133, 134, 135, 137, 138, *144, 147*
Gregory, N. W., 129, *148*
Grey, J., 109, 134, *145,* 153, 195, 197, *200, 201*
Grey, J. T., 195, *199*
Griem, H. R., 194, *200*
Griffith, D. R., 119, *148*
Grimley, R. T., 43, 44, 45, 57, 58, *77*
Grosse, A. V., 180, 181, *200*
Grosser, A. E., 118, *144*
Gruber, S., 155, *201*
Gruen, D. M., 97, *104*
Gspann, J., 132, *142, 144*
Guenebaut, H., 58, *76*
Gunn, S. R., 22, 23, 53, *77*
Gurry, R. W., 47, *76*
Gurvich, G. V., 40, 51, 69, *77, 82*
Gurvich, L. V., 40, 47, *77, 83*
Gusarov, A. V., 44, 51, *77*
Gutbier, H. B., 222, *233*
Gutman, D., 126, *144*
Guy, R. W., 121, *149*
Gyarmathy, G., 131, *143*

H

Haar, L., 9, 13, *77*
Habermann, C. E., 17, 45, 57, *77*

Haeusler, C., 10, *80*
Hagena, O., 111, 112, 114, 118, 131, 132, 133, 134, *141, 142, 144, 145*
Hahn, Jr., W. C., 23, 57, *77*
Hall, D. W., 130, *144*
Hall, R. T., 10, *77*
Hall, W. D., 115, 136, *145*
Hall, W. F., 17, *77*
Halow, I., 17, 18, 22, 23, 24, 25, 37, 38, 42, 46, 48, 49, 55, 56, 59, 62, 63, *83*
Hamamura, T., 61, *77*
Hamel, B. B., 112, *145*
Hampson, Jr., R. F., 17, 18, 51, 60, 61, 68, *77*
Haneman, D., 222, *233*
Hanes, G. R., 114, *145*
Happel, J., 177, *200*
Haranath, P. B. V., 66, *77*
Harbour, P. J., 134, *145*
Harnish, H., 155, 170, *200*
Harper, J. G., 222, *233*
Harrison, H., 116, 119, 123, *145*
Hay, A. J., 126, *144*
Hayashi, H., 142, *261*
Hayes, E. F., 96, *104*
Hayhurst, A. N., 125, 128, *145, 146*
Hazlett, R. N., 170, 197, *198*
Head, E. L., 22, 23, 24, 25, 45, 50, 53, 62, 69, 72, 73, *78*
Headman, M. L., 182, *201*
Heald, Jr., J. H., 114, *142*
Heath, G. A., 125, 127, *150*
Heggie, J. C., 114, *150*
Henkes, W., 112, 121, 130, 132, 133, 134, *141, 142, 144, 145*
Hennely, E. F., 39, *75*
Hepler, L. G., 23, *77*
Herm, R. R., 137, 139, *144, 145*
Herman, L., 38, *74, 77*
Herman, R., 38, 62, *77*
Hermann, G., 39, *77*
Herschbach, D. R., 107, 137, 139, *144, 145*
Hertz, H., 205, *233*
Herzberg, G., 9, 12, 13, 14, 18, 41, 47, 56, 58, *75, 77,* 137, 139, *145*
Heymer, G., 155, 170, *200*
Hildenbrand, D. L., 17, 40, 41, *77, 82,* 97, *104,* 116, 119, 124, *142, 145*
Hines, R. C., 130, *143*
Hinnov, E., 45, *77*
Hinten, B. B., 128, *144*
Hirayama, C., 170, 177, 186, *200*
Hirota, E., 10, *74*
Hirth, J. P., 230, *234*
Hiteshue, R. W., 178, *200*
Hoeberling, R. F., 115, *145*
Hoenig, C. L., 17, 48, 64, *77, 78*
Hollander, Tj., 40, *79*
Holley, Jr., C. E., 22, 23, 24, 25, 45, 46, 48, 49, 50, 52, 53, 54, 57, 62, 64, 67, 69, 72, 73, *74, 76, 78*
Holt, B., 126, *150*
Homann, K. H., 125, 129, *142, 145*
Hooper, G. W., 59, *74*
Hörbe, R., 59, *78*
Hostettler, H. V., 114, 118, *145*
Hougen, J. T., 97, *104*
Hove, J. E., 215, *233*
Howell, H. G., 9, 47, 51, 69, *78*
Hubbard, W. N., 63, *83*
Huber, Jr., E. J., 22, 23, 24, 25, 45, 46, 48, 49, 50, 52, 53, 62, 64, 67, 69, 72, 73, *78*
Hudson, D. E., 17, 18, 45, 57, *79, 82*
Hudson, R. L., 116, 117, 124, 125, 126, *143*
Huffman, E. H., 23, *82*
Hughes, V. W., 114, 139, *146*
Hulburt, H. M., 171, *200*
Huldt, L., 40, 47, 55, 57, *78, 79*
Hull, R. J., 128, *142*
Hultgren, R., 17, 18, 37, 38, 40, 44, 45, 46, 47, 48, 49, 51, 52, 53, 54, 55, 56, 57, 59, 60, 62, 64, 67, 69, 71, 72, 73, *78*
Hultin, M., 6, 9, *78*
Hummer, D., 116, 119, *145*
Humphrey, G. L., 23, *78*
Hundhausen, E., 114, 123, *144, 145*
Huo, W. H., 5, *78*
Hurlbut, F. C., 111, 119, 131, *140, 142*
Hurle, I. R., 125, *145*
Hurst, H. J., 11, 15, 66, *75*

I

Iczkowski, R. P., 118, *144*
Iglesias, L., 14, *78*
Inami, Y. H., 116, 119, *142*
Incropera, F., 164, *200*
Inghram, M. G., 11, 18, 37, 39, 42, 43, 44, 50, 52, 54, 55, 57, 58, 65, 66, 67, 68, 70, 71, 73, *74, 75, 76, 77, 78, 79, 81,* 108, *145*
Ino, T., 96, *103*
Inyard, R. E., 22, *74*
Isakova, E. P., 59, *80*
Israel, G. W., 129, *145*
Ito, H., 243, 246, *261*

J

Jacobs, P. F., 153, 169, 197, *200*
Jacox, M. E., 85, 97, *104*
Jaffe, I., 22, 23, *81*
James, C. G., 40, *79*
James, T. C., 14, *79,* 97, *104*
Janik, G. S., 128, *145*
Jarvinen, P. O., 129, *145*
Jelinek, F. J., 17, 46, *77*
Jen, C. K., 87, *104*
Jepson, D. W., 216, 219, 221, *233, 234*
Jernakoff, G., 117, *146*
Jeunehomme, M., 40, 42, 65, *75,* 216, *233*
Jevons, W., 11, *79*
Johnson, J. R., 167, *199*
Johnson, R. G., 18, 57, *79*
Johnston, W. D., 116, 119, *145*
Jolly, W. L., 48, *79*
Jones, L. M., 128, *144*
Jones, M., 181, *200*
Jones, T. V., 126, *145*
Jorba, P. Y., 254, 256, 257, *261*
Joshi, K. C., 55, *79*
Joubert, F., 197, *199*
Ju, F., 124, *145*
Junkes, J., 9, 50, *77*
Justice, B. H., 48, 52, *79*

K

Kahrs, J., 127, *145*
Kaiser, E. W., 94, *104*
Kaiser, K. W., 139, *145*
Kalff, P. J., 40, *79*
Kamada, O., 243, *261*
Kana'an, A. S., 197, *200*
Kane, J. S., 204, 207, 227, *233*
Kantrowitz, A., 109, 134, *145*
Kasai, P. H., 87, 97, 98, 100, 101, 103, *104, 105*
Kataev, D. I., 39, *80*
Katz, S., 9, 10, 12, 47, 62, *81*
Kaufman, M., 6, 11, 40, *75, 79, 83,* 89, 96, *104,* 115, 139, 140, *145, 150*
Kawa, W., 178, *200*
Kawana, Y., 178, *200*
Kebarle, P., 128, *145*
Keenan, P. C., 87, 92, *104*
Keith, M. L., 257, *262*
Keller, D. L., 60, *81*
Keller, R. A., 130, *144*
Kelley, K. K., 17, 18, 23, 24, 37, 38, 39, 40, 44, 45, 46, 47, 48, 49, 51, 52, 53, 54, 55, 56, 57, 59, 60, 61, 62, 64, 67, 68, 69, 70, 71, 72, 73, *78, 79, 80, 81*
Kelly, A. J., 193, *200*
Kern, R. D., 126, *144*
Kessler, R., 128, *146*
Khait, Y. L., 167, 170, 184, 186, *199, 200*
Kieffer, L. J., 140, *146*
Kienst, P. F., 177, *200*
Kimura, T., 178, *200*
King, E. G., 22, 23, 39, 44, 45, 48, 52, 53, 57, 61, 62, 64, 68, 72, 73, *75, 79, 80, 81, 83*
King, I. R., 125, *146*
King, J. G., 116, 119, *145*
King, R. B., 140, *141*
Kinsey, J. L., 118, *146*
Kirillin, V. A., 71, *79*
Kirk, R. E., 170, 171, 172, 173, 174, 182, *200*
Kistiakowsky, G. B., 126, *146*
Kitchener, J. A., 43, *77*
Kleman, B., 87, 89, *104*
Klemperer, W., 5, 6, 9, 11, 40, 52, *75, 78, 79, 83,* 89, 94, 96, 97, *103, 104, 105,* 115, 119, 139, 140, *141, 142, 143, 145, 150*

Klingelhöfer, R., 132, *141, 146*
Knacke, O., 59, *78, 79*
Knewstubb, O. F., 125, *146*
Knof, H., 130, *146*
Knudsen, M., 205, *233*
Knuth, E. L., 109, 114, 121, 123, 128, 134, 135, *143, 146, 150*
Koehler, G., 127, *148*
Koglin, B., 128, *146*
Kohn, H., 45, *77*
Kolesov, V. P., 53, *79*
Korman, S., 155, 177, 180, *200, 201*
Kornilov, A. N., 23, 25, 73, *79*
Kostryukov, V. N., 59, *79*
Kovtunenko, P. V., 22, 24, 39, *76*
Kozuka, T., 240, 243, 244, 246, 248, 250, 260, *261*
Kramer, L., 177, *200*
Kraner, H. M., 45, *80*
Krauss, L., 58, *80*
Krieg, G., 132, *144*
Krikorian, O. H., 65, 66, *79*
Krishnamurty, S. G., 10, 11, *76, 82*
Kruesi, P., 180, *200*
Krukonis, V., 178, 186, *198, 200*
Kubaschewski, O., 24, 70, *81*
Kuluva, N. M., 128, *146*
Kupferman, S. L., 87, *104*
Kusch, P., 114, 136, 139, *146, 147*
Kusenko, F. G., 23, 56, *77, 79*
Kydd, P. H., 126, *146*

L

Ladner, W. R., 162, 177, *198*
Lagerqvist, A., 6, 9, 10, 11, 15, 39, 40, 47, 55, 57, *78, 79, 80,* 87, 89, 92, 103, *104*
Laksham, S. V. J., 10, 15, 62, *80*
Lamar, L. E., 60, *80*
Lamb, J. D., 195, *200*
Langmuir, I., 205, *233*
Larson, L. P., 9, 46, *74*
Laszlo, T. S., 239, 242, 243, 259, *261*
Latimer, W. M., 48, *79*
Laurmann, J. A., 109, *146*
Lavrovskaya, G. K., 125, *149*
Leckenby, R. E., 112, 122, 129, 130, 131, 133, *146, 148*
Lee, L. P., 121, *149*
Lee, P. K., 109, *148,* 204, 207, *233*
Leece, J., 125, *141*
Leferer, R. A., 182, *200*
Lefkowitz, B., 134, *146*
Legay, F., 10, 12, 13, *76*
Leigh, C. H., 128, *146*
Leite, R. J., 128, *144*
Lemley, J., 227, *233*
Lenev, L. M., 22, 44, *81*
Leonas, V. B., 109, 114, *146*
Leonidov, V. Ya., 23, *79*
Leroi, G. E., 97, *104*
LeRoy, R. L., 112, 120, *144, 146, 150*
Lester, J. E., 204, 205, 206, 207, 209, 219, 227, 230, *233, 234*
Leutner, H. W., 155, 187, *200*
Levethal, M., 119, *146*
Levine, M. C., 222, *233*
Levine, S., 22, 23, *81*
Levy, A., 54, *74,* 127, *146*
Levy, D. H., 9, *75*
Lewis, G. N., 17, 22, 23, 24, 39, 49, 53, 57, 68, 73, *80*
Li, S. U., 154, 163, 194, *199*
Lichtenstein, M., 38, *75*
Lide, Jr., D. R., 10, *81*
Liebhafsky, H. A., 39, *75*
Liedmeier, F., 23, 56, *82*
Liljeqvist, B., 87, 89, *104*
Lincoln, K. A., 125, 126, *146*
Lind, E., 9, *80*
Lineberger, W. C., 128, *146*
Linevsky, M. J., 10, 52, 53, *83,* 92, 93, 94, 95, 96, *104, 105*
Linnett, J. W., 125, *141*
Lip, H. K., 222, *233*
Lippert, G., 164, *200*
Liquornik, D. L., 126, *148*
Littke, W., 252, *260*
Locke, J. W., 119, *143*
Loh, E., 252, *260*
Lohse, P., 132, *146*
Longfield, J. E., 165, 190, *199*
Lou, K., 222, 224, *233*
Lyon, H. B., 204, 209, 218, *234*

M

McAffee, Jr., K. B., 128, *146*
McArdle, R. T., 114, *149*
McCollum, W. A., 67, *76*
McConnell, G. I. T., 162, 177, *198*
McDaniel, E. W., 128, *147*
McDowell, C. A., 87, *104*
McElligott, P. E., 116, 117, *146, 148*
McEwan, M. J., 56, *80*
McFee, J., 136, *146*
McGarvey, J. J., 62, *80*
McGonigal, P. J., 180, 181, *200*
McGrath, W. D., 62, *80*
McGregor, W. K., 195, *200*
Mack, E., 45, *80*
Mack, L. I., 130, *143*
Mackay, D., 191, *201*
McKinley, J. D., 97, *104*, 126, *146*
McLaren, L., 130, *144*
MacLean, D. I., 125, *145*
McLeod, Jr., D., 2, 9, 11, *83*, 87, 93, 94, 100, *105*
McMichael, G. E., 112, *146*
McMullen, E. T., 127, *148*
Macur, G. J., 9, 10, 12, 47, 62, *80, 81*
Madip, R. J., 117, *149*
Maecker, H., 153, 154, 162, 164, *199, 200*
Mah, A. D., 22, 23, 24, 25, 40, 44, 45, 71, *80*
Maise, G., 109, 110, 112, 131, 134, *141*
Maiwald, B., 130, *146*
Makino, M., 178, *200*
Mal'tsev, A. A., 39, *80*
Maniero, D. A., 170, 177, 186, *200*
Mann, D. E., 10, 52, 53, *83*, 92, 94, 95, 96, *104, 105*
Mann, J. B., 61, *80*
Mann, K., 243, *261*
Marcus, P. M., 136, *146*
Margrave, J. L., 17, 18, 22, 23, 51, 52, 53, 55, 60, 61, 62, 63, *74, 76, 77, 81, 83*, 108, 118, *144, 150*, 183, 197, *198, 200*, 252, *262*
Marrone, P. V., 134, *146*
Marsden, D. G., 110, 132, *140*
Marsters, G. F., 126, *146*
Martin, D. W., 128, *147*
Martin, E. V., 62, *80*
Mason, C. J., 128, *144*
Masse, J. L., 49, *80*
Masse-Baerlocher, M., 49, *80*
Mastick, D. F., 2, 47, 55, 57, *75*
Matthews, C. O., 64, *78*
Maycock, J. N., 127, *150*
Mayer, H., 132, *141*
Mayer, J., 210, *234*
Mayer, W., 119, *149*
Meierkord, E. H., 24, 67, *78*
Meinel, H., 58, *80*
Melamed, V. G., 167, 184, 186, *199, 200*
Melville, H., 206, *233*
Mentzer, C. C., 187, 190, *199*
Merryman, E. L., 127, *146*
Meschi, D. J., 139, *142*
Messier, D. R., 47, 48, *80*
Meyer, B., 38, *80*
Meyer, C., 10, *80*
Meyer, G., 64, *81*
Meyer, H., 175, *200*
Meyer, R. T., 126, *146*
Michael, J. V., 126, *146*
Miescher, E., 58, *76*
Miller, D. P., 222, *233*
Miller, D. R., 112, 114, 121, 128, 132, 134, *140, 146*
Miller, R. C., 114, 136, *147*
Miller, T. M., 128, *147*
Miller, W. J., 125, *147*
Milligan, D. E., 85, 97, *104*
Milne, T. A., 109, 113, 116, 117, 119, 121, 122, 123, 124, 125, 127, 129, 130, 131, 132, 133, 134, 135, 137, 138, *144, 147*
Minten, A., 114, *147*
Minturn, R. E., 114, *143*
Mizuno, M., 242, 243, 247, 250, 251, 252, 254, 255, 256, 257, *261*
Mobley, R. C., 130, *143*
Modica, A. B., 126, *147*
Moller, C. K., 42, *76*
Moonan, J. F., 112, 134, *141*
Moore, C. E., 2, 12, 13, 14, 15, 16, *80*
Morabito, J. M., 209, 229, *233*
Moran, J. P., 119, *147*

Morgan, J. E., 9, 12, 38, *75*
Morgenstern, R., 130, *141*
Mori, S., 125, *147*
Morino, Y., 10, *74*
Morozova, G. K., 59, *79*
Morozova, M. P., 23, 56, *80*
Morton, H. S., 111, *144*
Moseley, J. T., 128, *147*
Moser, C. M., 3, 6, *75, 81,* 100, *104*
Mosharrafa, M., 128, *147*
Moss, M., 182, *200*
Moulton, D. M. L., 126, *143*
Moyer, J., 210, *234*
Muan, A., 23, 57, *77*
Muckermann, J. T., 107, *142*
Muenter, J., 89, 96, *104,* 139, *145*
Muenter, J. S., 94, *104,* 139, *145*
Mukhtarova, T. A., 184, 186, *200*
Mulford, R. N. R., 60, *80,* 109, *150*
Mullaly, J. R., 17, 46, *77*
Muller, R. H., 209, 229, *233*
Munir, Z. A., 17, 47, *80*
Muntz, E. P., 112, 119, *147*
Murad, E., 116, 119, 124, *142, 145*
Murphy, W. J., 180, 181, *200*
Muster, J., 251, *261*
Myers, M. N., 130, *144*

N

Nakayama, N., 242, *261*
Narasimham, N. A., 9, *76*
Narcisi, R. S., 128, *147*
Nesbet, R. K., 3, 69, *75*
Nesmeyanov, A. N., 59, *80*
Netusil, V. F., 127, *147*
Neuhaus, A., 216, *233*
Newbury, R. S., 39, 40, *80*
Niki, H., 125, *148*
Nikogosyan, K. S., 257, *261*
Nikonov, B. P., 39, *80*
Nilsson, N. E. L., 9, 11, 15, 39, *80*
Ninigradova, N. V., 257, *261*
Ninomiya, M., 9, *80*
Noguchi, T., 240, 242, 243, 244, 246, 247, 248, 249, 250, 251, 252, 254, 255, 256, 257, 260, *261*
Nordine, P. C., 17, 48, *78*
Norman, J. H., 51, 59, 60, 61, *80, 81,* 124, *148*
Norrish, R. G. W., 10, 15, 62, *81*
Novikov, M. M., 47, *77*
Novokhatskii, L. A., 22, 44, *81*
Nutt, C. W., 113, 126, *148*
Nutt, G. B., 125, *145*

O

Odell, F., 255, *260*
Oetting, F. L., 23, *81*
Ogden, H. S., 54, *74*
Ogden, J. S., 18, 70, *74*
O'Halloren, G. J., 128, *148,* 197, *200*
Ohse, R. W., 70, *81*
Okubo, T., 251, 254, *261*
Oldenberg, O., 49, *76*
Oldershaw, G. A., 10, 15, 62, *81*
Olds, L. E., 257, *261*
Olette, M., 63, *74*
Olsen, H. N., 160, *200*
Oman, R. A., 126, *148*
Omrod, G., 117, *148*
Orr, R. L., 17, 18, 37, 38, 40, 44, 45, 46, 47, 48, 49, 51, 52, 53, 54, 55, 56, 57, 59, 60, 62, 64, 67, 69, 71, 72, 73, *78*
Osberghaus, O., 114, *147*
Oskam, H. J., 128, *147*
Osterhof, G. G., 45, *80*
Osyka, V. S., 132, *150*
Othmer, D. F., 170, 171, 172, 173, 174, 182, *200*
Otmakhova, N. G., 39, *80*
Otto, E. M., 22, 23, *81*
Otto, H. E., 257, *261*
Ovchinnikov, K. V., 61, *82*
Ozier, I., 139, *142*

P

Padley, P. J., 54, *81,* 128, *145*
Palmberg, P. W., 209, *233*
Panish, M. B., 17, 18, 47, 49, 51, 58, 61, 62, 69, 72, *81*

Pankratz, L. B., 22, 23, 45, 48, 52, 54, 62, 72, *79, 80, 81*
Papousek, D., 13, *81*
Pardue, W. M., 18, 60, 70, *74, 81*
Parker, J. A., 126, *148*
Parker, V. B., 17, 18, 22, 23, 24, 25, 37, 38, 42, 46, 48, 49, 55, 56, 59, 62, 63, *83*
Parkinson, F. L., 182, *201*
Pasechnova, R. A., 40, *77*
Passchier, A. A., 129, *148*
Patterson, W., 117, *148*
Pattoret, A., 18, 70, *76*
Paule, R. C., 17, 18, 51, 62, *81*
Pauly, H., 107, 114, 115, 118, *144, 148*
Pavone, D., 23, 57, *76*
Pearson, G. S., 125, 127, *150*
Peeters, J., 125, *148*
Peiser, H. S., 102, 103, *104*
Peleg, M., 47, 48, 49, 67, 72, *74*
Pellett, G. L., 127, *148*
Pellett, G. W., 125, *148*
Peng, T. C., 126, *148*
Penmore, G. W., 113, *148*
Perry, T. R., 222, *233*
Pertel, R., 125, *148*
Peters, T., 154, 171, *199, 201*
Petrov, V. M., 71, *79*
Pfender, E., 155, 165, 197, *201*
Phillips, J. G., 2, 3, 11, *81*
Phillips, L. F., 50, 56, *80, 81*
Phillips, Jr., W. L., 61, *81*
Phipps, J. A., 110, 113, 119, *148, 149*
Phipps, T. E., 60, *81*
Pipkin, F. M., 87, *104*
Pitzer, K. S., 17, 22, 23, 24, 39, 49, 53, 57, 68, 73, *80,* 94, *105*
Plante, E. R., 18, 58, 62, 71, *75, 81, 82*
Platteeuw, J. C., 64, *81*
Polak, L. S., 167, 177, 178, 184, 186, *199, 200, 201, 202*
Polak, M., 113, *148*
Pope, R. B., 126, *148*
Porter, G., 9, 39, *81*
Porter, R. F., 52, 54, *75, 79, 81,* 95, 96, *103, 104,* 124, *149*
Potter, A. E., 125, *150*
Powell, F. X., 10, *81*
Preece, E. R., 130, *143*
Prescher, K. E., 59, *79*
Pressley, Jr., G. A., 124, *149*
Proffit, R. L., 127, *147*
Puckett, L. J., 128, *146*
Purgalis, P., 127, *148*

R

Rai, D. K., 39, *82*
Raisen, E., 128, *142*
Raithby, G. D., 197, *201*
Rambidi, N. G., *140, 150*
Ramsey, D. A., 9, 10, 12, 13, 42, 43, 50, *76*
Ramsey, N. F., 107, 114, 137, 139, *148*
Rand, M. H., 24, 60, 70, *74, 81*
Randall, M., 17, 22, 23, 24, 39, 49, 53, 57, 68, 73, *80*
Rank, D. H., 9, *81*
Rao, K. S., 10, 14, *81*
Rao, P. T., 14, 58, *82*
Rao, R. T., 66, *77*
Rao, V. V., 191, *201*
Rat'kovskii, I. A., 69, *82*
Rauh, E. G., 67, 72, *74*
Ray, A. B., 127, *148*
Raziunas, V., 9, 10, 12, 47, 62, *81*
Rebrov, A. K., 112, *148*
Redington, R. L., 94, *104*
Reed, J. W., 257, *261*
Reed, T. B., 152, 160, 176, 177, 184, *201*
Reese, R. M., 46, *76*
Reif, L., 17, 18, 49, 51, 58, 61, 62, *81*
Reinhard, R. R., 9, 46, *74*
Reis, V. H., 198, *201*
Resler, F. L., 126, *146*
Retting, W., 216, *233*
Rhodin, T. N., 209, *233*
Richards, A. W., 59, *81*
Richards, D., 9, 12, 42, *74,* 87, 92, 100, *104*
Richards, W. G., 5, 6, 41, *81, 83*
Richardson, F. D., 59, 63, *74, 81*
Rieman, C. W., 46, *74*
Robben, F., 112, 134, *148*

Robbins, E. J., 112, 122, 129, 130, 131, 133, *146, 148*
Roberts, R. W., 116, 117, *146, 148*
Robinson, E. J., 137, *141*
Robiscoe, R. T., 119, *146*
Roseberry, T. J., 182, *201*
Rosen, B., 9, 10, 50, *77, 81*
Rosenbaum, D. M., 18, *82*
Rosenblatt, G. M., 1, 2, 55, 67, 68, 70, 71, *75,* 109, *148,* 204, 207, *233*
Rosner, D. E., 131, *143*
Ross, J., 107, *148*
Rossini, F. D., 22, 23, *81*
Roth, R. S., 256, 257, *261*
Rothe, D. E., 121, *148*
Rouanet, A., 250, 256, 257, *260, 261*
Rouault, M., 114, 132, *141*
Routly, P. M., 140, *141*
Rowlinson, H. C., 9, 48, 63, 64, *74, 82*
Roy, R., 257, *262*
Royère, C., 252, *261*
Rubin, L., 126, *148*
Rudinger, G., 129, *148*
Runstadler, P. W., 154, *199*
Rusk, J. R., 140, *148*
Russell, G. J., 222, *233*
Russo, A. L., 134, *148*
Rutgaizer, Yu. S., 116, *148*
Ryabov, A. N., 24, *82*
Ryabova, V. G., 40, 47, *77, 82*
Ryason, P. R., 126, *148*
Ryhage, R., 130, *148*

S

Salpeter, E. W., 9, 50, *77*
Salser, G. E., 127, *148*
Salvemini, A., 183, *198, 201*
Salzmann, H., 197, *201*
Santharam, C. V., 14, 58, *82*
Saunders, A. R., 127, *148*
Savage, W. R., 17, 18, 45, 57, *79, 82*
Sawyer, R. F., 125, 127, *148*
Scarborough, J., 128, *145*
Scari, O., 42, *82*
Schaefer, E. J., 128, *144*
Schäfer, H., 23, 24, 56, *82*
Schawlow, A. L., 137, 139, *150*
Scheer, M. D., 56, *82*
Scheurich, J. T., 125, *146*
Schissel, P. O., 126, *148*
Schlick, S., 94, *104*
Schmeltekopf, A. L., 128, *143*
Schmude, K., 128, *142*
Schneider, S. J., 243, 247, 257, *260, 261*
Schneidereit, G., 24, *82*
Schnell, G., 194, *199*
Schnepp, O., 94, *104*
Schofield, K., 6, 40, 42, 54, 65, *82*
Schoonmaker, R. C., 95, *104,* 124, *149,* 227, *233*
Schort, Jr., D. R., 113, *149*
Schreiber, P. W., 195, *200*
Schroeder, L. W., 87, 92, *104*
Schügerl, K., 121, *149*
Schullus, E., 155, 170, *200*
Schumm, R. H., 17, 18, 22, 23, 24, 25, 37, 42, 46, 48, 49, 55, 56, 59, 62, 63, *83*
Schwarz, J. A., 117, *149*
Scoles, G., 112, 114, *144, 149*
Scott, J. E., 119, *148*
Scott, Jr., J. E., 110, 113, *149*
Scott, P. B., 113, 119, *149*
Searcy, A. W., 17, 39, 40, 47, 64, 73, *74, 77, 80,* 207, 209, *233*
Sears, G. W., 60, *81*
Sebacher, D. I., 121, *149*
Seehan, P. J., 243, 259, *261*
Segui, M., 251, *261*
Selin, L. E., 11, *76, 79,* 87, 89, 92, 103, *104*
Semenov, G. A., 39, 47, 56, 61, 62, 69, *82*
Sen Gupta, A. K., 55, 62, *82*
Seshadri, K. S., 10, 52, 53, *83,* 92, 94, 95, 96, *104, 105*
Seymour, E. V., 194, *201*
Shahin, M. M., 128, *149*
Shambon, A., 10, 69, *83*
Sharafutdinov, R. G., 112, *148*
Shardanand, 132, *149*
Shchukarev, S. A., 24, 39, 47, 56, 69, *82*
Sheer, C., 155, 180, *200, 201*
Sheindlin, A. E., 71, *79*

Sherman, F. S., 110, 111, 112, 121, 131, *141, 142, 149*
Sherman, M. P., 153, 169, *201*
Sherrill, II, W. A., 113, *149*
Sherwood, E. M., 18, *82*
Shetlar, M. D., 12, *82*
Shickell, W. D., 17, 46, *77*
Shimizu, H., 180, *202*
Shin-Piaw, C., 66, 67, *82*
Shomate, C. H., 23, *82*
Shteiner, E. V., 178, *199*
Simpson, O. C., 60, *81*
Sinanoglu, O., 4, *74*
Sinclair, T. F., 121, *149*
Singh, N. L., 10, 14, *82*
Singh, R. B., 39, *82*
Sinke, G. C., 18, 43, 44, 48, 53, 58, 70, 73, *82*
Sipler, D., 128, *146*
Sivaramamurty, V., 66, *77*
Skibenko, E. I., 112, *142*
Skinner, G. T., 119, 123, 126, *149, 150*
Skinner, H. A., 96, 97, *104*
Skofronick, J. G., 114, *149*
Skrivan, J. F., 153, 166, 167, 169, 177, 184, 193, *199, 201*
Skuratov, S. M., 23, 25, 53, 73, *79*
Smetana, F. O., 113, *149*
Smith, D. K., 247, *261*
Smith, D. L., 182, *200*
Smith, I. W. M., 56, *75*
Smith, K. F., 107, 139, *149*
Smith, W. T., 24, *75*
Smith, W. V., 137, *144*
Smity, J. A., 125, *150*
Smoes, S., 18, 38, 43, 52, 63, 68, 70, 71, *75, 76, 82*
Snelson, A., 94, 95, 96, 101, *104, 105*
Snyder, P. E., 22, 40, *75*
Somayajulu, G. R., 2, *75*
Sommer, A., 94, *105*
Somorjai, G. A., 204, 205, 206, 207, 209, 216, 218, 219, 221, 222, 224, 227, 229, 230, 231, *233, 234*
Sonin, A. A., 128, *149*
Spedding, F. H., 17, 18, 45, 46, 57, *77, 79, 82*
Sperling, R. L., 128, *149*
Spiridonov, V. P., *140, 150*
Sprouse, J. A., 193, *201*
Stafford, F. E., 40, *82*, 124, *149*
Stair, A. T., 114, *149*
Staley, H. G., 51, 59, 60, 61, *80, 81*, 124, *148*
Stark, D., 18, 61, *82*
Stauffer, J. L., 38, 52, *75*, 94, 96, *103*, 139, *141, 142, 143*
Stecura, S., *261*
Steenbeck, M., 162, *202*
Steiger, R. F., 209, 229, *233*
Stemple, N. R., 207, 216, *234*
Stern, S. A., 121, *149*
Stickney, R. E., 107, 116, *149, 150*
Stoecker, U., 43, *77*
Stokes, C. S., 169, 180, 181, 183, *200, 201*
Stolyarova, T. A., 23, 56, *80*
Storms, E. K., 18, 70, *82*
St. Pierre, A. G., 9, *81*
Stout, N. D., 17, 48, *78*
Stranski, I. N., 215, *234*
Strassmair, H., 18, 61, *82*
Streng, L. A., 183, *201*
Strigl, A., 175, *201*
Strunin, V. P., 125, *149*
Stuart, R. V., 119, *149*
Stubblefield, C. T., 23, *82*
Studier, M. H., 61, *82*
Studniarz, S. A., 128, *149*
Stull, D. R., 18, 43, 44, 48, 53, 58, 70, 73, *82*
Sugden, T. M., 50, 54, *81*, 125, *145*
Summers, W. H., 126, *149*
Swaminathan, T. M., 10, *82*
Sy, A., 194, *199*
Symons, M. C. R., 98, *103*
Szwarc, R., 18, 71, *81, 82*

T

Tagami, M., 22, 24, *74*
Takada, S., 246, *261*
Takezaki, Y., 125, *147*
Talbot, L., 109, 112, 134, *148, 149*
Talrose, V. L., 125, *149*

Tanaka, Y., 132, *150*
Tang, S., 112, 134, *141*
Tatevskii, V. M., 39, *80*
Taylor, E. H., 107, 114, *143*
Taylor, W. J., 111, *150*
Telford, N. R., 125, *145, 146*
Theard, L. P., 40, 41, *82,* 124, *145*
Thibault, R. J., 14, *79*
Thomann, H., 129, *150*
Thomas, A., 87, *103*
Thompson, J. C., 183, *198*
Thompson, K. R., 96, 97, *105*
Thorn, R. J., 43, 67, 72, *74,* 126, *150*
Thorpe, M. L., 175, *201*
Tiknomerov, V. B., 154, 164, *201*
Tilloca, G., 254, *261*
Timmins, R. S., 170, 177, 179, 186, 191, 197, *198, 201*
Toennies, J. P., 107, 115, 118, *148*
Toerring, T., 9, 10, *82*
Tolnas, E. L., 51, 72, *75*
Tomita, G., 61, *77*
Torello, F., 112, 114, *144, 149*
Toropov, N. A., 257, *261*
Torriano, G., 254, *260*
Townes, C. H., 137, 139, *150*
Trambarulo, R. F., 137, *144*
Trass, O., 191, *201*
Traverse, J. P., 252, 257, *260*
Travis, D. N., 10, *74*
Trevalion, P. A., 130, *146*
Trilling, L., 109, 113, 119, *147, 148, 149, 150*
Trischka, J. W., 114, 139, *150*
Trombe, F., 240, 251, 252, *261, 262*
Trulson, O. C., 126, *148*
Tuan Anh, N., 58, *76*
Turnbull, A. H., 121, 130, *143, 150*
Tyke, D. C., 37, *82*

U

Uhler, U., 9, 10, 11, 15, 37, 57, *80, 83,* 89, *105*
Unsöld, A., 194, *201*
Unsworth, P. J., 139, *149*
Ushakova, I. M., 25, 73, *79*

V

Vaidya, V. M., 50, *75*
Vandegrift, A. E., 129, 131, *144, 147*
Van Tiggelen, A., 125, *143, 148*
Varlamov, I. V., 154, 164, *201*
Varma, A. K., 118, *145*
Veenemans, C. F., 39, *75*
Veits, I. V., 40, 47, 51, 69, *77, 83*
Verhaegen, G., 5, 6, 41, 42, 48, 52, 54, 63, 64, 65, *75, 76, 81, 82, 83*
Verkhovtseva, E. T., 132, *150*
Verneker, V. R. P., 127, *150*
Vesselovskii, V. K., 64, *83*
Vickroy, D. G., 183, *198*
Vidal, R. J., 119, *150*
Vilkov, L. V., 140, *150*
Vinckier, C., 125, *148*
Von Engel, A., 162, *202*
Von Jaskowsky, W., 153, 154, 163, 166, 193, 194, *199, 201*
Von Zahn, U., 128, *150*
Vriens, L., 125, *150*
Vuillard, G., 250, *262*
Vurzel, F. B., 178, *202*
Vyse, R., 114, *150*

W

Wachman, H. Y., 109, 113, 119, *147, 149, 150*
Wadsö, I., 22, 52, *76*
Wagman, D. D., 17, 18, 22, 23, 24, 25, 37, 38, 42, 46, 48, 49, 55, 56, 59, 62, 63, *81, 83*
Wagner, H. G., 125, 129, *142, 145*
Wahlbeck, P. G., 47, 67, 68, *77, 80, 83,* 109, 113, *150*
Walker, R. F., 17, 18, 51, 58, 60, 61, 62, *75, 77,* 259, *260*
Walsh, A. D., 94, 96, *105*
Walsh, P. N., 17, 43, 45, 46, 47, 48, 50, 51, 52, 53, 54, 57, 59, 62, 63, 64, 66, 69, 72, 73, *74, 77, 83*
Walsh R. M., 140, *142*
Walter, T. A., 108, *142*
Walters, M., 128, *144*
Wang, C. P., 126, *150*

Wang, J. C. F., 121, *150*
Wang, K., 252, *262*
Wang, K. C., 109, 113, *150*
Wang, K. I., 140, *143*
Wang, T. C., 114, *144*
Ward, J. W., 109, *150*
Waring, J. L., 257, *261*
Warkehr, E., 58, *75*
Warn, J. W., 94, *105*
Warren, I. H., 180, *202*
Warshaw, I., 257, *262*
Washburn, C. A., 46, *83*
Waterman, P. C., 121, *149*
Watson, W. W., 10, 69, *83*
Webb, L. E., 59, *81*
Wedemeyer, R., 115, 116, *141*
Wegener, P. P., 131, *150*
Weidman, R., 155, 175, 1*99*
Weigner, C., 38, *77*
Weinstock, B., 125, *148*
Weiser, C. H., 126, *148*
Weiss, R., 115, *150*
Weller, W. W., 22, 45, 52, 62, *79, 80, 83*
Weltner, W., Jr., 2, 9, 11, *83*, 85, 87, 89, 93, 94, 97, 98, 100, 101, *104, 105*
Westenberg, A. A., 121, 124, 125, *150*
Westrum, E. F., Jr., 48, 52, *79*
Wharton, L., 6, 9, 11, 40, 52, *75, 79, 83,* 94, 96, *103, 105,* 115, 139, 140, *141, 142, 150*
White, D., 10, 17, 43, 45, 46, 47, 48, 50, 51,, 52, 53, 54, 57, 59, 62, 63, 64, 66, 69, 72, 73, *74, 77, 83,* 92, 94 95, 96, *104, 105*
Whitfield, D. L., 130, *144*
Whipple, E. B., 97, 101, *104*
Whyman, D. 162,, *202*
Wigartz, K., 9, 11, 15, 39, *80*
Wiggins, T. A., 9, *81*
Williams, G. T., 125, 127, *150*
Willis, D. R., 112, *145*
Willis, P., 131, 133, *148*
Winicur, D. H., 123, 128, *150*
Winterbottom, W. L., 109, *150,* 206, 230, *234*
Wise, S. S., 63, *83*
Witt, A., 183, *202*
Wolff, E. G., 67, *83*
Wong, E. L., 125, *150*
Wood, R. H., 22, 23, *76*
Woodbury, H. H., 216, 217, *234*
Woods, R., 116, *150*
Wutzke, S. A., 197, *201, 202*

Y

Yamamoto, S., 116, *150*
Yamamoto, T., 250, *261*
Yealland, R. M., 120, *150*
Yoder, H. S., 257, *262*
Yonemochi, O., 251, 254, *261*
Yoshida, M., 250, *261*
Yuferov, V. B., 112, *142*

Z

Zaikin, I. D., 53, *79*
Zapata, R. N., 114, *150*
Zimm, B., 210, *234*
Zmbov, K. F., 108, *150*
Zorn, J. C., 107, 115, 136, *145, 150*

Subject Index

A

Acetylene synthesis, streaming thermal plasmas in, 171–172, 177–179
Activation energy of vaporization, 207
Alkaline earth diatomic gases, low-lying electronic states of, 7
Aluminum dioxide refractory systems, high temperature phase diagrams, 257, 259
Aluminum fluoride, matrix-isolation studies on, 94–96
Aluminum monoxide, dissociation energy and free energy functions of, 37–38
Antimony monoxide, dissociation energy of, 62
Argon monoxide, dissociation energy of, 38
Arsenic monoxide, dissociation energy of, 38

B

Barium monoxide, dissociation energies and free energy functions of, 38–40
Beams, molecular, *see* Molecular beams
Beryllium monoxide, free energy functions and dissociation energies of, 40–41
Bismuth monoxide, dissociation energy of, 42
Boron monoxide, dissociation energy and free energy functions of, 38–39
matrix-isolation techniques for, 94
Bromine monoxide, dissociation energy of, 42

C

Cadmium monoxide, dissociation energy of, 43
Cadmium sulfide crystals, sublimation-mechanism studies on, 216–221
Calcium monoxide, dissociation energy of, 42–43
Carbon black, from streaming thermal plasma device, 183
Carbon monoxide, dissociation energy of, 42
Centrifugal plasma furnace, 180–181
Ceramic oxides, freezing point data on, 246
Cerium monoxide, dissociation energy of, 43
Cesium monoxide, dissociation energy of, 44
Chlorine oxide, dissociation energy of, 43
Chromium monoxide, dissociation energy of, 44
Coal, upgrading of, streaming thermal plasmas in, 177–178
Cobalt monoxide, dissociation energy of, 43–44
Condensation, rate of, equation for, 205
Condensed oxides, heats of formation of, 22–25
Cupric oxide, dissociation energy of, 45

D

Diatomic oxides (gaseous)
dissociation energies of, 19–20
enthalpy data for, 7
heats of formation of, 21
rotational and vibrational constants of, 9–11
Dihalides, matrix-isolation studies on, 96–97
Dissociation energies of gaseous monoxides, 1–83
Dysprosium oxide, dissociation energy of, 45

E

Electron-paramagnetic-resonance spectroscopy, of high-temperature molecules, 98–101
Elements, heats of atomization of, 17–18
Erbium monoxide, free energy function and dissociation energy of, 45–46

Europium monoxide, free energy function and dissociation energy of, 46
Evaporation rate, equation for, 205

F

Ferrous monoxide, dissociation energy of, 46–47
Fluorine monoxide, dissociation energy of, 46
Free energy functions of gaseous monoxides, 1–83
calculation of, 1
electronic contribution to partition function, 2
electronic energy levels and degeneracies used in computation of, 12–16
electronic partition function Q_{el}, 3–4
number of electronic states and, 2

G

Gadolinium monoxide, dissociation energy of, 47–48
Gallium arsenide crystals, sublimation-mechanism studies, 221–228
Gallium monoxide, dissociation energy of, 47
Gaseous mixtures, expansion of, 112
Gaseous monoxides, dissociation energies and free energy functions of, 1–83
sources and evaluation of data, 37–73
Gases
monatomic, expansion of, 112
polyatomic, expansion of, 112
Germanium monoxide, dissociation energy of, 48

H

Hafnium monoxide,
dissociation energy of, 49–50
electronic partition function of, 2–4
Halides, matrix-isolation studies on, 94–97
Halogen chemistry, in streaming thermal plasmas, 182–183
Heats of atomization of elements, 17–18
Heats of formation
of condensed oxides, 22–25
of gaseous diatomic oxides, 21
Helium monoxide, dissociation energy of, 49
High temperature chemistry, molecular beams in, 107–150
High temperature equilibria data, evaluation of, 7
High-temperature molecules
EPR studies on, 98–101
magnetic properties of, 98
matrix-isolation technique applied to, 85–105
High temperature phase studies with solar furnace, 235–262
freezing point determination of high melting refractory oxides, 244–248
phase studies at high temperatures, 250–259
high temperature reactions, 250–251
phase diagrams, 252
vaporization and rate studies, 251
solar furnaces for
attainable temperatures, 239–240
flux density of, 239
irradiation techniques, 240–243
nature of, 236–237
optical system of, 236
structure and types, 238–239
temperature and emissivity measurement, 243–244
by centrifugal blackbody furnace, 244
specular reflection method, 243–244
Holmium monoxide, dissociation energy of, 50
Hydrogen cyanide, synthesis of, streaming thermal plasma studies, 187–191
Hydrogen monoxide, dissociation energy of, 49

I

Indium monoxide, dissociation energy of, 50–51

Iodine monoxide, dissociation energy of, 50
Iridium monoxide, dissociation energy of, 51

K

Krypton monoxide, dissociation energy of, 51

L

Lanthanum monoxide, dissociation energy of, 51–52
Lead monoxide, dissociation energy of, 59
Lime, dissociation energy of, 42–43
Liquidus curve measurement with solar furnace, 248–250
Lithium fluoride, matrix-isolation studies on, 94–96
Lithium monoxide, dissociation energy of, 52–53
Lithium oxides, matrix-isolation techniques for, 94
Lutecium oxide, dissociation energy of, 53–54

M

Mach number as a function of distance downstream, 110
Magnesium monoxide, dissociation energy of, 54
Manganese monoxide, dissociation energy of, 54–54
Matrix-isolation technique applied to high temperature molecules, 85–105
  description of, 85
  experimental procedures for, 86–87
  optical spectroscopy in, 87–97
  orientation of molecules in matrices, 101–103
  vanadium monoxide in studies of, 87–89
Metal dihalides, matrix infrared spectroscopy, 96–97
Metal dioxides, properties, from matrix spectra, 93
Metal halides, matrix infrared spectroscopy of, 94–97
Metal oxides, matrix-isolation techniques for, 89–94
  vibrational frequencies, 97
Methane,
  acetylene and olefins from, 177
  decomposition of by streaming thermal plasma studies, 184–186
Molecular beams in high temperature chemistry, 107–150
  applications of, 120–140
    direct sampling at high pressures, 120–130
    electric and magnetic deflection experiments, 135–139
    flames, 123
    to high temperature electron diffraction, 140
    to high temperature spetroscopy, 139
    homogeneous nucleation in free jets, 121
    interaction of the system and probe, 122–123
    to ionization cross sections, 140
    metastability of clusters, 122
    to microwave spectroscopy, 140
    to sampling, 124–130
      ablation and beam-surface reaction, 126
      auto engines, 127
      free evaporation-laser heated surfaces, 125
      particle sampling, 129
      plasmas, 127–128
      propellant burning, 126–127
      shock tubes, 126
      transpiration and Knudsen cell vaporization, 124
      upper atmosphere composition, 128
      weakly bound species, 129
    studies using high pressure beam properties, 130–135
      collisions at low temperatures, 135
      free-jet isolation spectroscopy, 133–135
      high-energy molecules, 132

nucleation, 130–132
properties of clustered neutrals and ions, 132–133
temperature dependence of fragmentation patterns, 122
beam composition, 120–121
beam formation from high pressures, 108–113
ideal free-jet expansion, 109–111
phenomena involved in, 111–113
beam modulation, 116–117
beam sources, 113–114
effusive beams, 114
nozzle beams, 114
detectors for, 115–116
ionization methods in, 115
techniques, 113–120
transition from molecular flow to continuum flow, 113
velocity selection and analysis in, 117–120
phase shift velocity, 119–120
time-of-flight techniques, 118–119
Monoxides, gaseous type, *see* Gaseous monoxides
Molybdenum monoxide, dissociation energy of, 55

N

Neodymium monoxide, dissociation energy of, 56–57
Neon monoxide, dissociation energy of, 57
Nickel oxide, dissociation energy of, 57
Niobium monoxide, dissociation energy of, 56–57
Nitric oxide, dissociation energy of, 55–56

O

Osmium monoxide, dissociation energy of, 58
Oxides, condensed type, *see* Condensed oxides
Oxygen (molecular), dissociation energy of, 58

P

Petroleum fractions, upgrading of, streaming thermal processes in, 178–179
Phosphorous monoxide, dissociation energy of, 58
Platinium monoxide, dissociation energy of, 60
Plutonium monoxide, dissociation energy of, 60
Potassium oxide, dissociation energy of, 51
Praseodymium monoxide, dissociation energy of, 59–60

R

Refractory oxides, freezing point determination of, 244–248
Rhenium monoxide, dissociation energy of, 60–61
Rhodium monoxide, dissociation energy of, 61
Rubidium monoxide, dissociation energy of, 60
Ruthenium monoxide, dissociation energy of, 61–62

S

Samarium monoxide, dissociation energy of, 64
Scandium monoxide, dissociation energy of, 62–63
Selenium monoxide, dissociation energy of, 63
Silicon monoxide, dissociation energy of, 63
Silver, low energy electron diffraction studies of vaporizing surface, 228–232
Silver oxide, dissociation energy of, 37
Sodium chloride, sublimation-mechanism studies on, 209–216
Sodium oxide, dissociation energy of, 56
Solar furnace, high temperature phase studies using, 235–262

Solids, sublimation mechanism of, 203–237
Streaming thermal plasmas,
  chemical research in, 151–202
  description of, 152–153
  generation of, 153–162
    by consumable-electrode devices, 155, 158
    by large volume arcs, 160–162
    by plasma jet, 154–155, 156
    by transpiration cooled devices, 155, 159
  by welding arc, 159–160
  industrial applications, 170–175
    acetylene synthesis, 171–172
    spheroidizing, 175
    surface modification, 173–175
  properties of, 162–170
    arc process, 163–165
    intermediate section, 165–169
      heat flow, 166–167
      mixing, 167, 169
      supersonic jet, 169
    quenching section, 169–170
  recent research on, 175–198
    diagnostics, 191–198
      arc diagnostics, 197
      average properties, 192–194
      chemical diagnostics, 197–198
      radial profile measurements, 194–197
    mechanism studies, 183–191
      HCN synthesis, 187–191
      methane decomposition, 184–186
    product-oriented type, 176–183
      acetylene synthesis, 177–179
      in metallurgy, 180–182
      synthesis of oxide materials, 179–180
      surface modification, 182
Strontium monoxide, dissociation energy of, 65
Sublimation mechanism of solids, 203–237
  cadmium sulfide single crystals, 216–221
    doping effects, 216–218
    light effects, 219
    surface concentration effects, 219–221
    temperature dependence, 216
  charge transfer in, 232
  experimental aspects, 207–209
  gallium arsenide single crystals, 221–228
    activation energy of, 226
    liquid gallium effects, 226
    mass spectrometric studies, 224–226
    microbalance studies, 222–224
    vacuum evaporation rates, 227
  impurities in, 232
  low energy electron diffraction study of vaporizing silver surface, 228–231
  sodium chloride single crystals, 209–216
    dislocation effects, 210–212
    divalent impurity effects, 212–214
    monomer-to-dimer ratios, 214–215
    steady state vacuum vaporization, 214
    vapor flux composition, 211
    vaporization characteristics, 215–216
  surface composition in, 231–232
  vapor-composition invariance, 232
  vaporization studies, 205–207
    mechanisms, 232
Sulfur monoxide, dissociation energy of, 62

T

Tantalum monoxide, dissociation energy of, 65–66
Tellurium monoxide, dissociation energy of, 66–67
Terbium monoxide, dissociation energy of, 66
Thallium monoxide, dissociation energy of, 69
Thorium monoxide, dissociation energy of, 67
Thulium monoxide, dissociation energy of, 69
Tin monoxide, dissociation energy of, 64–65

Titanium monoxide, dissociation energy and free energy functions of, 67–69
Titanium oxide, electronic partition function of, 2–4
Transition-metal oxides, matrix-isolation techniques for, 89–94
Tungsten monoxide, dissociation energy of, 71

U

Uranium monoxide, dissociation energy of, 70

V

Vanadium monoxide,
  absorption spectrum of, 87
  absorption bands of, 90–91
  dissociation energy of, 70–71
  electronic spectrum of, 88
Vaporization, principles of, 205–207

X

Xenon monoxide, dissociation energy of, 72

Y

Ytterbium monoxide, dissociation energy of, 72–73
Yttrium monoxide, dissociation energy of, 72

Z

Zinc monoxide, dissociation energy of, 73
Zirconium dioxide refractory systems, high temperature phase diagrams, 254–257
Zirconium monoxide,
  electronic partition function of, 2–4
  diffusion energy of, 73